The VHDL Handbook

The VHDL Handbook

by

David R. Coelho
Vantage Analysis Systems, Inc.

Kluwer Academic Publishers
Boston/Dordrecht/London

Distributors for North America:
Kluwer Academic Publishers
101 Philip Drive
Assinippi Park
Norwell, MA 02061, USA

Distributors for all other countries:
Kluwer Academic Publishers Group
Distribution Centre
Post Office Box 322
3300 AH Dordrecht, THE NETHERLANDS

Consulting Editor: Jonathan Allen, Massachusetts Institute of Technology

Library of Congress Cataloging-in-Publication Data

Coelho, David R.
  The VHDL handbook.

  Bibliography: p.
  Includes index.
  1. VHDL (Computer program language) I. Title.
  TK7874.C6 1989   621.39'2   89-11084
  ISBN 0-7923-9031-8

Copyright © 1989 by Kluwer Academic Publishers **Sixth Printing, 1992.**

All rights reserved. No part of this publication may be reproduced, stored in a retrieval system or transmitted in any form or by any means, mechanical, photocopying, recording, or otherwise, without the prior written permission of the publisher, Kluwer Academic Publishers, 101 Philip Drive, Assinippi Park, Norwell, Massachusetts 02061.

Printed in the United States of America.

Contents

1 Introduction   **1**
    1.1 Introduction to the VHDL Language 1
        1.1.1 History of VHDL 1
        1.1.2 DOD Requirements and VHDL 2
        1.1.3 VHDL As a Design Tool 3
    1.2 Multi-Level Design 5
    1.3 The Model Accuracy Continuum 7

2 Anatomy of a VHDL Model   **11**
    2.1 Describing Electronic Hardware in VHDL 11
    2.2 A VHDL File 14
    2.3 The Standard Logic Package 20
    2.4 User Defined Packages 21
    2.5 VHDL Models and the Accuracy Continuum 25
        2.5.1 2-Value Unit-Delay Approach 25
        2.5.2 46-Value Unit-Delay Approach 28
        2.5.3 Fixed-Delay Approach 30
        2.5.4 Variable-Delay Approach 31
        2.5.5 Generic Variable-Delay Approach 33
        2.5.6 Full-Delay Approach 36
        2.5.7 Error Checking and Model Structure 38
    2.6 Handling Timing Using Configurations 42
    2.7 Using VHDL as a Stimulus Language 46
    2.8 Standardized VHDL Modelling Conventions 48
        2.8.1 Generic Parameters 50
        2.8.2 Naming Conventions 53
        2.8.3 Constraints 54
        2.8.4 Unknown Handling 54

3 Combinational Devices — 57
3.1 Simple Gates — 57
- 3.1.1 2-Input Positive-Nand Gate — 57
- 3.1.2 2-Input Positive-Nand with Open-Collector Outputs — 60
- 3.1.3 2-Input Positive-Nor Gate — 62
- 3.1.4 Inverter — 64
- 3.1.5 Inverter with Open-Collector Outputs — 66
- 3.1.6 3-Input Positive-And Gate — 68
- 3.1.7 3-Input Positive-Nand Gate — 71
- 3.1.8 2-Input Positive-Or Gate — 73
- 3.1.9 2-Input Positive-Xor Gate — 76

3.2 Selectors/Multiplexers — 78
- 3.2.1 3 to 8 Decoder/Multiplexer — 79
- 3.2.2 2 to 4 Decoder/Multiplexer — 84
- 3.2.3 1 of 8 Selector/Multiplexer — 87
- 3.2.4 1 of 4 Selector/Multiplexer — 92
- 3.2.5 1 of 2 Selector/Multiplexer — 96

3.3 Switch Level Devices — 99
- 3.3.1 Switch Modelling Utilities — 100
- 3.3.2 Bidirectional Transmission Element — 104
- 3.3.3 Basic Complementary Transmission Gate — 107
- 3.3.4 Basic Transmission Gate — 107

3.4 Simple ALU's — 108
- 3.4.1 ALU/Function Generator — 109

3.5 One Shots — 119
- 3.5.1 Monostable Multivibrator — 119

3.6 Comparators — 123
- 3.6.1 4 Bit Magnitude Comparator — 124

3.7 Parity Generators/Checkers — 129
- 3.7.1 9 bit Odd/Even Parity Generator/Checker — 129

4 Sequential Devices — 135
4.1 Flip-Flops — 135
- 4.1.1 D-Type Positive-Edge Triggered Flip-Flop with Preset/Clear — 136
- 4.1.2 JK Pos-Edge Triggered Flip-Flop with Preset/Clear — 140
- 4.1.3 JK Neg-Edge Triggered Flip-Flop with Preset/Clear — 146
- 4.1.4 JK Negative-Edge Triggered Flip-Flop with Preset — 152

4.2 Registers — 157
- 4.2.1 4-Bit Parallel-Access Shift Register — 158
- 4.2.2 3 to 8 Decoder/Demultiplexer with Register — 165
- 4.2.3 3 to 8 Decoder/Demultiplexer with Latch — 171

		4.2.4	8 Bit Parallel-Out Serial Shift Register	177
		4.2.5	Parallel Load 8 Bit Shift Register	182
		4.2.6	Parallel Load 8 Bit Shift Register with Clear	188
	4.3	Counters		195
		4.3.1	Synchronous 4 Bit Decade Counter with Asynchronous Clear	196
		4.3.2	Synchronous 4 Bit Binary Counter with Asynchronous Clear	202
		4.3.3	Synchronous 4 Bit Decade Counter	208
		4.3.4	Synchronous 4 Bit Binary Counter	214
		4.3.5	Synchronous Up/Down 4-Bit Decade Counter	220
5	**Memory Devices**			**227**
	5.1	Memory Initialization		227
	5.2	Read Only Memories		231
		5.2.1	1024 bit (256 by 4) ROM	231
		5.2.2	16,384 bit (4096 by 4) register PROM	236
	5.3	Random Access Memories		242
		5.3.1	64 bit RAM	243
	5.4	PALs, PLDs		251
		5.4.1	Calculating Products	252
		5.4.2	10 input, 2 output, 6 I/O PAL	255
		5.4.3	8 input, 2 I/O, 6 clocked output PAL	259
		5.4.4	8 input, 8 clocked output PAL	264
6	**Complex Devices**			**269**
	6.1	Getting Started		269
		6.1.1	Partial versus Full Functional Models	270
		6.1.2	Architecture	271
		6.1.3	Behavior	278
	6.2	The Timing Model		288
		6.2.1	Device Speeds	288
		6.2.2	Min/Max Timing	290
		6.2.3	Drive/Loading Dependencies	290
		6.2.4	A Uniform Approach to Device Dependent Data . .	291
	6.3	Error Handling		293
		6.3.1	Unknowns	293
		6.3.2	Setup / Hold Time Techniques	295
		6.3.3	Waveform Checking	296
	6.4	Techniques for Modeling		299
		6.4.1	Bus Handlers	299
		6.4.2	Instruction Decoders	300

		6.4.3	Sequencers	301
		6.4.4	Instruction Sets	301
	6.5	Quality Assurance		301
		6.5.1	Developing a Test Plan	302
		6.5.2	Validation of the Model	304

7 The Standard Logic Package 309

	7.1	Using the Standard Logic Package		310
	7.2	The Logic Value System		311
	7.3	Technology Rules		315
		7.3.1	ECL - Emitter Coupled Logic	322
		7.3.2	CMOS - Complementary MOS	324
		7.3.3	NMOS - n-Channel MOS	326
		7.3.4	TTL - Transistor transistor logic	328
		7.3.5	TTLOC - Open-collector TTL	330
	7.4	Bus Resolution		332
	7.5	Logic Manipulation		337
		7.5.1	Overloaded Comparison Operators	337
		7.5.2	State/Strength Lookup Tables	339
		7.5.3	Logic Lookup Tables	352
	7.6	Timing Utilities		357
	7.7	Integer Data Utilities		364

Bibliography 377

Index 381

List of Figures

1.1	DOD/VHDL Documentation Requirements	3
1.2	Accuracy/Speed/Abstraction Continuum	7
2.1	Structure of Design as Schematic	12
2.2	Circuit Simulation Results	15
2.3	RSFF Structure	18
2.4	RSFF Simulation Results	19
2.5	Circuit Initialization Problems	26
2.6	Circuit Behavior and Timing	27
2.7	Simulation of Variable Delay	32
2.8	Schematic with Generic Parameter Values	35
2.9	Delay Model	37
2.10	Input Delays	38
2.11	Configurations and Timing	46
2.12	VHDL Test Bench	48
2.13	RSFF Test Bench	49
3.1	Logic Diagram 2-Input Positive-Nand	58
3.2	Logic Diagram Open Collector 2-Input Positive Nand	60
3.3	Logic Diagram 2-Input Positive-Nor	62
3.4	Logic Diagram Inverter	65
3.5	Logic Diagram Open-Collector Inverter	67
3.6	Logic Diagram 3-Input Positive-And	68
3.7	Logic Diagram 3-Input Positive-Nand	71
3.8	Logic Diagram 2-Input Positive-Or	74
3.9	Logic Diagram 2-Input Positive-Xor	76
3.10	Logic Diagram 3 to 8 Decoder/Multiplexer	80
3.11	Logic Diagram 2 to 4 Decoder/Multiplexer	84
3.12	Logic Diagram 1 of 8 Selector/Multiplexer	87
3.13	Logic Diagram 1 of 4 Selector/Multiplexer	92
3.14	Logic Diagram 1 of 2 Selector/Multiplexer	96

3.15	Resistor strength drop	102
3.16	Capacitor application	103
3.17	Logic Diagram ALU/Function Generator	109
3.18	Logic Diagram One Shot	120
3.19	Logic Diagram 4 Bit Magnitude Comparator	124
3.20	Logic Diagram 9 Bit Odd/Even Parity Generator/Checker	130
4.1	Logic Diagram D Pos Edge Flip-Flop with Preset/Clear	136
4.2	Logic Diagram JK Pos Edge Flip-Flop with Preset/Clear	141
4.3	Logic Diagram JK Neg Edge Flip-Flop with Preset/Clear	147
4.4	Logic Diagram JK Neg Edge Flip-Flop with Preset	153
4.5	Logic Diagram 4 Bit Parallel-Access Shift Register	159
4.6	Logic Diagram 3 to 8 Decoder/Demultiplexer	165
4.7	Logic Diagram 3 to 8 Decoder/Demultiplexer with Latch	171
4.8	Logic Diagram 8 Bit Parallel-Out Serial Shift Register	177
4.9	Logic Diagram 8 Bit Parallel Load Shift Register	183
4.10	Logic Diagram 8 Bit Parallel Load Shift Register with Clear	189
4.11	Logic Diagram 4 Bit Decade Counter with Async Clear	196
4.12	Logic Diagram 4 Bit Binary Counter with Async Clear	202
4.13	Logic Diagram 4 Bit Decade Counter	209
4.14	Logic Diagram 4 Bit Sync Binary Counter	215
4.15	Logic Diagram Bit Up/Down Sync Decade Counter	221
5.1	Logic Diagram 1024 Bit ROM	232
5.2	Logic Diagram 16384 Bit PROM	236
5.3	Logic Diagram 64 Bit RAM	243
5.4	PAL16L8 Block Diagram	258
5.5	PAL16R6 Block Diagram	262
5.6	PAL16R8 Block Diagram	267
6.1	Bus Functional Model	270
6.2	TDC1028 Block Diagram	272
6.3	Capacitive Loading for Fujitsu MB86901	291
6.4	Model Development Time	304
6.5	Setup/Hold Time	306
6.6	Hardware Model Testing	307

Preface

This book is intended to be a working reference for electronic hardware designers who are interested in writing VHDL models. A handbook/cookbook approach is taken, with many complete examples used to illustrate the features of the VHDL language and to provide insight into how particular classes of hardware devices can be modelled in VHDL. It is possible to use these models directly or to adapt them to similar problems with minimal effort.

This book is not intended to be a complete reference manual for the VHDL language. It is possible to begin writing VHDL models with little background in VHDL by copying examples from the book and adapting them to particular problems. Some exposure to the VHDL language prior to using this book is recommended. The reader is assumed to have a solid hardware design background, preferably with some simulation experience. For the reader who is interested in getting a complete overview of the VHDL language, the following publications are recommended reading:

- *An Introduction to VHDL: Hardware Description and Design* [LIP89]
- *IEEE Standard VHDL Language Reference Manual* [IEEE87]
- *Chip-Level Behavioral Modelling* [ARMS88]
- *Multi-Level Simulation of VLSI Systems* [COEL87]

Other references of interest are [USG88], [DOD88] and [CLSI87].

Use of the Book

If the reader is familiar with VHDL, the models described in chapters 3 through 7 can be applied directly to design problems. A review of chapters 1 and 2 will give a clear overview of the methodology and intended scope of the models described in this book. All other chapters can be used in a reference fashion, reading only those sections which discuss the specific type of device for which a VHDL model is desired.

Chapter 1 describes the history and background of the VHDL language and provides recommendations on how to get started quickly with VHDL.

Chapter 2 discusses the overall format of a VHDL file and the particular methodology used in models described in this book.

Chapter 3 describes the VHDL models for a number of combinational devices. The reader is encouraged to review this chapter before studying other more complex device models.

Chapter 4 describes the VHDL models code for a wide variety of sequential devices including registers, flip-flops, counters and others.

Chapter 5 discusses memory devices and includes the VHDL source code for devices ranging from RAMs and ROMS to PLAs.

Chapter 6 discusses the VHDL modelling of complex devices such as microprocessors. Examples are given to highlight modelling techniques. Recommendations regarding proper methodologies for complex devices are given.

Finally, chapter 7 gives the complete VHDL source code for the standard logic package and discusses how to build a logic value system, bus resolution functions, logic modelling utilities and timing utilities.

Getting Started Quickly with VHDL

All of the examples in this book are written using strictly standard VHDL which adheres to the IEEE standard 1076-1987 [IEEE87]. The modelling features and approaches used in the VHDL code in this book have been tested against both the Vantage and Intermetrics tools. Other tools which adhere to the VHDL standard should handle the examples without difficulty (assuming that the required VHDL features are available). Virtually all of the examples in this book rely on the availability of the Standard Logic Package described in chapter 7. The easiest and fastest way to get started writing and simulating with VHDL models discussed in this book is to get a copy of this standard package. This package is in the public domain and is not specific to any particular VHDL environment. Once compiled in your VHDL environment, you can reference this package from your models and begin creating models which follow the conventions and approaches discussed exactly as shown or with your own modifications incorporated. By following this approach it should be possible to begin writing VHDL models which simulate effectively within an afternoon. In order to facilitate this process, all examples including the Standard Logic Package discussed in this book are available on IBM PC floppy format for MS-DOS. If you are interested in getting on-line VHDL source code for the examples in this book send your request to Coelho Publications, 43000 Christy Street, Fremont, CA 94538-3167.

Background on the Writing of this Book

All examples cited in this book have been validated using the Vantage-Spreadsheet simulation environment [VANU89] [VANT89] developed and marketed by Vantage Analysis Systems, Inc. The text for this book was prepared using the TEX text formatting system [LAM86] [KNU86] developed by Donald Knuth at Stanford University running on an Apollo workstation with camera ready copy produced on an Apple Laserwriter. Graphics figures were created using the Vantage environment for simulation results, and AutoCAD [AUT88] for schematic and free form graphics.

Acknowledgments

Bill Billowitch provided insightful input into techniques for modelling complex devices and is largely responsible for the discussions regarding the modelling of microprocessors. Bill also contributed heavily in review of the manuscript. I am greatful for Bill's substantial contribution.

I am indebted to Ken Scott, Doug Perry, Andy Tsay and Alec Stanculescu all of whom contributed to the development and evolution of the standard logic package and the switch level modelling techniques. Thanks also goes to others who reviewed the manuscript including Rick Lazansky, Paul Krol, Tom Miller, Al Dewey, John Hines, John Van Tassel, Karen Serafino, John Evans and David Hemmendinger.

The VHDL Handbook

Chapter 1

Introduction

1.1 Introduction to the VHDL Language

VHDL (VHSIC Hardware Description Language) is rapidly emerging as one of the most important electronic design languages in both the commercial and military electronic design areas. This section will summarize the history of the language and discuss its applicability to hardware design problems.

1.1.1 History of VHDL

Initiated in the early 1980's by the Department of Defense, version 7.2 of the language was released for public review in August 1985. The primary requirement for this first version of the language was to provide a vehicle for improved documentation of electronic systems delivered to the Government. Current estimates indicate that over 50 percent of all procurement dollars spent on electronic systems are allocated to on-going life cycle costs including maintenance, second sourcing and upgrading of equipment. Projections indicate that by the early 90's this percentage will grow to 75 percent [COELA88]. VHDL's primary design requirement was to provide a tool which will reduce the on-going costs of operating military electronic systems. In the context of the billions of dollars being spent upgrading and maintaining existing electronic systems, the $20 million the government has spent to date developing and promoting VHDL will pay for itself rapidly. More importantly, there are clear economic benefits to the government for spending more funds up-front during the design of electronic systems in order to get VHDL documentation which will facilitate reduced on-going life cycle costs once an electronic system is installed. Many electronic systems have a life in excess of 20 years, making this motivation even more compelling.

VHDL has been compared to ADA and in fact has many of the same goals as the ADA language. VHDL evolution however has differed from ADA and as a result is establishing itself as an industry standard language much more rapidly. From the inception of the VHDL program, the DOD acknowledged the weaknesses in the ADA standardization approach and took steps which have resulted in a strong language. VHDL has benefited from substantial industry review and participation throughout its development. In February 1986 all rights to the VHDL language were transferred to the IEEE as part of its industry wide hardware description language standardization effort. As a result, the original version 7.2 language went through a phase of substantial changes driven primarily by industry representation from CAE vendors as well as users. The goals of the IEEE in developing a standard language were broader in scope. Driven by the need for a production quality language suitable as a design tool to be used throughout the design cycle, the IEEE enhanced the language substantially. This effort resulted in the creation of a world-class hardware description language which was ratified in December 1987 as the first industry standard hardware description language with overwhelming support from industry. One important reason for the strong support VHDL has received throughout industry is that unlike ADA, there are no other standard languages which VHDL must compete against, while at the same time industry has been demanding industry standards as evidenced by the growing support for other standards such as Unix and EDIF (Electronic Design Interchange Format).

1.1.2 DOD Requirements and VHDL

The IEEE VHDL standard, designated officially as IEEE 1076-1987 is a required deliverable for all military electronics contracts which include ASICs. The specific government requirements are spelled out in the VHDL Data Item Description (DID) which is referenced by military standard 454. This standard requires that every physical implementation level of an electronic system be documented both structurally (netlist) and behaviorally. Although the current military requirements applies primarily to ASIC's, future CALS requirements will apply to boards and entire systems. In other words, a schematic and a high-level behavioral model will be required for the entire system, for each board in the system, for every chip in the system, and possibly for each macro used in chips in the system. In addition, each VHDL entity (netlist or behavioral model) must have an associated test bench, a VHDL description which thoroughly tests the VHDL in order to insure proper simulation of the design as shown in figure 1.1. This top-down approach to documenting electronic systems has significant

1.1. INTRODUCTION TO THE VHDL LANGUAGE

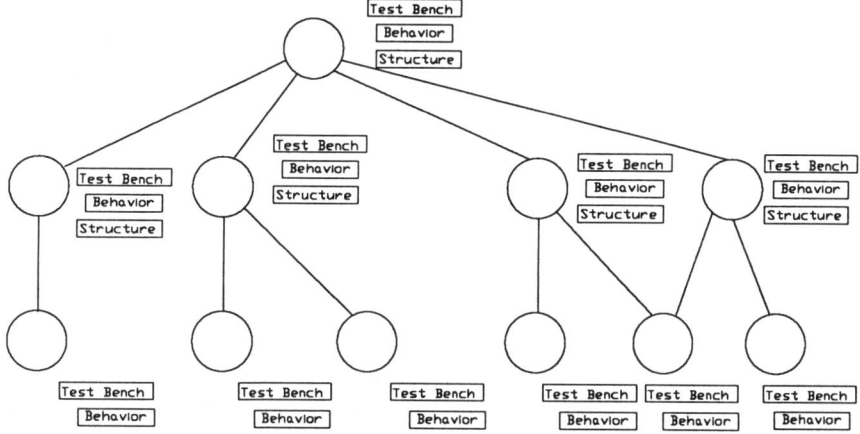

Figure 1.1: DOD/VHDL Documentation Requirements

implications to the design methodology currently used by many military contractors. It will be no longer acceptable to provide the government with low-level netlists or schematics of electronic systems. High-level behavioral descriptions which capture the intent of the design in a technology independent fashion will be required. In most cases, the most cost effective way to address this new requirement will be to incorporate VHDL into the design cycle from the start.

Recently, the number of contracts with VHDL requirements have increased dramatically. Total contracts with VHDL which are awarded or in the procurement phase currently exceed $3 billion. Over time, military contractors can expect VHDL to be used even during the procurement process. The DOD has expressed clear intent to solicit VHDL descriptions as part of contract bids, and to perform simulation on alternative solutions in order to evaluate the best technical approaches prior to contract award. In addition, the Defense Electronics Supply Center (DESC), traditionally responsible for controlling military spec devices, is gearing up for management of military standard VHDL descriptions for devices which are used in military systems.

1.1.3 VHDL As a Design Tool

One of the primary criticisms of the VHDL language is the lack of tools, especially production quality simulators. In the past year the VHDL language has moved from having only limited tool support to being by far the most heavily supported language from a multi-vendor standpoint of any language available. Virtually every major simulator vendor in the CAE business is evaluating their VHDL strategy, and a number have announced their intentions to support the language. More importantly, a wide variety of other tools will also support the language including the coming generation of hardware synthesis tools.

An important gauge of the success of a hardware description language is its effectiveness as a design tool. Of prime importance is the fact that the VHDL language definition is independent of the tools which support it. Virtually every other language available today is tied closely with the tool which supports it, and in fact in most cases the definitions for these languages are ambiguous without the tool being available for interpretation. VHDL supports a wide spectrum of tools from simulation to synthesis to layout. The language has a concise specification and is controlled by the IEEE.

VHDL provides a wide range of abstraction levels from the architecture level down to the gate level. More importantly, VHDL supports the easy mixing of widely varying levels of abstraction during simulation, making it possible to adopt a true top-down design style. In addition, the VHDL language has exceptional facilities for modelling hardware timing. Min/max timing, setup, hold and spike detection facilities are all supported. In addition, VHDL provides a mechanism for building generic models which can be back-annotated with timing for specific devices. From a library standpoint, this is a significant advantage since it allows a single model (such as an AND gate) to represent any manufacturers device with no change to the model. This makes construction of VHDL model libraries significantly easier than other languages.

Another key feature in the VHDL language is its layered characteristic. Unlike ADA, which is a difficult language to learn because of its complexity, VHDL can be used initially in a subset fashion resulting in a rapid learning curve for the language. Experience with VHDL indicates that a hardware designer can come up to speed with the most important features of the language and be effectively modelling a wide variety of electronic devices within 3 weeks of opening the manual. Over time, the designer can expand his use of the more advanced VHDL features by peeling the VHDL feature onion. The beauty of the language is that it allows a hardware designer to use an easy to learn subset without being forced to learn the more complex (and lesser used) features of the language. At the same time, the language has sufficient power to solve even the most complex modelling problems including describing VLSI chips or entire electronic systems. In this context, VHDL provides the best of the characteristics of the simple but easy to learn HDL's, while still allowing the advanced modeller the flexibility he needs to solve the most demanding problems.

In summary, VHDL is clearly one of the most important hardware description languages to emerge in the past decade. As a design tool, VHDL has significant merit and because of its industry standard status provides a much lower risk alternative to proprietary languages. Of prime importance is the rapid emergence of a wide variety of production quality tools which support VHDL from many different vendors. For the military contractor, VHDL is unquestionably the most important HDL. Military/aerospace companies that move quickly in adopting VHDL for design work have an excellent opportunity to establish a significant competitive advantage over other companies in bidding on military procurements; companies that ignore the VHDL language will rapidly find that they are at a severe disadvantage competitively in contract bids.

1.2 Multi-Level Design

Traditionally, electronic design has been performed at the gate-level using standard off the shelf components. The complexity of hardware designs has grown significantly. The building blocks in typical systems are now microprocessors or ASICs which represent thousands of gates in complexity. As a result, the traditional bottom-up design methodologies are giving way to hierarchical design practices which make management of this complexity possible. An effective approach is to incorporate a hierarchical hardware description language such as VHDL into the design process. By utilizing VHDL as a specification tool it is possible to begin simulation of complex systems before details regarding the implementation are fully specified. In addition, VHDL facilitates the top-down design process where a higher level specification is developed first, debugged and ultimately used to judge the correctness of the next lower level implementation.

Pure top-down design methodologies are rarely followed due to the pressures of limited resources and schedules and because in many cases a pure top-down approach is less effective than a mixture of design styles. A common approach is for a design group to employ several high-level system architects who code the overall system specifications in VHDL and provide these VHDL models to the logic designers in the team for logic or standard component level implementation. This approach can be viewed as a combination of top-down methodology (through the use of VHDL specifications) and bottom-up methodology (during the implementation phase for each major subsystem in the design) and is effective. Other methodologies are possible including a middle-out approach where prototype approaches are used to drive both the high-level specification of a system and to aid in low-level design. In a real design situation, all of these approaches are utilized.

In a typical design environment, VHDL is used in three distinct fashions as summarized here:

- **High-level specification**; use of abstract data types for signals; combined use of schematics and VHDL

- **Logic/standard component level design**; use of a specific logic level data type for signals; primary design representation is the schematic diagram

- **Standard component model library support**; use of a specific logic level data type for signals; all models represented in VHDL

During the high-level specification phase, VHDL is used as an architectural tool to aid in abstract analysis and design of a system. In this process, virtually all of the features of VHDL including the data abstraction mechanisms (such as user defined signal data types) will generally be utilized. This level of design is most commonly performed by system architects and is often performed without detailed knowledge or concern for specific implementation details.

During the logic/standard component level of design, the actual implementation of a design is determined. Generally this level of design is performed in a structural manner, most commonly through the use of schematic entry tools. In many environments, this level of design will not utilize VHDL except through the utilization of pre-built VHDL model libraries which correspond to the devices used in the schematics.

Finally, the basic building blocks of a design are coded in VHDL. Generally, these models are provided to the design team in a pre-built and pre-compiled format by the simulation environment. Alternatively, some design groups will have a captive library development group which builds additional VHDL model libraries. Also, many semiconductor manufacturers are beginning to develop VHDL model libraries which correspond to devices which they sell.

This book is primarily concerned with the development of models which represent standard components and devices. Many of the aspects of high-level design and specification of a system using VHDL are beyond the scope of this book. In particular, most of the examples in this book deal primarily with a single standard logic value system which represents signals at a level which is fairly close to the actual implementation level. As a result, this book provides techniques and examples of models which cover a wide range of devices from SSI through VLSI, but modelled in a fashion as to be plug-compatible at the pin level with the actual device. The intent of this book is to provide the user with sufficient knowledge to allow the creation of accurate VHDL models which represent the functioning of an actual standard component or hardware device.

For the reader interested in higher level design with VHDL, the reader is referred to several excellent publications discussed in the introductory section of this book.

1.3 The Model Accuracy Continuum

Electronic simulation is just that, *simulation*. What this means is that a simulator will never predict the exact behavior of an electronic design, nor would this be desirable. The value which a simulator provides to the designer is the ability to provide an abstract prediction of how an electronic system will behave. The abstraction is important because the designer can easily be swamped with data which is not relevant to the design process or the debugging of his design.

One of the most difficult aspects of electronic simulation for a hardware designer to grasp is the fact that a continuum of accuracy exists and often the choice of how much accuracy the simulator will provide is under the control of the designer. Figure 1.2 shows the continuum over which simulation results can be produced.

Figure 1.2: Accuracy/Speed/Abstraction Continuum

On the one extreme, an abstract simulation which provides limited detail regarding the functioning of the hardware is possible. For this level of simulation, software simulator run-time speeds are fast and often exceed those of dedicated hardware accelerators running at the gate level. This type of simulation is often associated with high-level architectural work. For example, the system architect may choose to model the data bus in his design as an integer value. Clearly, this abstraction has no correspondence to the actual hardware (until he specifies how many bits and what order those bits will be used to represent the data value) and merely represents a higher level conceptual view of the system, but this may be just what the designer wishes in the early stage of a design where architectural choices are being studied.

On the other extreme, an analog level simulation can be performed which models the current and charge storage effects of the design with a high degree of accuracy. The designer is flooded with detailed information, and the simulator performance is 100 to 10,000 times slower than more abstract levels of simulation. In fact, at this level only simple circuits are cost-effective for simulation.

Although less obvious, even during gate level simulation a continuum exists as shown below

Logic Modelling Accuracy Continuum

Less Accurate	More Accurate
unit delay	technology dependent delay
0/1 states	0/1/X/U states, strengths
no error checking	setup, hold, spike checking

The simplest approach to modelling a logic device is to view it as a pure binary transform with no variations in timing. In this view, false and true values are propagated with unit delay through each model. In addition, the simplest form of this type of model would have no error checking. For many design applications, this view of a design can be adequate and in fact many hardware acceleration products utilize approaches which are similar to this. The benefit to the user of this level of simplicity is that the data fed to the simulator and the data generated by the simulator is easy to interpret. Simulation will tend to be efficient as well.

Most simulators and logic models tend to have more sophisticated modelling schemes. Designers generally require feedback regarding the timing problems in a design and for this reason models must process technology dependent delays which can include proper handling of load related delays, layout specific delays derived from wire-lengths and their associated capacitive and resistive properties, and propagation delays which can be a function of the load and the state transition. In order to adequately handle charge storage effects in transistor level circuits a strength system is required in addition to the conventional logic states. Both an unknown

1.3. THE MODEL ACCURACY CONTINUUM

and an uninitialized state are required to effectively provide circuit startup and accurately propagate ambiguous data through a circuit. Finally, it is highly desirable for logic models to detect other failures or errors in a design including setup constraint violations, hold constraint violations and spike detection.

In the examples which will follow (see section 2.5), several alternative models will be given for each device type, each covering a different point on the accuracy continuum for logic models. No one example is universally the best model. The designer needs to evaluate his particular needs and choose the model which is most appropriate for his application. If in doubt, often choosing a middle ground which provides the bulk of the error checking, unknown and uninitialized handling and propagation delays can be effective. For ASIC design, accurate timing models will be required with special emphasis on load dependent and technology dependent timing.

Chapter 2

Anatomy of a VHDL Model

2.1 Describing Electronic Hardware in VHDL

In order to simulate electronic hardware designs, the following information is required:

- Structural description of the design (netlist or schematic)

- Behavioral model for each device in the design (VHDL source or model library)

- Stimulus for the design (test vectors)

- Design configuration information (specify which version of each device model to use during simulation)

Each of these pieces of information will be highlighted in the following example.

Figure 2.1 shows the schematic diagram for an electronic circuit. Three nand gates perform the and-or operation for four inputs as shown. This schematic describes the *structure* by specifying the building blocks the design utilizes, and the interconnection between these building blocks and the outside world. The most common way of representing design structure is through the use of schematic diagrams, and typically these diagrams would be entered into a simulation environment via a schematic editor. An alternative is to express this information in the VHDL language using the structural description facilities available in VHDL as shown below

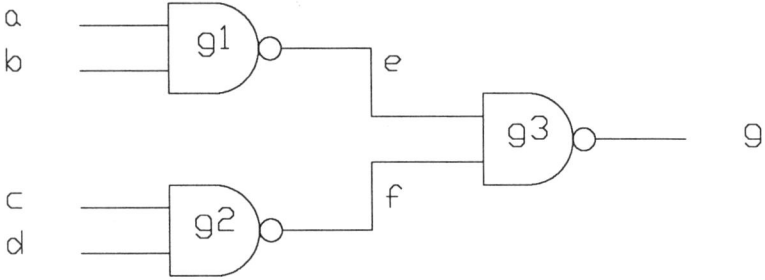

Figure 2.1: Structure of Design as Schematic

Structure of Design in VHDL

```
ENTITY simple IS
    PORT    (a,b,c,d: IN BIT; g: OUT BIT);
END simple;

ARCHITECTURE structural OF simple IS
    COMPONENT nand_gate PORT (a,b:IN BIT; y: OUT BIT); END COMPONENT;
    SIGNAL e,f: BIT;
BEGIN
    g1:nand_gate PORT MAP(a,b,e);
    g2:nand_gate PORT MAP(c,d,f);
    g3:nand_gate PORT MAP(e,f,g);
END structural;
```

The external ports of this circuit are defined in the entity declaration and indicate four input ports called **a**, **b**, **c** and **d** and one output port called **g**.

The architecture section describes the VHDL structural information where three gates **g1**, **g2** and **g3** are declared, and the interconnection between the ports on these gates and the external ports of the **simple** circuit and the internal signals **e** and **f** is specified. This VHDL code corresponds exactly to the schematic representation given in figure 2.1. Often this type of input is referred to as a netlist. Most of the examples in this book assume that the structure for circuits will be entered via a schematic editor and in most cases the schematic diagram will be shown in an associated figure.

The structure shown in figure 2.1 is not sufficient to begin simulation. Equally important, each building block in the design must have a *behavior* associated with it. A behavioral description can be thought of as information that tells the simulator how a given building block in a design should react to all possible inputs that it sees. Each black box in a design can be thought of as a transfer function, with inputs and state determining the

2.1. DESCRIBING ELECTRONIC HARDWARE IN VHDL

outputs. The following shows the VHDL code for the nand gate found in the schematic diagram shown in figure 2.1.

Behavior of Design

```
ENTITY nand_gate IS
    PORT    (a,b: IN BIT; y: OUT BIT);
END nand_gate;

ARCHITECTURE behavioral OF nand_gate IS
BEGIN
    y <= a NAND b AFTER 1 ns;
END behavioral;
```

In this case, the entity declares the ports of the device as **a**, **b** (inputs) and **y** (output). The architecture defines the behavior of the nand gate. The behavior of a device in a circuit will always be described using the VHDL language. The bulk of this book presents examples and approaches to writing these behavioral VHDL models for a wide variety of devices.

In order to perform simulation, stimulus to the circuit must be specified. Typically stimulus will be described as a time based sequence of true and false signal values applied to the various input ports of the circuit. The format of stimulus varies from one simulation environment to another. The following shows an example of a stimulus model specified in VHDL suitable for stimulating the circuit shown in figure 2.1.

Circuit Stimulus

```
ENTITY nand_stim IS
    PORT    (a,b,c,d: OUT BIT; g: IN BIT);
END nand_stim;

ARCHITECTURE behavioral OF nand_stim IS
BEGIN
    a <= '0' AFTER 0 ns,    '1' AFTER 80 ns;

    b <= '0' AFTER 0 ns,    '1' AFTER 40 ns,
         '0' AFTER 80 ns,   '1' AFTER 120 ns;

    c <= '0' AFTER 0 ns,    '1' AFTER 20 ns,
         '0' AFTER 40 ns,   '1' AFTER 60 ns,
         '0' AFTER 80 ns,   '1' AFTER 100 ns,
         '0' AFTER 120 ns,  '1' AFTER 140 ns;

    d <= '0' AFTER 0 ns,    '1' AFTER 10 ns,
         '0' AFTER 20 ns,   '1' AFTER 30 ns,
         '0' AFTER 40 ns,   '1' AFTER 50 ns,
```

```
            '0' AFTER  60 ns,  '1' AFTER  70 ns,
            '0' AFTER  80 ns,  '1' AFTER  90 ns,
            '0' AFTER 100 ns,  '1' AFTER 110 ns,
            '0' AFTER 120 ns,  '1' AFTER 130 ns,
            '0' AFTER 140 ns,  '1' AFTER 150 ns;
END behavioral;
```

For each input port on the **simple** circuit, a stimulus pattern is specified. For example, for port **a** the value false is applied at time 0 and the value true is applied at time 80. All stimulus and simulation results are shown graphically in the following examples.

Finally, VHDL has built-in facilities for specifying design management information. In particular, VHDL allows the designer to specify the configuration of the circuit he wants to simulate. This configuration information determines what level of hierarchy will be simulated, and if more than one alternative version or style of model is available for a given device, then the configuration information determines which will be used during simulation. The following shows the configuration definition for the **simple** circuit of figure 2.1.

Design Configuration

```
CONFIGURATION parts OF simple IS
FOR structural
    FOR g1,g2,g3:nand_gate USE ENTITY work.nand_gate(behavioral);
    END FOR;
END FOR;
END parts;
```

In this example, the three gates **g1**, **g2** and **g3** are associated with the **nand_gate** model and the specific version named **behavioral** is used.

When all of the above inputs are specified, the user can proceed with simulation[1] Figure 2.2 shows the simulations results for the circuit of figure 2.1 when stimulated with the data shown above.

2.2 A VHDL File

Each of the different pieces of information discussed in section 2.1 must be combined into a VHDL file or files and compiled in proper sequence in order

---

[1] This is not totally true in this example since a top level test bench circuit is required to connect the circuit under study with the stimulus generator. See section 2.7 for details on how to create the required descriptions.

2.2. A VHDL FILE

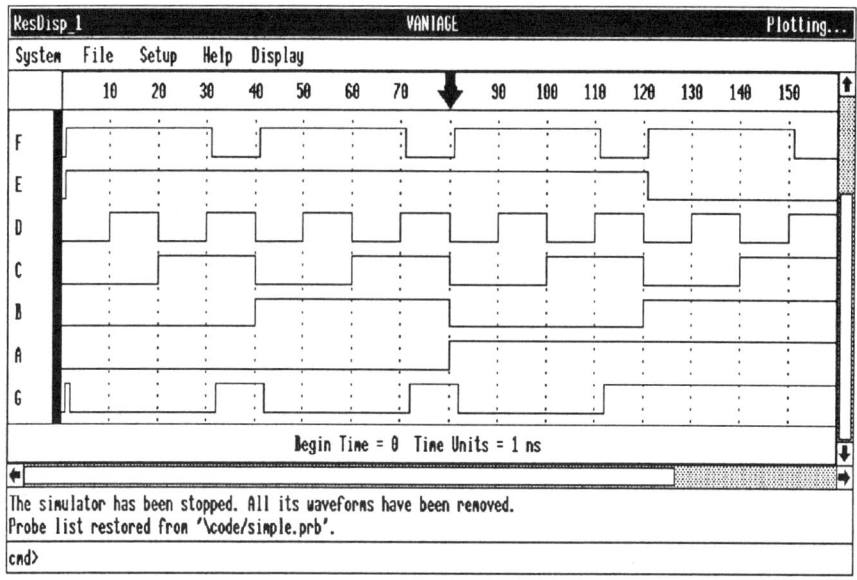

Figure 2.2: Circuit Simulation Results

to perform simulation. This section will outline the sequence and overall format of a VHDL file.

The following shows the declarations for a VHDL file that describes a complete circuit suitable for simulation.

VHDL File Example

```
-- ***** nand gate model *****
-- external ports
ENTITY nand_gate IS
    PORT    (a,b: IN BIT; y: OUT BIT);
END nand_gate;

-- internal behavior
ARCHITECTURE behavioral OF nand_gate IS
BEGIN
    y <= a NAND b AFTER 1 ns;
END behavioral;

-- ***** buffer model *****
-- external ports
ENTITY buf_gate IS
    PORT    (a: IN BIT; y: OUT BIT);
END buf_gate;

-- internal behavior
ARCHITECTURE behavioral OF buf_gate IS
BEGIN
    y <= a AFTER 1 ns;
END behavioral;

-- ***** rs flip flop model *****
-- external ports
ENTITY rsff IS
    PORT    (r,s: IN BIT; q: OUT BIT);
END rsff;

-- internal structure
ARCHITECTURE structural OF rsff IS
    -- component types to use
    COMPONENT nand_gate
        PORT (a,b:IN BIT; y: OUT BIT); END COMPONENT;
    COMPONENT buf_gate
        PORT (a:IN BIT; y: OUT BIT); END COMPONENT;

    -- internal signals
    SIGNAL qn,qb: BIT;
BEGIN
    -- 1 buffer named b1, connected to signals qb and q
    b1:buf_gate PORT MAP(qb,q);
```

2.2. A VHDL FILE

```
    -- 2 gates, named g1 and g2
    -- g1 connected to signals s, qn and qb
    g1:nand_gate PORT MAP(s,qn,qb);
    -- g2 connected to signals qb, r and qn
    g2:nand_gate PORT MAP(qb,r,qn);
END structural;

-- design management/configuration
CONFIGURATION parts OF rsff IS
FOR structural
    -- use behavioral architecture for gates g1 and g2
    FOR g1,g2:nand_gate
        USE ENTITY work.nand_gate(behavioral); END FOR;

    -- use behavioral architecture for buffer b1
    FOR b1:buf_gate
        USE ENTITY work.buf_gate(behavioral); END FOR;
END FOR;
END parts;
```

The first part of the file specifies the **nand_gate** device. Both the interface specification and the internal behavior are shown. The entity declaration for this device defines two input ports **a** and **b** and one output port **y**. The architecture declaration defines a behavior named **behavioral** that has a transfer function operation in which the nand function of the two inputs is propagated to the output with a delay of 1 nanosecond. Typically, each device will have one entity declaration, and can have many architecture definitions. At least one architecture definition is required in order for simulation to proceed. In VHDL the following summarizes how each of these types of descriptions is used:

- Entity declaration - defines the black box (outside) characteristics of a device

- Architecture declaration - defines the operational (internal) characteristics of a device

- Configuration declaration - used for configuration management; extremely useful for transforming generic models into specific manufacturer specific models, and also for back-annotation of detailed timing data from layout.

Some implementations of VHDL require a specific order of compilation for these declarations. It is best to order them as described above.

Similar information for the **buf_gate** device is shown, with both the entity declaration and architecture definition being specified.

The actual circuit under study is specified in the **rsff** model. This is a structural model of a set/reset flip-flop. Figure 2.3 shows the corresponding

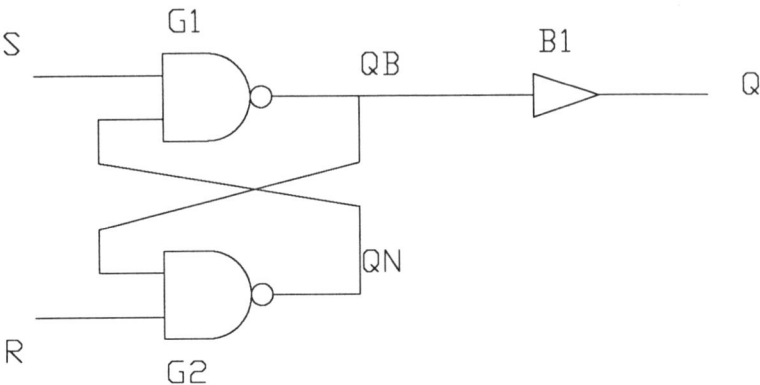

Figure 2.3: RSFF Structure

schematic diagram for this model. The interface to this circuit is specified in the entity declaration with two input ports **r** and **s** and one output port **q**. The architecture for this circuit is called **structural** and identifies two devices **nand_gate** and **buf_gate**. Two internal signals **qn** and **qb** are declared. Finally, the circuit structure is shown with 3 components **b1** (a buf_gate), **g1** and **g2** (nand_gate devices). Each component is connected to the internal signals and the external ports of the circuit as indicated in the port map declarations. The **b1** component is connected to signal **qb** on port **a** and to signal **q** on port **y**. Similarly, component **g1** is connected to signal **s** on port **a**, to signal **qn** on port **b** and to signal **qb** on port **y**.

Finally, the configuration body of the **rsff** circuit with the name **parts** is specified. The components **g1** and **g2** are associated with the **behavioral** architecture of entity **nand_gate** and the **b1** component is associated with the **behavioral** architecture of entity **buf_gate**. Architectures allow specification of a given version of the model. This facility becomes essential when more than one architecture exists for a given entity or when technology dependent timing is associated with a generic model. In certain cases, some VHDL implementations provide for automatic generation of default configuration bodies. In these cases, it is possible to skip the step of creating a configuration for a given model.

Figure 2.4 shows the simulations results for this circuit using the stimulus as shown below:

RSFF Stimulus

```
ENTITY rsff_stim IS
    PORT   (r,s: OUT BIT; q: IN BIT);
```

2.2. A VHDL FILE

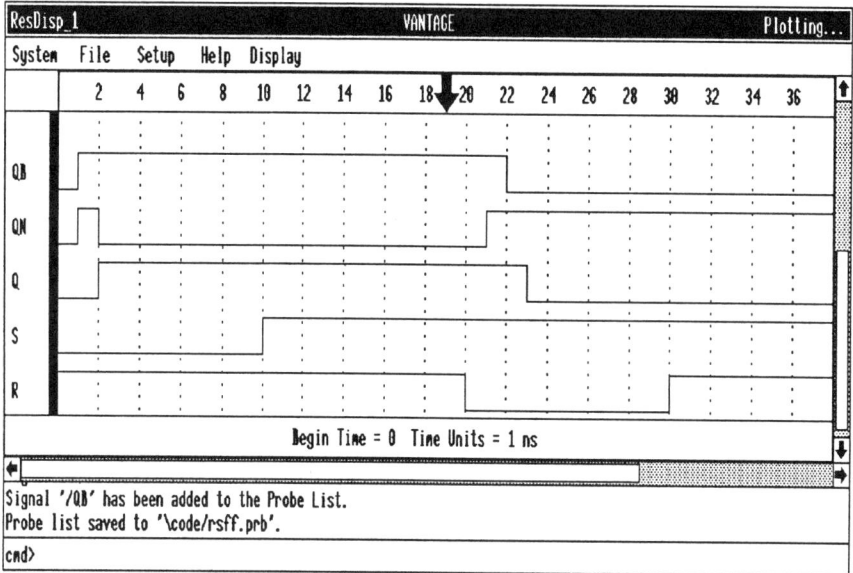

Figure 2.4: RSFF Simulation Results

```
END rsff_stim;

ARCHITECTURE behavioral OF rsff_stim IS
BEGIN
    r <= '1' AFTER 0 ns,
         '1' AFTER 10 ns,
         '0' AFTER 20 ns,
         '1' AFTER 30 ns;

    s <= '0' AFTER 0 ns,
         '1' AFTER 10 ns,
         '1' AFTER 20 ns,
         '1' AFTER 30 ns;
END behavioral;
```

as described in VHDL. A test bench was created to connect the stimulus model **rsff_stim** shown above with the circuit **rsff** shown above. This test bench is shown below

RSFF Test Bench

```
ENTITY rsff_bench IS
END rsff_bench;
```

```
ARCHITECTURE structural OF rsff_bench IS
    COMPONENT rsff_stim PORT (r,s: OUT BIT; q: IN BIT); END COMPONENT;
    COMPONENT rsff PORT (r,s: IN BIT; q: OUT BIT); END COMPONENT;
    SIGNAL r,s,q: BIT;
BEGIN
    generator:rsff_stim PORT MAP(r,s,q);
    circuit:rsff PORT MAP(r,s,q);
END structural;

CONFIGURATION parts OF rsff_bench IS
FOR structural
    FOR generator:rsff_stim USE ENTITY work.rsff_stim(behavioral);
    END FOR;
    FOR circuit:rsff USE ENTITY work.rsff(structural);
      FOR structural
        FOR g1,g2:nand_gate USE ENTITY work.nand_gate(behavioral); END FOR;
        FOR b1:buf_gate USE ENTITY work.buf_gate(behavioral); END FOR;
      END FOR;
    END FOR;
END FOR;
END parts;
```

with no external ports. Three internal signals **r**, **s** and **q** are used to connect the test bench **generator** pin for pin to the test circuit **circuit**. The configuration body specifies which architectures of the two components to use. In this case the **behavioral** architecture of the **generator** component (see entity **rsff_stim** is used. The **structural** architecture of the **circuit** component for entity **rsff** is used. Both of the **nand_gate** components g1 and g2 utilize the **behavioral** architectures. And finally, the **buf_gate** component **b1** utilizes the **behavioral** architecture.

2.3 The Standard Logic Package

Chapter 7 shows the complete definition for the standard logic package. This package gives the hardware designer a pre-defined modelling environment that makes the modelling of hardware devices easy and convenient. In order to access this package, the use statement must be provided. The following shows the use of the standard logic package for a VHDL model entity declaration.

Using the Standard Logic Package

```
USE std.std_logic.all;
USE std.std_ttl.all;

ENTITY jkff IS
```

```
    PORT    (j,k,clr,clk: IN t_wlogic; q,qb: INOUT t_wlogic);
END jkff;
```

In this case, the first line brings in the technology independent aspects of the standard logic package and the second line brings in the TTL technology dependent aspects of the package. The entity declaration for **jkff** specifies a number of input and output ports all of type **t_wlogic** (the standard logic 46 value system as described in section 7.2). In most cases, the complexity of this value system is completely hidden from the user because of the features of the logic package, but in general the full power of the 46 value system is available making it possible for the hardware designer to handle a wide variety of special case modelling situations including circuit initialization, wired gate processing, modelling of busses, and switch level device modelling with charge storage effects.

The bulk of the examples given in this book assume the availability of the standard logic package as defined in chapter 7. Where ambiguous the use statement will be incorporated in the VHDL examples given, but in most cases the inclusion of this declaration is assumed. For a complete overview of the facilities and operation of the standard logic package, the reader should review chapter 7.

2.4 User Defined Packages

One of the most powerful features of the VHDL language is the facility it provides for building separately compiled packages of predefined functions and declarations.

This facility makes it possible for the hardware designer to extend the VHDL modelling environment. As a result, VHDL models are easier to write and are more readable. Pre-built facilities can be created and supported by users who are experts in VHDL and provided to less experienced VHDL users without requiring that the less experienced users know the details of the packages. The following shows a VHDL file that declares a package.

A User Defined VHDL Package

```
USE STD.std_ttl.ALL;
PACKAGE utility IS

    type t_nibble is array ( 0 to 3 ) of t_wlogic;

    FUNCTION f_logictoint(
             SIGNAL s : IN t_nibble )
```

```
        RETURN integer;
END utility;

USE STD.std_ttl.ALL;
PACKAGE BODY utility IS
    FUNCTION f_logictoint(
                SIGNAL s : IN t_nibble )
    RETURN integer IS
        VARIABLE work : integer := 0;
    BEGIN
        FOR i IN t_nibble'RANGE LOOP
            IF s(i) = '1' THEN
                work := work + 2 ** i;
            END IF;
        END LOOP;
        RETURN work;
    END;
END utility;
```

The first part of this file shows the package declaration. The second part of this file shows the package body. This package defines a data type **t_nibble** which is an array with four elements each of the standard logic data type. A function **f_logictoint** is defined that takes as an input four logic values and returns the integer value represented.

The following describes the entity declaration for the **check_edge** component.

Entity declaration for check_edge

```
USE std.std_logic.all;
USE std.std_ttl.all;
USE work.utility.all;

ENTITY check_edge IS
    PORT    ( clk:          IN  t_wlogic;
              a, b, t:      IN  t_nibble;
              edge, dark:   OUT t_wlogic );
END check_edge;
```

This device is part of a pattern recognition system. Three four-bit integer values are fed to the device, **a**, **b** and **t**. The first two of these are incoming data representing digital values corresponding to brightness detected by attached circuitry. The third value is a tolerance value that determines the difference required between the two measured values over which a light to dark or a dark to light transition is detected. The device is driven by a clock signal **clk** and generates outputs **edge** and **dark**. The ports, represented at a somewhat higher level than may be customary for logic design, have

2.4. USER DEFINED PACKAGES

two distinct types, **t_nibble**, and **t_wlogic**. VHDL simplifies the process of design by allowing the designer to form descriptions which are shorter and more concise than is otherwise possible by describing new types of data, rather than '0' or '1' that flows through the circuit. In this case, the data type **t_nibble** is a collection of four bits, which represents a value from 0 to 15, with 0 being the darkest value that a cell takes on, and 15 the lightest. By using the 4 bits together, as a single unit, the possibility of inadvertently switching a pair of signals is eliminated. VHDL helps the designer to discover errors in complex designs which might otherwise be overlooked, by requiring signals, or wires between ports, and ports to conform in the type of data they carry.

VHDL accomplishes sharing data types and common functionality through the use of design packages. Packages are collections of datatypes and operations that are commonly used and shared either through a design or throughout an organization. The data types **t_nibble** and **t_wlogic** in the entity declarations are used throughout the design of the edge detector circuit. The type **t_wlogic** is defined as part of the standard logic package described in chapter 7. **t_wlogic** represents a single bit, incorporating a 46 value logic modeling system (as described in section 7.2), which is a refinement of the more common 3 state/4 strength systems used in some logic simulators. This approach makes possible accurate technology dependent gate level simulation.

Both structure and behavior are described by means of VHDL architectures. An architecture describes the implementation of an entity. An architectural description of the **check_edge** component is shown below

An architecture of the check_edge entity

```
ARCHITECTURE behavioral OF check_edge IS
BEGIN
    PROCESS (clk)
        VARIABLE avalue, bvalue, tvalue : integer;
    BEGIN
        -- Watch clock
        IF (clk = '1') THEN
            -- Pickup integer values for inputs

            avalue := f_logictoint( a );
            bvalue := f_logictoint( b );
            tvalue := f_logictoint( t );

            -- Check for light to dark edge transition
            IF avalue - bvalue > tvalue THEN
                edge <= F1 AFTER 0 NS, F0 AFTER 5 NS;
                dark <= F1;
```

```
            -- Check for dark to light transition
            ELSIF bvalue - avalue > tvalue THEN
                edge <= F1 AFTER 0 NS, F0 AFTER 5 NS;
                dark <= F0;

            -- No transition
            ELSE
                edge <= F0;
            END IF;
        END IF;
    END PROCESS;
END behavioral;
```

The reader is referred to chapter 7 for a detailed discussion of the **F0** and **F1** values shown here. For the purposes of this discussion, these values represent the logic values false and true respectively.

This architecture describes the behavior of a single cell edge checker. check_edge has a single process statement (see the keyword **PROCESS**), sensitive only to events (change of value) on the port **clk**. Whenever **clk** is toggled, the body of the process statement is activated by the simulator.

Whenever **clk** is at a rising edge (note that an unknown to true transition is also treated as a rising edge), as checked by (**clk** = '1') the inputs **a** and **b** and threshold **t** are converted to integer values and compared. If the difference exceeds the threshold, a 5 nanosecond pulse is sent to the output **edge** and the current transition from light to dark or dark to light is output on **dark**.

There are several interesting points to note about the assignment of values to signals. In this example, the signal assignment, represented by **edge <= new value** is a waveform assignment. Unlike many other hardware description languages, VHDL allows the assignment of many values in series for a signal or port. This facility allows more economical and precise specification of the behavior. In the case where no transition occurs in the above example, the value of the port **dark** remains unchanged.

For more complex descriptions, VHDL allows multiple process statements to describe concurrent behavior. Furthermore, a small number of easily mastered language features permits the description of synchronous behavior, such as the wait statement and guarded assignments in which assignments are conditional on the values of their guards.

In this example, the package defined above described the data type t_nibble and the function f_logictoint both of which are referenced by the check_edge model. By packaging the logic to integer conversion facilities in this fashion, the model development was facilitated. In addition, this package is reusable, and can be referenced in other models as well.

2.5. VHDL MODELS AND THE ACCURACY CONTINUUM

Since the **utility**, **std_ttl** and **std_logic** packages were referenced with the use statement in the entity declaration for this model, their facilities are automatically available in the associated architecture.

The package concept in VHDL is one of the most powerful features of the language. Packages provide a way for VHDL experts to extend the language in a fashion which hides the complexity of detailed language features from the model developer. A good example of this is the standard logic package discussed in this book and described in chapter 7. By utilizing the operator overloading facilities and by providing a set of pre-defined functions the modeller can leverage the powerful features of the standard logic package without needing to know the details of the 46 value logic system (see section 7.2) upon which it is based.

2.5 VHDL Models and the Accuracy Continuum

In section 1.3 the concept of a continuum of modelling accuracy was discussed. The following examples will highlight models which range from simple to complex and will discuss the relationship between model complexity and simulation accuracy.

The versatility of the VHDL language allows the description of models which can have widely varying levels of accuracy. Clearly a tradeoff exists between the accuracy of a VHDL model and the performance of the model during simulation. A more complex model with more accurate timing will operate more slowly than a simple model, and will require more memory. In addition, a more accurate model will generate more complex simulation results which require a more detailed review by the hardware designer. No single model is the right model for every application. In certain circumstances the simplest model will be sufficient and most desirable. In the final stages of verifying a design with layout driven timing, the most sophisticated model shown here may be required.

2.5.1 2-Value Unit-Delay Approach

The following shows the entity declaration and architecture for a NAND gate.

NAND Model

```
ENTITY nand_gate IS
    PORT    (a,b: IN BIT; y: OUT BIT);
```

```
END nand_gate;

ARCHITECTURE behavioral OF nand_gate IS
BEGIN
    y <= a NAND b AFTER 1 ns;
END behavioral;
```

This model has the following characteristics:

- Continually watches the **a** and **b** ports for input signal value changes.
- Whenever either input port changes value, the output port **y** will receive the new NAND value after 1 nanoseconds of delay.

This example correctly models the NAND device, but lacks the ability to handle typical hardware requirements such as technology dependent timing. This is the simplest recommended approach to modelling a device such as a gate.

The BIT value system represents logic as follows:

- '0' - logic false
- '1' - logic true

For most synchronous designs where all circuit storage is gated by a clock this value system is adequate provided that the circuit also anticipates circuit power-up adequately. If circuits with asynchronous characteristics are simulated (or power-up characteristics are complex), often the two value approach will not work. In these types of circuits the initialization phase of the simulation can lock in invalid signal values. Figure 2.5 is a circuit which

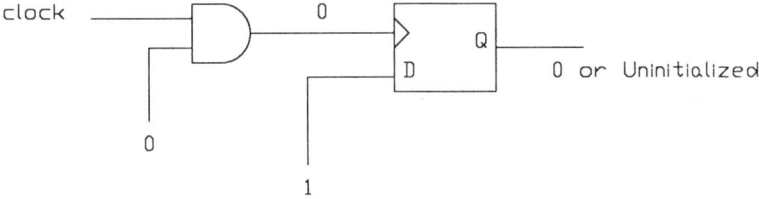

Figure 2.5: Circuit Initialization Problems

highlights this difficulty. Assume that the designer in this circuit made an error in which one of the inputs to the and gate starts initially as false, disabling the clocking of the D-latch. In this case, the D-latch will never receive the data value true and as a result will have an indeterminate value. Using the BIT value system, the modeller would have to choose to default the D-latch state to either 0 or 1. As a result, although the simulator will indicate that all signals have reached a stable initial value, the results of

2.5. VHDL MODELS AND THE ACCURACY CONTINUUM

the simulation are inaccurate and don't match the actual behavior of the circuit (which could power-up with either a 0 or 1 value in the D-latch).

By introducing an uninitialized value (as will be discussed in the next section) the simulator can accurately predict that the initial value of the flip-flop is not known indicating a possible design failure. In this case, the D-latch state would start up as uninitialized and would stay in this state. The designer can quickly identify this error and insure that the anded clock signal is cycled at least once with valid input data.

The model above takes a simple approach to timing with all transitions having an associated small amount of delay (often unit delay simulation utilizes delta delay, although the use of 1 nanosecond or smaller is safer). This approach is commonly referred to as unit-delay simulation, since all devices in the circuit have a fixed single unit of delay associated with them. For synchronous designs, this approach to modelling timing can often be the most effective approach. Simulation can be efficient, and the designer can ignore any complexities introduced in the simulation results due to timing skews. The problem with this approach is that in designs which have any asynchronous characteristics, timing can have a significant effect on the behavior of the circuit. Figure 2.6 shows a circuit in which the data

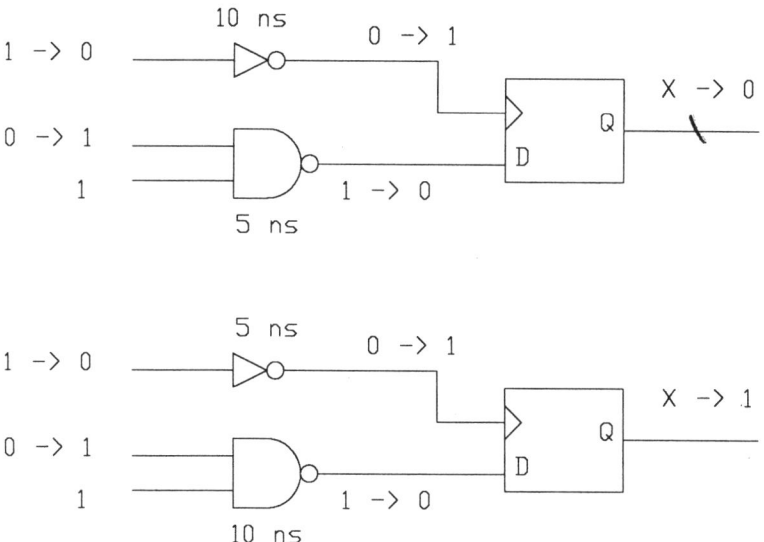

Figure 2.6: Circuit Behavior and Timing

value which gets latched into the D-latch can be altered by the timing of the nand and inverter gates which drive it. Two different timing options are shown. In the upper scenario, the clock transition appears ahead of the data transition and the false value is clocked into the flip-flop. In the bottom scenario, the clock transition appears after the data transition and the true value is clocked into the flip-flop. Clearly a unit-delay model of timing would not be appropriate for simulation of this type of circuit. Later examples will illustrate more sophisticated ways to model timing in a VHDL model.

2.5.2 46-Value Unit-Delay Approach

The following shows a unit delay model similar to that discussed in the previous section, but with a more sophisticated underlying value system.

46-Value Unit-Delay Nand Model

```
USE std.std_logic.ALL;
USE std.std_ttl.ALL;
ENTITY nand_simple IS
    PORT    (a,b: IN t_wlogic; y: OUT t_wlogic);
END nand_simple;

ARCHITECTURE behavioral OF nand_simple IS
BEGIN
    y <= a NAND b AFTER 1 ns;
END behavioral;
```

In this example, the *standard logic* package is referenced, with all signals declared as type **t_wlogic** which represents a 46 value logic system as described in chapter 7. The important aspects of using the 46 value logic system can be summarized here:

- Initially, the signal values will default to uninitialized. This applies to both the input as well as output ports for this model. (Default occurs because the value system is defined with **t_wlogic'LOW** equal to the uninitialized value, and VHDL uses **t_wlogic'LOW** as the default value for all signals of this type).
- Propagation of unknown values from input to output ports.
- Signal strength processing.

The **NAND** operator in this model is a VHDL overloaded operator. This VHDL facility causes the nand operator to in essence be replaced by VHDL code which is defined in the standard logic package. The reader is referred to chapter 7 for a detailed discussion on how the overloaded operators are defined. The **a** and **b** are passed into the overloaded NAND expression

2.5. VHDL MODELS AND THE ACCURACY CONTINUUM

which returns the nand result. The overloaded facility in VHDL improves the readability of the model while providing the full 46 value functionality required for this model. Without the use of the VHDL overloaded operator NAND, the model would look like that shown below

46-Value Unit-Delay Nand Model/No Overloading

```
ARCHITECTURE behavioral OF nand_simple IS
BEGIN
    y <= f_ttl(f_nand(f_state(a))(f_state(b)));
END behavioral;
```

In this model, the input port values **a** and **b** are converted to a signal state with the f_state table, the nand table f_nand returns the proper output state, and the f_ttl table returns the TTL specific technology value (which incorporates state and strength). These tables are discussed in detail in chapter 7. The truth table for the nand operation is summarized below

Nand truth table

input 1	input 2	output
0	0	1
0	1	1
0	X	1
1	0	1
1	1	0
1	X	X
X	0	1
X	1	X
X	X	X

From this table, it is clear that the device anticipates not only true and false inputs, but also uninitialized and unknown inputs (uninitialized is treated identically to the unknown value in this table). In general, the modeller should take care to propagate pessimistic rather than optimistic results. For example, input of a true and unknown value is dependent on the unknown value and therefore an unknown should be generated on the output. If an optimistic value of false were generated, then the user might incorrectly believe that his circuit operates correctly, when in fact there is a fifty percent chance that it might operate incorrectly. This technique is commonly referred to as unknown propagation, and is an important aspect of accurate device modelling since if in doubt, the simulator will always generate an unknown signal value. Occasionally, sufficient information exists for the model to intelligently suppress propagation of unknowns as in the case where the inputs are unknown and false respectively. In this case, regardless of the value of the unknown input, we know that the output will always be true and we therefore propagate this value. This is also an important decision since although an unknown output could be correct from a simulation standpoint, it is of little use to the designer in helping him debug his circuit.

The uninitialized value is useful to the hardware designer in that it indicates a signal that has never been given a definitive value since circuit power-up. This signals a possibly important design fault which will occur during power-up, since without a deterministic final value being achieved the operation of the circuit is not dependable. Uninitialized differs from unknown in the sense that an uninitialized signal has never been driven to a value of true or false, whereas an unknown signal has been given a value of true or false but we cannot determine which. During circuit power-up, ideally all uninitialized signal values will eventually be filtered out. Unknown values however, can occur at any time during simulation, often appearing if a device is operated incorrectly (for example when a flip-flop is given inputs which force it into an unstable state).

Finally, the following shows the table used by the nand model for establishing the output strengths.

TTL strength output

state	value
0	F0
1	F1
X	FX

The device always generates a forced false output, a forced true output, and a forced unknown output. These selections of output strength when combined with switch-level device models accurately reflect the operation of TTL devices. For further details on overloaded operators, the use of value strengths and the table lookup functions see chapter 7.

2.5.3 Fixed-Delay Approach

The following shows a model which utilizes a fixed delay rather than the unit delay approach seen before.

2.5. VHDL MODELS AND THE ACCURACY CONTINUUM

Fixed-Delay Nand Model

```
USE std.std_logic.ALL;
USE std.std_ttl.ALL;
ENTITY nand_onedelay IS
    PORT    (a,b: IN t_wlogic; y: OUT t_wlogic);
END nand_onedelay;

ARCHITECTURE behavioral OF nand_onedelay IS
BEGIN
    y <= a NAND b AFTER 5 ns;
END behavioral;
```

This is the most limited type of timing which can be introduced into a device model and represents at best an approximation of the behavior of a device. Often, choosing the average or nominal delay for all transitions of a device is a good technique, and in many cases represents sufficient timing detail for the purposes of the hardware designer. This technique is most effective for relatively high level modelling tasks, such as a VLSI block in which the detailed timing characteristics of the device have been abstracted and the circuit utilizing such a device is sufficiently tolerant that expected variations in the timing of the actual device won't affect circuit behavior as compared to the simulated behavior.

2.5.4 Variable-Delay Approach

The following shows a model which utilizes a delay which is a function of the output transition state of the device.

Variable Delay Nand Model

```
USE std.std_logic.ALL;
USE std.std_ttl.ALL;
ENTITY nand_twodelay IS
    PORT    (a,b: IN t_wlogic; y: OUT t_wlogic);
END nand_twodelay;

ARCHITECTURE behavioral OF nand_twodelay IS
BEGIN
    PROCESS (a,b)
        VARIABLE nextstate : t_wlogic;
    BEGIN
        nextstate := a NAND b;
        CASE f_state(nextstate) IS
            WHEN '0' =>         y <= nextstate AFTER 7 ns;
            WHEN '1' =>         y <= nextstate AFTER 4 ns;
            WHEN 'X' | 'U' =>   y <= nextstate AFTER 7 ns;
```

```
        END CASE;
     END PROCESS;
 END behavioral;
```

In this case, the delays are summarized as follows:

- **true to false** - 7 nanoseconds
- **false to true** - 4 nanoseconds
- **to unknown/uninitialized** - 7 nanoseconds

Figure 2.7 shows the simulation results for this model. The delay for the

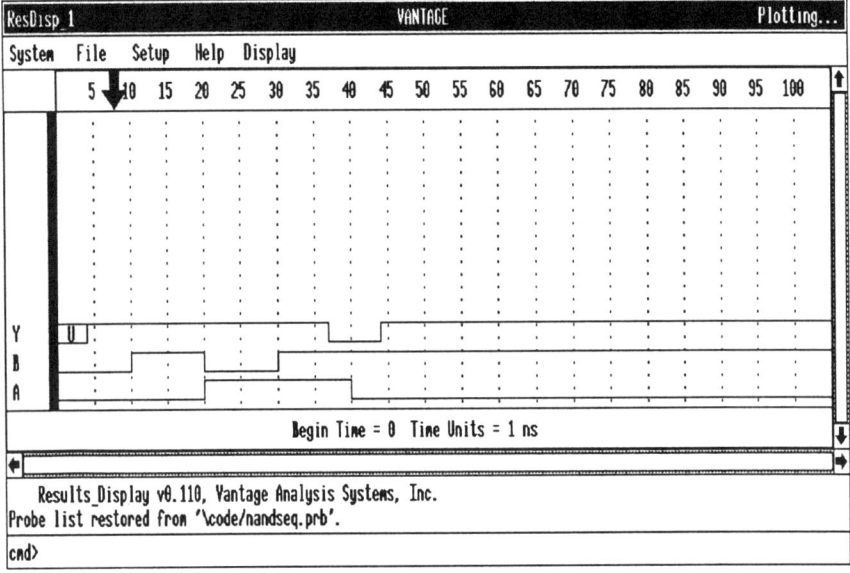

Figure 2.7: Simulation of Variable Delay

high to low transition of the **y** output occurs 7 nanoseconds (at time 37) after the two inputs **a** and **b** both become high (at time 30). The low to high transition of **y** occurs 4 nanoseconds (at time 44) after **a** drops low (at time 40).

Often the specification for a device will indicate different delays to be associated with a specific state transition, and this model demonstrates how these differences can factor into a model. The unknown and uninitialized timing has been chosen to represent the most pessimistic view of the device. An alternative approach would be to place the average delay

2.5. VHDL MODELS AND THE ACCURACY CONTINUUM

for either transition direction. Neither approach is ideal, and often the designer will have a specific requirement in his expectations in simulation which will drive the choice of timing values. For example, if the designer is concerned about whether his circuit will continue to operate with extreme variations in device delays then the end condition values and a pessimistic approach to unknown transitions would be utilized. On the other hand, the designer may be more concerned with the nominal timing characteristics of the circuit, and here an averaging approach might make more sense.

2.5.5 Generic Variable-Delay Approach

A desirable characteristic of a VHDL model is technology independence. Many devices can have a wide range of technology and manufacturer specific characteristics but common functional characteristics. The best example is the nand gate. Several hundred versions of the nand gate exist, but all have the same functionality. The primary difference between nand gates from different technologies and manufacturers is the propagation delay. It is possible using the techniques described in this section to write a single nand model which can accurately simulate every one of these many versions of the nand gate without requiring source code changes to the model. In this case, the model library developer has leveraged his VHDL investment by several hundred fold, a substantial savings.

A model which can be utilized in a wide range of applications without modification is commonly referred to as a generic model (although related, this should not be confused with the term, generic parameter which is a specific VHDL concept). The following rules provide a basis for developing generic VHDL models:

- All technology dependent information is passed to the model through generic parameters. This information is generally restricted to timing data only.

- All delay calculations are performed outside of the VHDL environment, and only the actual timing values are passed to the generic model.

- The device behavior is modelled in a technology independent fashion.

The following shows a nand model which utilizes generic parameters for establishing propagation delays.

Generic Variable Delay Nand Model

```
USE std.std_logic.ALL;
USE std.std_ttl.ALL;
```

```
ENTITY nand_twogeneric IS
    GENERIC (tplh,tphl : TIME := 7 ns);
    PORT    (a,b: IN t_wlogic; y: OUT t_wlogic);
END nand_twogeneric;

ARCHITECTURE behavioral OF nand_twogeneric IS
BEGIN
    y <= a NAND b AFTER f_delay(a NAND b, tplh, tphl);
END behavioral;
```

In this model, two generic parameters **tplh** and **tphl** are declared, both with the same default value of 7 nanoseconds. These parameters represent the propagation delay for the low to high and high to low transitions respectively of the device. In addition, this model utilizes a standard logic package function **f_delay** which facilitates handling of variable delay in models. The first parameter to this function is the new state which will be assigned to the output port, the second parameter is the low to high transition delay, and the third parameter is the high to low transition delay. The function returns either the second or third parameter values dependent on the new state (and thus the transition delay value) of the device. Although a little less readable, an alternative which avoids the duplicate calculation of the device's next state (and is slightly more efficient as a result) is shown below

                Alternate Generic Variable Delay Nand Model

```
ARCHITECTURE behavioral_other OF nand_twogeneric IS
BEGIN
    PROCESS (a,b)
        VARIABLE newstate : t_wlogic;
        BEGIN
        -- calculate new state
        newstate := a NAND b;

        -- assign new state with proper timing
        y <= newstate AFTER f_delay(newstate, tplh, tphl);
    END PROCESS;
END behavioral_other;
```

This example also illustrates the use of the process statement as an alternative to the concurrent signal assignment used in most of the examples in this section. The process is sensitive to the two input ports **a** and **b**, and whenever either changes value, the new state of the device is calculated and saved in the local variable **newstate**. This value is then used in the output assignment to port **y** utilizing the **f_delay** function to determine which of the transition delays to use.

2.5. VHDL MODELS AND THE ACCURACY CONTINUUM

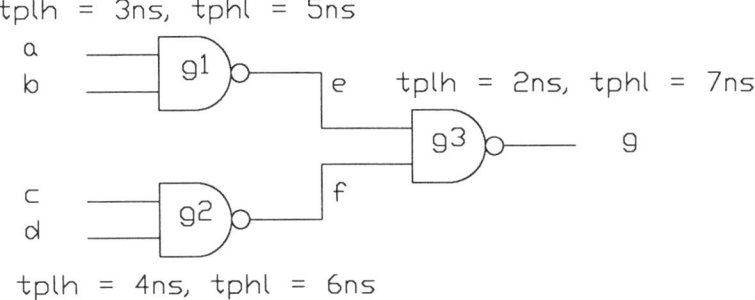

Figure 2.8: Schematic with Generic Parameter Values

Figure 2.8 shows a schematic diagram which references the model discussed above. The specific delays associated with each instance of the nand gate are shown on the diagram. This example illustrates the advantage of using a generic VHDL model. Although each instance of the nand gate has different timing, only one VHDL model is required to accurately simulate the circuit. The following shows the equivalent VHDL structural description which is analogous to the schematic diagram shown in figure 2.8.

Generic Parameter Instantiation

```
USE std.std_logic.ALL;
USE std.std_ttl.ALL;
ENTITY sample_twogeneric IS
    PORT    (a,b,c,d: IN t_wlogic; g: OUT t_wlogic);
END sample_twogeneric;

ARCHITECTURE structural OF sample_twogeneric IS
    COMPONENT nand_twogeneric
        GENERIC (tplh,tphl : TIME);
        PORT (a,b:IN t_wlogic; y: OUT t_wlogic);
    END COMPONENT;
    SIGNAL e,f: t_wlogic;
BEGIN
    g1:nand_twogeneric GENERIC MAP (3 ns,5 ns) PORT MAP(a,b,e);
    g2:nand_twogeneric GENERIC MAP (4 ns,6 ns) PORT MAP(c,d,f);
    g3:nand_twogeneric GENERIC MAP (2 ns,7 ns) PORT MAP(e,f,g);
END structural;

CONFIGURATION parts OF sample_twogeneric IS
FOR structural
```

```
        FOR g1,g2,g3:nand_twogeneric USE ENTITY work.nand_twogeneric(behavioral);
        END FOR;
    END FOR;
    END parts;
```

For each of the gates **g1**, **g2** and **g3** the generic map construct is used to specify the actual delays to be used by the device. During simulation the output transition of the **g1** gate onto signal **e** occurs at time 3. The similar transition of the **g2** gate onto signal **f** occurs at time 4. Both of these transitions correspond to the low to high delay times specified in the parameter substitutions shown above. Further analysis indicates that each gate has different delays as summarized here:

- **g1**: 3 ns low to high, 5 ns high to low
- **g2**: 4 ns low to high, 6 ns high to low
- **g3**: 2 ns low to high, 7 ns high to low

2.5.6 Full-Delay Approach

Previous models in this section have anticipated only the propagation delay associated with the device. These models did not reflect the delay which might be associated with wiring between devices. In ASIC design especially, the assumption of no wiring delay can be a dangerous one since many designs of this type experience an important if not dominant timing effect from wires. The following shows a VHDL model for the nand device which provides an accurate view of the timing associated not only with the device but also with the delay associated with the input wiring to the device.

Full Timing Nand Model

```
USE std.std_logic.ALL;
USE std.std_ttl.ALL;
ENTITY nand_indelay IS
    GENERIC (a_iO1,a_i10,b_iO1,b_i10,y_oO1,y_o10 : TIME := 1 ns);
    PORT    (a,b: IN t_wlogic; y: OUT t_wlogic);
END nand_indelay;

ARCHITECTURE behavioral OF nand_indelay IS
    SIGNAL adelay, bdelay : t_wlogic;
BEGIN
    -- input delay processing
    adelay <= a AFTER f_delay(a,a_iO1,a_i10);
    bdelay <= b AFTER f_delay(b,b_iO1,b_i10);

    -- gate function
    y <= adelay NAND bdelay
```

2.5. VHDL MODELS AND THE ACCURACY CONTINUUM

```
        AFTER f_delay(adelay NAND bdelay, y_o01, y_o10);
END behavioral;
```

Figure 2.9 shows an abstract view of a VHDL model. The model

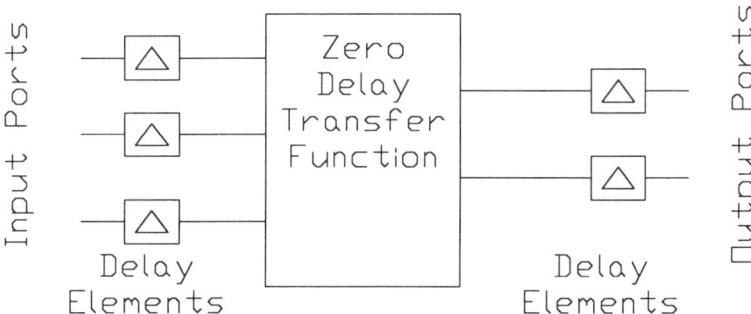

Figure 2.9: Delay Model

has a set of structural delay elements attached to each input port, a transfer function which performs the behavior of the device, and output delay elements. In the model shown above, the signals **adelay** and **bdelay** utilize the **f_delay** function to incorporate the appropriate input delays passed as generic parameters. In this case the functioning of the device is straightforward enough that the nand expression is incorporated directly in the output delay processing. In this case the nand expression is passed to the **f_delay** function for calculation of the appropriate output delay. The output signal **y** receives the expression value with this calculated delay. Notice that the nand expression utilizes the internal input delayed signals **adelay** and **bdelay** and not the primary inputs **a** and **b**.

Figure 2.10 shows a circuit in which the detailed delays associated with signal segments are known. Since VHDL does not have a mechanism for handling delays on signal segments, the delays are placed on input ports of the devices which are attached to the signal. Note that the path must be analyzed in order to determine the correct input port delay. For example, the upper path is composed of a 1.0 nanosecond delay in the common segment plus an additional 0.5 nanosecond delay in the upper segment; thus a total delay of 1.5 nanoseconds is passed as a generic parameter value to the upper nand gate. Similarly, the lower nand gate has a delay of 1.3 nanoseconds passed as a generic parameter for its input port. By following the above prescribed methodology for incorporating input port delays as

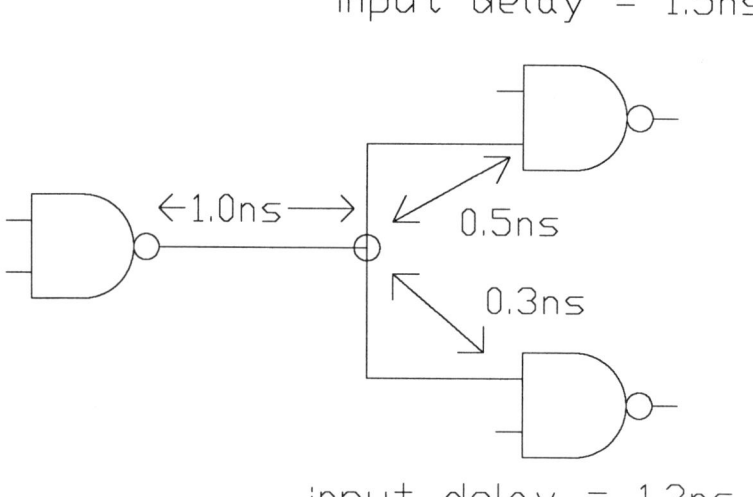

Figure 2.10: Input Delays

delay elements in a VHDL model, and by providing generic parameters for these delays, it is possible to handle the vast majority of timing problems related to various technologies. In particular, this mechanism will work effectively in support of back-annotation of delay values from the circuit layout.

An effective means in specifying the actual instance values for the generic parameters is to use VHDL configuration bodies. These configuration bodies can be generated automatically from a delay calculation program (often this type of tool is available from the ASIC manufacturer).

2.5.7 Error Checking and Model Structure

Models in this book are structured as follows:

Structure of VHDL Model

```
-- package use statements
ENTITY devicename IS
    -- generic parameters
    -- port declarations
END devicename;
```

2.5. VHDL MODELS AND THE ACCURACY CONTINUUM 39

```
ARCHITECTURE full OF devicename IS
    -- internal input delay signals
BEGIN
    -- error checking processes

    -- input delay processing

    -- device operation process
END full;
```

By following this approach the modeller is assured of uniformity in his models and a structure which insures that the key aspects of model operation are anticipated. Every model should have an error checking section. The most effective way to handle error checking is to separate it from the input delay processing and from the device operation by utilizing separate processes.

The following shows the complete model for a D flip-flop.

VHDL Model with Error Checks

```
USE std.std_logic.ALL;
USE std.std_ttl.ALL;
ENTITY dff IS
    GENERIC (preset_i01,preset_i10,
             clear_i01,clear_i10,
             clock_i01,clock_i10,
             d_i01,d_i10,
             q_o01,q_o10,
             qb_o01,qb_o10,
             clock_min, preset_min, clear_min,
             d_setup,d_hold : TIME := 2 ns);
    PORT    (preset,clear,clock,d: IN t_wlogic; q,qb: OUT t_wlogic);
END dff;

ARCHITECTURE full OF dff IS
    SIGNAL presetdelay, cleardelay, clockdelay, ddelay : t_wlogic;
BEGIN
    -- check clock frequency/spike detection
    PROCESS (clock)
        VARIABLE clocklastev : TIME := 0 ns;
        BEGIN
        ASSERT (NOW = 0 ns) OR ((NOW - clocklastev) >= clock_min)
            REPORT "Spike detected on clock" SEVERITY warning;
        clocklastev := NOW;
    END PROCESS;

    -- spike detection
    PROCESS (preset)
        VARIABLE presetlastev : TIME := 0 ns;
        BEGIN
```

```vhdl
         ASSERT (NOW = 0 ns) OR ((NOW - presetlastev) >= preset_min)
            REPORT "Spike detected on preset" SEVERITY warning;
         presetlastev := NOW;
      END PROCESS;

   PROCESS (clear)
      VARIABLE clearlastev : TIME := 0 ns;
      BEGIN
         ASSERT (NOW = 0 ns) OR ((NOW - clearlastev) >= clear_min)
            REPORT "Spike detected on clear" SEVERITY warning;
         clearlastev := NOW;
      END PROCESS;

   -- check for setup/hold violations
   PROCESS (clock,d)
      VARIABLE dlastev : TIME := 0 ns;
      VARIABLE clocklastev : TIME := 0 ns;
      BEGIN
         -- Hold check
         IF d'EVENT THEN
            ASSERT (NOW = 0 ns) OR ((NOW - clocklastev) >= d_hold)
               REPORT "Hold error" SEVERITY warning;
            dlastev := NOW;
         END IF;

         -- Setup check
         IF (clock'EVENT) AND (clock = '1') THEN
            ASSERT (NOW = 0 ns) OR ((NOW - dlastev) >= d_setup)
               REPORT "Setup error" SEVERITY warning;
            clocklastev := NOW;
         END IF;
      END PROCESS;

   -- check for invalid control
   ASSERT NOT ( (preset = '0') AND (clear = '0') )
      REPORT "Preset and clear both active"
      SEVERITY warning;

   -- input delay processing
   presetdelay <= preset AFTER f_delay(preset,preset_i01,preset_i10);
   cleardelay <= clear AFTER f_delay(clear,clear_i01,clear_i10);
   clockdelay <= clock AFTER f_delay(clock,clock_i01,clock_i10);
   ddelay <= d AFTER f_delay(d,d_i01,d_i10);

   -- flip-flop operation
   PROCESS (presetdelay,cleardelay,clockdelay)
      BEGIN
         -- Check for preset
         IF (presetdelay = '0') AND (cleardelay = '1') THEN
            q <= f_ttl('1') AFTER f_delay(f_ttl('1'), q_o01, q_o10);
            qb <= f_ttl('0') AFTER f_delay(f_ttl('0'), qb_o01, qb_o10);
```

2.5. VHDL MODELS AND THE ACCURACY CONTINUUM

```
          -- Check for clear
          ELSIF (presetdelay = '1') AND (cleardelay = '0') THEN
              q <= f_ttl('0') AFTER f_delay(f_ttl('0'), q_o01, q_o10);
              qb <= f_ttl('1') AFTER f_delay(f_ttl('1'), qb_o01, qb_o10);

          -- Check for control error
          ELSIF (presetdelay = '0') AND (cleardelay = '0') THEN
              q <= f_ttl('X') AFTER f_delay(f_ttl('X'), q_o01, q_o10);
              qb <= f_ttl('X') AFTER f_delay(f_ttl('X'), qb_o01, qb_o10);

          -- Check for unknown controls
          ELSIF (presetdelay = 'X') OR (cleardelay = 'X') THEN
              q <= f_ttl('X') AFTER f_delay(f_ttl('X'), q_o01, q_o10);
              qb <= f_ttl('X') AFTER f_delay(f_ttl('X'), qb_o01, qb_o10);

          -- Check for unknown clock
          ELSIF (clockdelay'EVENT) AND (clockdelay = 'X') THEN
              q <= f_ttl('X') AFTER f_delay(f_ttl('X'), q_o01, q_o10);
              qb <= f_ttl('X') AFTER f_delay(f_ttl('X'), qb_o01, qb_o10);

          -- Check for clocked data
          ELSIF (clockdelay'EVENT) AND (clockdelay = '1') THEN
              q <= d AFTER f_delay(d, q_o01, q_o10);
              qb <= NOT d AFTER f_delay(NOT d, qb_o01, qb_o10);
          END IF;
          END PROCESS;
END full;
```

In the model above, five separate processes check for the following errors:

- **Check clock frequency** - flag an error if clock pulses are shorter than required minimum
- **Spike detection on preset input** - this asynchronous input is continually monitored for spikes
- **Spike detection on clear input** - this asynchronous input is continually monitored for spikes
- **Setup and hold checks** - the clock and data inputs are monitored for setup or hold violations
- **Invalid control** - any invalid control input is flagged

Note that all error checking is performed on the input ports directly (not on the internal delayed signals) to insure that errors are reported at the expected time.

Four statements provide the internal delay processing for input ports to the signals **presetdelay**, **cleardelay**, **clockdelay** and **ddelay**. The **f_delay** function (described in more detail in section 7.6) provides the timing calculations as appropriate.

The main flip flop operation is a single process which is sensitive to the **presetdelay, cleardelay** and **clockdelay** signals. This process utilizes an if then else statement to provide the logic of the device. In this particular model, the output signal assignments are executed in-line in each if clause. An alternate approach used for more complex models is to store intermediate results in local variables and perform actual output at the of the process. This approach is more efficient and makes the model more readable for some devices.

2.6 Handling Timing Using Configurations

The configuration capabilities of VHDL are unique and are one of the most powerful and useful features of the language. The methodology used for modelling devices in this book follows this approach:

- All device models are written in a technology and timing independent fashion
- All timing data is passed to these models through generic parameters
- All technology dependent timing data is bound to the model instances through the use of configurations

The following shows a generic nand gate model.

Generic VHDL Model

```
USE std.std_logic.ALL;
USE std.std_ttl.ALL;
USE work.config.ALL;
ENTITY nand2generic IS
    GENERIC (lh_min,lh_typ,lh_max,hl_min,hl_typ,hl_max: time);
    PORT (a,b: IN t_wlogic;
          y: OUT t_wlogic);
END nand2generic;
--
ARCHITECTURE behavior OF nand2generic IS
BEGIN
    PROCESS(a,b)
        VARIABLE tlh,thl:time;
    BEGIN
        tlh := f_tchoice (s_t,lh_min,lh_typ,lh_max);
        thl := f_tchoice (s_t,hl_min,hl_typ,hl_max);
        y <= a NAND b AFTER f_delay(a NAND b,tlh,thl);
    END process;
END behavior;
--
CONFIGURATION c_nand2 OF nand2generic IS
  FOR behavior
```

2.6. HANDLING TIMING USING CONFIGURATIONS

```
END FOR;
END c_nand2;
```

The entity declaration indicates the device name as **nand2generic** with ports **a**, **b**, and **y**. Six generic parameters are passed to this model: **lh_min**, **lh_typ**, **lh_max**, **hl_min**, **hl_typ**, and **hl_max**. These parameters represent the low to high and high to low transitions times. Three values representing minimum, typical and maximum delay times are given. The architecture of the model shows a single process which is sensitive to the inputs **a** and **b**. Two local variables **tlh** and **thl** are used in the calculation of the actual delay the device will use. The function **f_tchoice** determines which of the minimum, typical or maximum delays will be used (see the package definition below) and utilizes a global constant **s_t** which is defined in the **config** package (see below). The **f_delay** function takes the output value and the calculated delays **tlh** and **thl** and returns the appropriate value depending on the transition. This function is described in more detail in chapter 7.

The following shows the source code for the **config** package which was referenced above and which defines the **f_tchoice** function and the **s_t** global constant.

Min/Typ/Max Timing Package

```
USE std.std_logic.ALL;
PACKAGE config IS
TYPE t_time IS (t_min, t_typ, t_max,no_time);
CONSTANT s_t:t_time := no_time;
FUNCTION f_tchoice (sim_time: t_time; vmin,vtyp,vmax: time) RETURN time;
END config;

USE std.std_logic.ALL;
PACKAGE BODY config IS
    FUNCTION f_tchoice (sim_time: t_time; vmin,vtyp,vmax: time)
        RETURN time IS
    BEGIN
        CASE sim_time IS
            WHEN no_time => RETURN 0 ns;
            WHEN t_min =>   RETURN vmin;
            WHEN t_typ =>   RETURN vtyp;
            WHEN t_max =>   RETURN vmax;
        END case;
    END f_tchoice;
END config;
```

A constant **s_t** of type **t_time** is declared which is referenced globally from each device model to determine whether minimum, maximum, typical or

44   CHAPTER 2. ANATOMY OF A VHDL MODEL

zero delay simulation will be used. The function f_tchoice is called from each device model when delays are being calculated in order to determine the proper delays to use (see the nand model above).

Once the generic device model is available, the device dependent versions of the device are created using structural hierarchy as shown below:

Device Dependent VHDL Models

```
USE std.std_logic.ALL;
USE work.config.ALL;
ENTITY v_nand2 IS
    PORT (a,b: IN t_wlogic;
          y: OUT t_wlogic);
END v_nand2;
--
ARCHITECTURE lib OF v_nand2 IS
COMPONENT nand2generic
    GENERIC (lh_min,lh_typ,lh_max,hl_min,hl_typ,hl_max: time);
    PORT (a,b: IN t_wlogic;
          y: OUT t_wlogic);
END COMPONENT;
BEGIN
  N1: nand2generic
      GENERIC MAP (0 ns, 0 ns, 0 ns, 0 ns, 0 ns, 0 ns)
      PORT MAP (a,b,y);
END lib;
--
CONFIGURATION sn5400 OF v_nand2 IS
   FOR lib
      FOR ALL: nand2generic USE CONFIGURATION work.c_nand2
      GENERIC MAP (6600 ps, 11 ns, 22 ns,
                   4200 ps, 7 ns, 15 ns)
      PORT MAP (a,b,y);
      END FOR;
   END FOR;
END SN5400;
--
CONFIGURATION sn7400 OF v_nand2 IS
   FOR lib
      FOR ALL: nand2generic USE CONFIGURATION work.c_nand2
      GENERIC MAP (6600 ps, 11 ns, 22 ns,
                   4200 ps, 7 ns, 15 ns)
      PORT MAP (a,b,y);
      END FOR;
   END FOR;
END SN7400;
```

A technology dependent entity v_nand2 is defined which has a number of architectures, one for each manufacturer specific part. In this case, three architectures are defined: lib, sn5400 and sn7400. The lib version is

2.6. HANDLING TIMING USING CONFIGURATIONS

a zero delay model. These structural architectures have an identical port correspondence to the generic model but attach technology dependent delays to the model.

Finally, the following circuit (a single nand gate) shows how the technology dependent model is specified.

Using Technology Dependent Models

```
USE std.std_logic.ALL;
USE std.std_ttl.ALL;
USE work.config.ALL;
ENTITY test_nand2 IS
SIGNAL ta,tb,ty: t_wlogic;
CONSTANT s_t: t_time := no_time;
END test_nand2;
--
ARCHITECTURE ar OF test_nand2 IS
COMPONENT v_nand2
    PORT (a,b: IN t_wlogic;
          y: OUT t_wlogic);
END COMPONENT;
BEGIN
  n1: v_nand2
    PORT MAP (ta,tb,ty);
END ar;
--
CONFIGURATION c OF test_nand2 IS
  FOR ar
    FOR ALL: v_nand2 USE CONFIGURATION work.sn5400
      PORT MAP (ta,tb,ty);
    END FOR;
  END FOR;
END c;
```

Three signals **ta**, **tb** and **ty** are connected to the nand gate. The structural architecture **ar** indicates a single instance **n1** of the nand gate. The configuration c of this circuit indicates the method used to reference the technology dependent model. In this case the **sn5400** version of the nand gate is used.

Figure 2.11 shows graphically the relationship between the various VHDL objects discussed above. At the top, the technology dependent entity **v_nand2** is defined. This entity contains a number of architectures, one for each manufacturers part such as **lib**, **sn5400**, **sn7400** and others as desired. Each of these refers structurally to the generic entity **nand2generic**. This entity has a single behavioral architecture **behavior** which defines the technology independent behavior of all of the nand devices.

46   CHAPTER 2. ANATOMY OF A VHDL MODEL

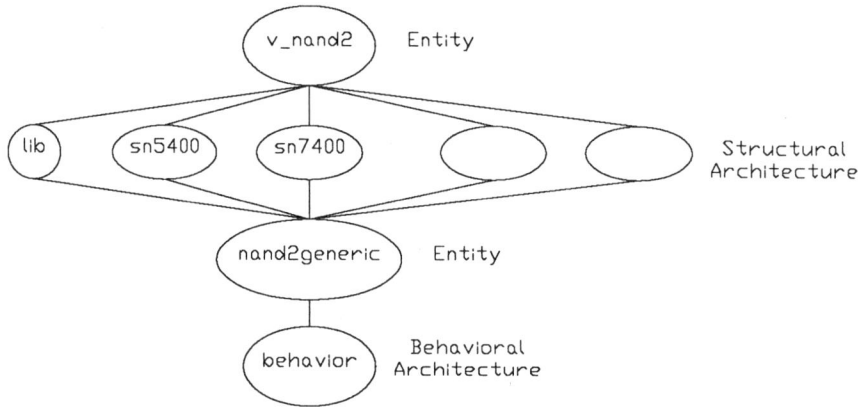

Figure 2.11: Configurations and Timing

By using the methodology discussed in this section, large libraries of modular VHDL models which encompass technology dependent timing characteristics while still maintaining a single generic behavioral architecture can be developed. For devices which are complex, this approach offers significant advantages over more brute forces methods since the timing shells are relatively easy to maintain, and only a single behavioral architecture needs to be maintained. Thus if a bug is found in the behavioral model, the fix is propagated to all versions of this model.

2.7 Using VHDL as a Stimulus Language

VHDL can be used effectively as a stimulus language. Constructs are available which make test vector information easy to code as follows

Test Vectors in VHDL

```
ENTITY rsff_stim IS
    PORT    (r,s: OUT BIT; q: IN BIT);
END rsff_stim;

ARCHITECTURE behavioral OF rsff_stim IS
```

2.7. USING VHDL AS A STIMULUS LANGUAGE

```
BEGIN
    r <= '1' AFTER 0 ns,
         '1' AFTER 10 ns,
         '0' AFTER 20 ns,
         '1' AFTER 30 ns;

    s <= '0' AFTER 0 ns,
         '1' AFTER 10 ns,
         '1' AFTER 20 ns,
         '1' AFTER 30 ns;
END behavioral;
```

which describes a stimulus entity **rsff_stim** with three ports r, s and q. In the behavioral body, the test vectors to be applied to the ports r and s are described. In this case, the data applied is summarized here

Test Vector Summary

time	r	s
0	1	0
10	1	1
20	0	1
30	1	1

The test bench approach is used in VHDL to connect the stimulus entity described above to the circuit under study. Figure 2.12 shows the basic approach which can be used in VHDL. A stimulus generator is created as described above in **rsff_stim** with ports which correspond exactly to the circuit under study as follows

- Each circuit input port has a corresponding output port on the generator
- Each circuit output port has a corresponding input port on the generator
- Each circuit input/output port has a corresponding input/output port on the generator

A top level test bench is created which connects each of the ports on the generator to the corresponding port on the circuit under study. When this test bench is simulated, the generator will provide stimulus to the circuit, and circuit will respond to this stimulus and will send results back to the generator. It is possible to include assertions in the generator to provide automatic tests for proper circuit operation. The following shows an example of the VHDL coding for a test bench which tests the **rsff** circuit described previously in section 2.2.

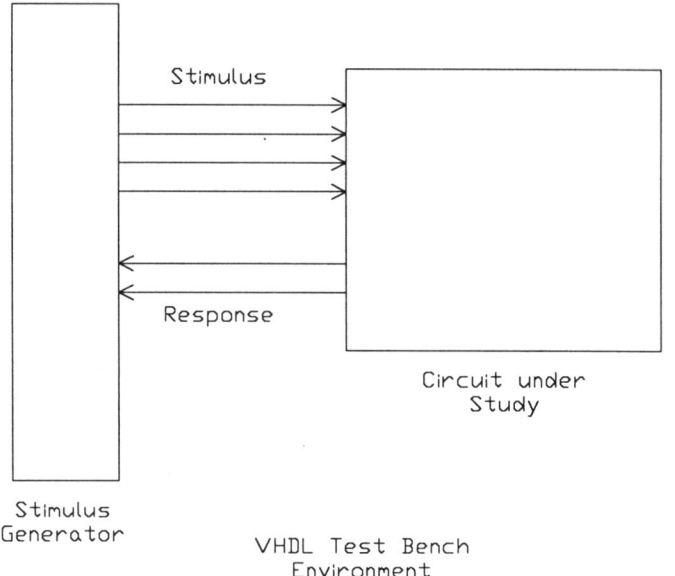

Figure 2.12: VHDL Test Bench

VHDL Test Bench

```
ENTITY rsff_bench IS
END rsff_bench;

ARCHITECTURE structural OF rsff_bench IS
    COMPONENT rsff_stim PORT (r,s: OUT BIT; q: IN BIT); END COMPONENT;
    COMPONENT rsff PORT (r,s: IN BIT; q: OUT BIT); END COMPONENT;
    SIGNAL r,s,q: BIT;
BEGIN
    generator:rsff_stim PORT MAP(r,s,q);
    circuit:rsff PORT MAP(r,s,q);
END structural;
```

The architecture **rsff_bench** describes a test bench as shown in figure 2.13.

2.8 Standardized VHDL Modelling Conventions

The scope of VHDL is large, allowing the description of systems ranging from the microcode and architectural levels down to the gate level. As a

2.8. STANDARDIZED VHDL MODELLING CONVENTIONS

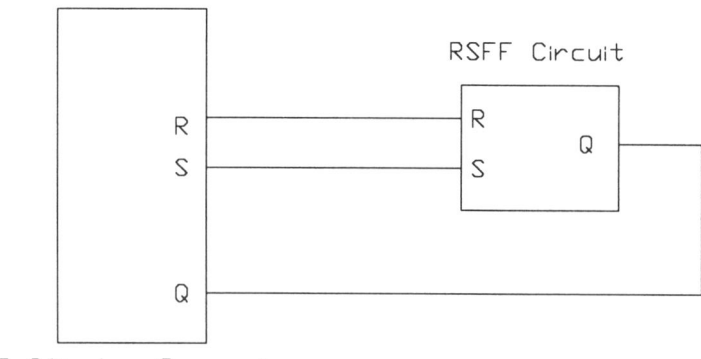

Figure 2.13: RSFF Test Bench

result of this multi-level capability, VHDL has considerably more flexibility and power than most other hardware description languages. For the vast majority of hardware designers, this flexibility and power is not required, and in fact can result in the creation of models that have severe compatibility problems when mixed during simulation [COELB88].

An analogy exists to VHDL in the area of programming languages. With a language such as C or Pascal, unless the use of these languages is restricted according to a strict methodology, a team of programmers will have severe difficulties during integration of a large programming effort. As a result, one of the key characteristics of an effective software engineering effort is the introduction of practices including:

- Code Sharing; Effective use of Libraries
- Standardized Code Formats
- Proper Documentation of Code
- Effective Modularization of Code
- Standardization of Coding Techniques including handling of errors and messages

These good engineering practices are well accepted in the software area, but are not as well understood or adopted in the hardware modelling area.

With a language such as VHDL, the use of structured software engineering practices is critical in order to insure proper compatibility of models during simulation.

Another motivation for the above practices relates to one of the primary requirements which influenced the development of the VHDL language, the need to effectively document hardware designs. The VHDL language in this context is useful as a mechanism to describe the operation of a hardware system, and to allow this information to pass from one designer to another, or from one organization to another. In this regard, many of the aspects mentioned above become even more important.

From this perspective, it is clear that the VHDL Language Reference Manual is not sufficient to insure the creation of compatible and documentation quality models. The remainder of this section will outline the specific problems which must be addressed in regard to VHDL modelling practices.

The approaches which have been incorporated in most of the examples given throughout the book will also be summarized.

Experience with VHDL and other related languages such as HHDL [COEL83,85] and ADLIB [COEL87] have clearly demonstrated that without a set of modelling guidelines and standard packages, a wide variance in coding quality and compatibility will exist from one designer to another. For this reason, a standard logic modelling package and associated guidelines is important and is described in detail is chapter 7 (other packages for other levels of design may also be useful). By leveraging existing knowledge in the development of behavioral models, a standard VHDL package such as the one shown in this book can be thought of as a **knowledge base** from which other designers can learn and leverage. Further, VHDL users will experience an increase in productivity by following standard practices and utilizing a standard package.

2.8.1 Generic Parameters

Generic parameters in VHDL allow instantiations of a model to have different characteristics. The most obvious use of this capability relates to timing characteristics of devices. In a typical design, although the basic function of a nand gate may remain the same for all instances of the gate, the timing can change from one instance to another. Generics allow the timing associated with each gate to be passed as a parameter. When many models are combined during simulation, a consistent methodology must be used in each model. If timing parameters are handled in an inconsistent fashion, it may not be possible to effectively back-annotate delays from the layout of a design. The following illustrates this problem.

2.8. STANDARDIZED VHDL MODELLING CONVENTIONS

Incompatibility between Generic Parameters

```
-- Nand gate interface with a single propagation
-- delay for the output port
ENTITY nand_gate IS
  GENERIC (y_prop : TIME);
  PORT (a, b : IN logic;
        y : OUT logic);
  USE standardlogic.ALL;
END nand_gate;

-- Nor gate interface with input port delays,
-- and different delays on output for rising
-- and falling values
ENTITY nor_gate IS
  GENERIC (a_in, b_in, y_f, y_t : TIME);
  PORT (a, b : IN logic;
        y : OUT logic);
  USE standardlogic.ALL;
END nand_gate;
```

In this example, a model of a nand gate and a nor gate are shown. Two different conventions were used in these models in regards to the information passed as generic parameters. In the nand gate, a single generic parameter **y_prop** which represents the propagation delay associated with the device in all cases was used. In the nor gate, several time values were passed; **a_in** and **b_in** represent input port delays, **y_t** and **y_f** represent rising and falling propagation delays associated with the output port. The problem in this situation is that a back-annotation program must provide information that varies depending on what model the information is passed to. This problem is compounded severely when more than two models are used during simulation. Other compatibility problems can occur when some models require more detailed information than others. For example, setup and hold time constraints are useful generic parameters for models. Unfortunately, if one designer ignores these values, and another designer includes these parameters, the resulting simulation will be limited in its effectiveness.

The methodology which is used throughout this book for handling generic parameters which represent timing is the following:

- Timing delays associated with wiring/metal from the actual layout are tied to input ports (see section 2.5.6 for a full discussion of this technique). Consider this delay to be **TW** for wire timing.

- Charge effects associated with input port value changes are tied to input ports. Consider this delay to be **TT** for transition timing.

- Propagation delays through the device can be associated with input ports, with input/output port pairs, or with output ports. Consider

this delay to be **TP** for propagation timing.
- Charge effects associated with output port value changes are tied to output ports. Consider this to be **TT** for transition timing.

The total delay associated with a particular path through the device is summarized here:

$$delay = IPort\_TW + IPort\_TT + IPort\_to\_OPort\_TP + OPort\_TT$$

In this equation, **IPort_TT** or **OPort_TT** can be substituted for

$IPort\_to\_OPort\_TT$

should the modeller not need the complete flexibility of handling every path through the device.

Naming conventions for generic parameters are summarized here:

- For a given input port named **portname**, the following options exist for specifying delays to be associated with the device:
    - **portname_I** - a single input port delay for the given input port.
    - **portname_Ixxzz** where **xx** and **zz** are the representations of a given logic value. The transition direction is from **xx** to **zz**. Examples are:
        * A_I - a single delay to be used for all transition types
        * A_IR1R0 - delay to be associated with the R1 to R0 transition
        * A_IR1F - delay to be associated with a transition from R1 to any forced value
        * A_IRF - delay to be associated with any resistive to forced value transition
        * A_I10 - delay to be associated with any true to false transition
        * A_I1 - delay to be associated with any true to other value transition
        * A_I_0 - delay to be associated with any to false transition. Notice the use of _ to signify a placeholder for the starting state/strength.

    Wire delays associated with individual input ports will be combined in an appropriate fashion to delays shown above by the back-annotation facility.

- For a given output port named **portname**, the following options exist for specifying delays to be associated with the device:

2.8. STANDARDIZED VHDL MODELLING CONVENTIONS

- **portname_O** - a single output port delay for the given output port.
- **portname_Oxxzz** where **xx** and **zz** are the two character representations of a given logic value. The transition direction is from **xx** to **yy**. Examples are:
  * A_O - a single delay to be used for all transition types
  * A_OR1R0 - delay to be associated with the R1 to R2 transition
  * A_OR1F - delay to be associated with a transition from R1 to any forced value
  * A_ORF - delay to be associated with any resistive to forced value transition
  * A_O10 - delay to be associated with any true to false transition
  * A_O1 - delay to be associated with any true to other value transition
  * A_O_0 - delay to be associated with any to false transition. Notice the use of _ to signify a placeholder for the starting state/strength.

- For a given input/output port pair, the following options exist:
  - **iport_oport_P** - the propagation delay associated with a signal path from the the given input to output port.
  - **oport_P** - the propagation delay associated with an output port only.
  - **iport_P** - the propagation delay associated with an input port only.

2.8.2 Naming Conventions

Good modelling practices dictate that consistent and uniform naming conventions be applied to the development of VHDL models. This becomes even more important when generic parameters are utilized, especially with respect to back-annotation of delays from layout. Without standardized naming conventions, it may not be possible for an automatic back-annotation facility to deposit values into the VHDL database.

Areas of concern are summarized here:

- Architectural body names
- Port names

- Generic parameter names

Clearly, maintaining a consistent mapping between schematic types and architectural bodies will make implementation of back-annotation easier. Consistency in generic parameter names makes automation of instantiation of generic parameter values possible. Since generic parameters often contain information which must ultimately be associated with port names, maintaining consistency between generic parameter names and port names will improve model readability and consistency.

2.8.3 Constraints

Behavioral simulators and hardware description languages have a significant advantage over older gate level technologies in supporting semantic checks as part of the behavioral model. The following summarizes some of the more important checks which can be performed effectively in VHDL:

- Setup limits
- Hold limits
- Spike detection
- Special timing requirements
- Invalid data
- Invalid control

It is important that models which are combined during simulation consistently handle constraint checks. If this rule is not followed, the effectiveness of the simulation will be reduced, and in the case of spike detection the simulation results can be erroneous. For example, if selected models suppress pass-through of spikes, but others don't, the results of the simulation can in fact be false. Although less severe, the results of a simulation can be misleading if only selected models check setup and hold constraints. The designer may falsely assume that he has no timing errors in his design, when in fact some of his models are just not checking for these constraints.

2.8.4 Unknown Handling

One of the most difficult but most important aspects of writing a behavioral model is the proper handling of unknown values during simulation. Experience with simulators has shown that the introduction of an unknown state is required in order to correctly handle the following situations:

- Circuit power-up and associated simulator initialization
- Recovery from improper device use, both in timing and function

2.8. STANDARDIZED VHDL MODELLING CONVENTIONS

During circuit power-up, uninitialized values are required in order to accurately predict the state of circuit after the power-up sequence is completed. Consider the case of a flip-flop; At the end of a power-up sequence, the simulator must choose to set the flip-flop state as either true or false. The state of the flip-flop is indeterminate, since the value of the flip-flop will lock in based on the current and voltage levels in the actual device which are highly dependent on the topology of the device. A more accurate reflection of the power-up sequence is to place the flip-flop initially into an uninitialized value, and only after a sequence of inputs to the device which force the device to a known state are received does the uninitialized value disappear. Proper handling of uninitialized values in models can be an effective tool in diagnosing hardware designs, especially for power-up conditions. Should a design fail to properly eliminate unknown values during this stage of simulation, the designer can expect to see indeterminate behavior in the actual circuit.

A secondary use of unknowns occurs during error recovery. Consider a device which has indeterministic behavior for a given set of inputs. A J-K flip-flop is a good example. If both the J and K inputs are held high, the state of the device can not be predicted. One approach to handling this situation would be to report an error to the user, and halt the simulation. This approach is not a good one, since it doesn't give the user the option of proceeding with simulation in order to observe other effects in the circuit, and does not allow error propagation effects to be observed. An alternative would be to arbitrarily choose a true or false value. Here, the user may be deceived into believing that the simulation results are correct when in fact they may not accurately model the behavior of the actual circuit. The best solution is to place an unknown state in the flip-flop which indicates to the user that the state of the flip-flop has an indeterminate value.

From the standpoint of model compatibility, it is critical that all models combined during a simulation use the conventions in regards to the handling of unknown. Clearly, if some models utilize unknowns, and others don't, improper simulation results are possible and at the minimum, any utility which might have been gained from the unknown state will be lost.

Chapter 3

Combinational Devices

3.1 Simple Gates

The following models describe various SSI device models. These models follow a standard format and all are presented with full timing functionality. It is possible to reduce the simulation and model complexity in stages as follows

- Remove input delay processing
  - Remove all input timing generic parameters
  - Remove all internal delay signals
  - Remove the input delay signal assignment statements
  - Rename all references in the main behavioral process from the delayed port names to primary input port names
- Remove error checks: delete the error checking processes which occur prior to the main behavioral process
- Use simpler output delay calculations: adjust the final output assignment statements

By methodically adjusting these major components of the models it is possible to build a wide range of models which fit into the model accuracy continuum discussed in section 2.5.

3.1.1 2-Input Positive-Nand Gate

This nand gate device performs the boolean functions $y = \overline{a \cup b}$ or $y = \overline{a} \cup \overline{b}$ in positive logic and corresponds to the Texas Instruments 00 series TTL

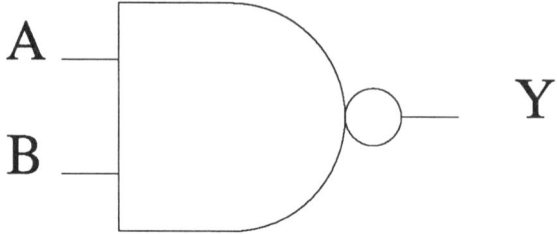

Figure 3.1: Logic Diagram 2-Input Positive-Nand

parts [TI86]. Figure 3.1 shows the symbolic representation of this device. The function table for the device is shown below

Function Table

inputs		output
a	b	y
H	H	L
L	X	H
X	L	H

The following shows the full model for a nand gate.

2-Input Positive-Nand Gate

```
USE std.std_logic.ALL;
USE std.std_ttl.ALL;
ENTITY pos2nand IS
    GENERIC (a_i01,a_i10,b_i01,b_i10,y_o01,y_o10,a_min,b_min : TIME := 2ns);
    PORT    (a,b: IN t_wlogic; y: OUT t_wlogic);
END pos2nand;

ARCHITECTURE full OF pos2nand IS
    SIGNAL adelay, bdelay : t_wlogic;
BEGIN
    -- check for spikes
    PROCESS (a)
        VARIABLE alastev : TIME := 0ns;
        BEGIN
        ASSERT (NOW = 0ns) OR ((NOW - alastev) >= a_min)
            REPORT "Spike detected" SEVERITY warning;
        alastev := NOW;
        END PROCESS;
```

3.1. SIMPLE GATES

```
    PROCESS (b)
       VARIABLE blastev : TIME := 0ns;
       BEGIN
       ASSERT (NOW = 0ns) OR ((NOW - blastev) >= b_min)
           REPORT "Spike detected" SEVERITY warning;
       blastev := NOW;
    END PROCESS;

    -- input delay processing
    adelay <= a AFTER f_delay(a,a_i01,a_i10);
    bdelay <= b AFTER f_delay(b,b_i01,b_i10);

    -- gate function
    y <= adelay NAND bdelay
        AFTER f_delay(adelay NAND bdelay, y_o01, y_o10);
END full;
```

This device has the characteristics
- Positive logic
- 2 inputs

The ports for this model are
- **a** - data input
- **b** - data input
- **y** - data output

Generic parameters to this model are summarized here
- **a_i01, a_i10** - low to high and high to low **a** input port delays
- **b_i01, b_i10** - low to high and high to low **b** input port delays
- **y_o01, y_o10** - low to high and high to low **y** output port delays
- Minimum **a** pulse width **a_min**
- Minimum **b** pulse width **b_min**

This model features two error checks
- **a** data spike detection
- **b** data spike detection

as shown in the error checking section of the architecture. Each of the error checks utilizes a separate process statement. Each of the spike detection processes uses a local variable to save the previous event time for the signal being checked. This is somewhat more efficient than using the delayed attribute since it does not require the simulator to create an additional signal.

The input delay processing section of the model incorporates the appropriate delay for each of the input ports. An internal signal for each input port is declared. The main device function process utilizes the delayed signals rather than the primary inputs.

The gate operation is handled by a single concurrent signal assignment statement which determines the output with the expression **adelay NAND bdelay**. Appropriate delay is introduced through the use of the **f_delay** function as discussed in section 7.6. This function utilizes the output rise and fall delays to choose the proper delay based on the new value.

3.1.2 2-Input Positive-Nand with Open-Collector Outputs

This nand gate device performs the boolean functions $y = \overline{a \cup b}$ or $y = \overline{a} \cup \overline{b}$ in positive logic and corresponds to the Texas Instruments 01 series TTL parts [TI86]. The open collector output requires a pull-up resistor to perform correctly. This device may be connected to other open-collector outputs to implement active-low wired-OR or active-high wired-AND functions. Figure 3.2 shows the symbolic representation of this device. The

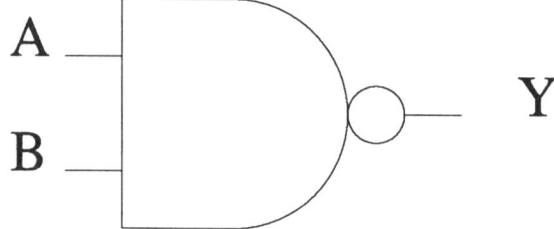

Figure 3.2: Logic Diagram Open Collector 2-Input Positive Nand

function table for the device is shown below

Function Table

inputs		output
a	b	y
H	H	L
L	X	H
X	L	H

The following shows the full model for a nand gate.

2-Input Positive-Nand with Open-Collector Outputs

3.1. SIMPLE GATES

```
USE std.std_logic.ALL;
USE std.std_ttloc.ALL;       -- note the use of TTLOC here!!
ENTITY pos2nandoc IS
    GENERIC (a_i01,a_i10,b_i01,b_i10,y_o01,y_o10,a_min,b_min : TIME := 2ns);
    PORT    (a,b: IN t_wlogic; y: OUT t_wlogic);
END pos2nandoc;

ARCHITECTURE full OF pos2nandoc IS
    SIGNAL adelay, bdelay : t_wlogic;
BEGIN
    -- check for spikes
    PROCESS (a)
        VARIABLE alastev : TIME := 0ns;
        BEGIN
        ASSERT (NOW = 0ns) OR ((NOW - alastev) >= a_min)
            REPORT "Spike detected" SEVERITY warning;
        alastev := NOW;
        END PROCESS;
    PROCESS (b)
        VARIABLE blastev : TIME := 0ns;
        BEGIN
        ASSERT (NOW = 0ns) OR ((NOW - blastev) >= b_min)
            REPORT "Spike detected" SEVERITY warning;
        blastev := NOW;
        END PROCESS;

    -- input delay processing
    adelay <= a AFTER f_delay(a,a_i01,a_i10);
    bdelay <= b AFTER f_delay(b,b_i01,b_i10);

    -- gate function
    y <= adelay NAND bdelay
        AFTER f_delay(adelay NAND bdelay, y_o01, y_o10);
END full;
```

This device has the characteristics
- Positive logic
- 2 inputs
- Open collector output

The ports for this model are
- **a** - data input
- **b** - data input
- **y** - data output

Generic parameters to this model are summarized here
- **a_i01, a_i10** - low to high and high to low **a** input port delays
- **b_i01, b_i10** - low to high and high to low **b** input port delays

- **y_o01, y_o10** - low to high and high to low **y** output port delays
- Minimum **a** pulse width **a_min**
- Minimum **b** pulse width **b_min**

This model features two error checks

- **a** data spike detection
- **b** data spike detection

as shown in the error checking section of the architecture. Each of the error checks utilizes a separate process statement. Each of the spike detection processes uses a local variable to save the previous event time for the signal being checked. This is somewhat more efficient than using the delayed attribute since it does not require the simulator to create an additional signal.

The input delay processing section of the model incorporates the appropriate delay for each of the input ports. An internal signal for each input port is declared. The main device function process utilizes the delayed signals rather than the primary inputs.

The gate operation is handled by a single concurrent signal assignment statement which calculates the output with the expression **adelay NAND bdelay**. Appropriate delay is introduced through the use of the **f_delay** function as discussed in section 7.6. This function utilizes the output rise and fall delays to choose the proper delay based on the new value.

3.1.3 2-Input Positive-Nor Gate

This nor gate device performs the boolean functions $y = \overline{a \cup b}$ or $y = \overline{a} \cap \overline{b}$ in positive logic and corresponds to the Texas Instruments 02 series TTL parts [TI86]. Figure 3.3 shows the symbolic representation of this device.

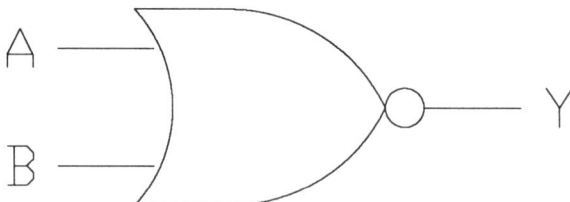

Figure 3.3: Logic Diagram 2-Input Positive-Nor

The function table for the device is shown below

3.1. SIMPLE GATES

Function Table

inputs		output
a	b	y
H	X	L
X	H	L
L	L	H

The following shows the full model for a nor gate.

2-Input Positive-Nor Gate

```
USE std.std_logic.ALL;
USE std.std_ttl.ALL;
ENTITY pos2nor IS
    GENERIC (a_i01,a_i10,b_i01,b_i10,y_o01,y_o10,a_min,b_min : TIME := 2ns);
    PORT    (a,b: IN t_wlogic; y: OUT t_wlogic);
END pos2nor;

ARCHITECTURE full OF pos2nor IS
    SIGNAL adelay, bdelay : t_wlogic;
BEGIN
    -- check for spikes
    PROCESS (a)
        VARIABLE alastev : TIME := 0ns;
        BEGIN
        ASSERT (NOW = 0ns) OR ((NOW - alastev) >= a_min)
            REPORT "Spike detected" SEVERITY warning;
        alastev := NOW;
        END PROCESS;
    PROCESS (b)
        VARIABLE blastev : TIME := 0ns;
        BEGIN
        ASSERT (NOW = 0ns) OR ((NOW - blastev) >= b_min)
            REPORT "Spike detected" SEVERITY warning;
        blastev := NOW;
        END PROCESS;

    -- input delay processing
    adelay <= a AFTER f_delay(a,a_i01,a_i10);
    bdelay <= b AFTER f_delay(b,b_i01,b_i10);

    -- gate function
    y <= adelay NOR bdelay
        AFTER f_delay(adelay NOR bdelay, y_o01, y_o10);
END full;
```

This device has the characteristics

- Positive logic

- 2 inputs

The ports for this model are

- **a** - data input
- **b** - data input
- **y** - data output

Generic parameters to this model are summarized here

- **a_i01, a_i10** - low to high and high to low **a** input port delays
- **b_i01, b_i10** - low to high and high to low **b** input port delays
- **y_o01, y_o10** - low to high and high to low **y** output port delays
- Minimum **a** pulse width **a_min**
- Minimum **b** pulse width **b_min**

This model features two error checks

- **a** data spike detection
- **b** data spike detection

as shown in the error checking section of the architecture. Each of the error checks utilizes a separate process statement. Each of the spike detection processes uses a local variable to save the previous event time for the signal being checked. This is somewhat more efficient than using the delayed attribute since it does not require the simulator to create an additional signal.

The input delay processing section of the model incorporates the appropriate delay for each of the input ports. An internal signal for each input port is declared. The main device function process utilizes the delayed signals rather than the primary inputs.

The gate operation is handled by a single concurrent signal assignment statement which calculates the output with the expression **adelay NOR bdelay**. Appropriate delay is introduced through the use of the **f_delay** function as discussed in section 7.6. This function utilizes the output rise and fall delays to choose the proper delay based on the new value.

3.1.4 Inverter

This inverter device performs the boolean function $y = \overline{a}$ and corresponds to the Texas Instruments 04 series TTL parts [TI86]. Figure 3.4 shows the symbolic representation of this device. The function table for the device is shown below

3.1. SIMPLE GATES

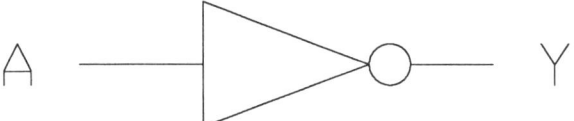

Figure 3.4: Logic Diagram Inverter

Function Table

input a	output y
H	L
L	H

The following shows the full model for an inverter gate.

Inverter

```
USE std.std_logic.ALL;
USE std.std_ttl.ALL;
ENTITY inverter IS
    GENERIC (a_i01,a_i10,y_o01,y_o10,a_min : TIME := 2ns);
    PORT    (a: IN t_wlogic; y: OUT t_wlogic);
END inverter;

ARCHITECTURE full OF inverter IS
    SIGNAL adelay : t_wlogic;
BEGIN
    -- check for spikes
    PROCESS (a)
        VARIABLE alastev : TIME := 0ns;
        BEGIN
        ASSERT (NOW = 0ns) OR ((NOW - alastev) >= a_min)
            REPORT "Spike detected" SEVERITY warning;
        alastev := NOW;
        END PROCESS;

    -- input delay processing
    adelay <= a AFTER f_delay(a,a_i01,a_i10);

    -- gate function
    y <= NOT adelay AFTER f_delay(NOT adelay, y_o01, y_o10);
END full;
```

This device has the characteristics
- Positive logic

The ports for this model are

- **a** - data input
- **y** - data output

Generic parameters to this model are summarized here

- **a_i01, a_i10** - low to high and high to low **a** input port delays
- **y_o01, y_o10** - low to high and high to low **y** output port delays
- Minimum **a** pulse width **a_min**

This model features one error check

- **a** data spike detection

as shown in the error checking section of the architecture. The error check utilizes a separate process statement. The spike detection process uses a local variable to save the previous event time for the signal being checked. This is somewhat more efficient than using the delayed attribute since it does not require the simulator to create an additional signal.

The input delay processing section of the model incorporates the appropriate delay for the input port. An internal signal for the input port is declared. The main device function process utilizes the delayed signals rather than the primary input.

The gate operation is handled by a single concurrent signal assignment statement which calculates the output with the expression **NOT adelay**. Appropriate delay is introduced through the use of the **f_delay** function as discussed in section 7.6. This function utilizes the output rise and fall delays to choose the proper delay based on the new value.

3.1.5 Inverter with Open-Collector Outputs

This inverter device performs the boolean function $y = \overline{a}$ and corresponds to the Texas Instruments 05 series TTL parts [TI86]. The open collector output requires a pull-up resistor to perform correctly. This device may be connected to other open-collector outputs to implement active-low wired-OR or active-high wired-AND functions. Figure 3.5 shows the symbolic representation of this device. The function table for the device is shown below

Function Table

input a	output y
H	L
L	H

3.1. SIMPLE GATES

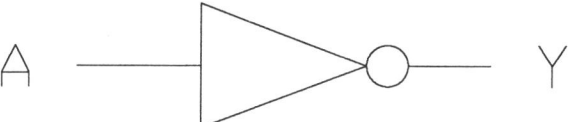

Figure 3.5: Logic Diagram Open-Collector Inverter

The following shows the full model for an inverter gate with the characteristics

- Positive logic
- Open collector output

Inverter with Open-Collector Outputs

```
USE std.std_logic.ALL;
USE std.std_ttloc.ALL;
ENTITY inverter IS
    GENERIC (a_i01,a_i10,y_o01,y_o10,a_min : TIME := 2ns);
    PORT    (a: IN t_wlogic; y: OUT t_wlogic);
END inverter;

ARCHITECTURE full OF inverter IS
    SIGNAL adelay : t_wlogic;
BEGIN
    -- check for spikes
    PROCESS (a)
        VARIABLE alastev : TIME := 0ns;
        BEGIN
        ASSERT (NOW = 0ns) OR ((NOW - alastev) >= a_min)
            REPORT "Spike detected" SEVERITY warning;
        alastev := NOW;
        END PROCESS;

    -- input delay processing
    adelay <= a AFTER f_delay(a,a_i01,a_i10);

    -- gate function
    y <= NOT adelay AFTER f_delay(NOT adelay, y_o01, y_o10);
END full;
```

The ports for this model are

- **a** - data input
- **y** - data output

Generic parameters to this model are summarized here

- **a_i01, a_i10** - low to high and high to low **a** input port delays
- **y_o01, y_o10** - low to high and high to low **y** output port delays
- Minimum **a** pulse width **a_min**

This model features one error check

- **a** data spike detection

as shown in the error checking section of the architecture. The error check utilizes a separate process statement. The spike detection process uses a local variable to save the previous event time for the signal being checked. This is somewhat more efficient than using the delayed attribute since it does not require the simulator to create an additional signal.

The input delay processing section of the model incorporates the appropriate delay for the input port. An internal signal for the input port is declared. The main device function process utilizes the delayed signals rather than the primary input.

The gate operation is handled by a single concurrent signal assignment statement which calculates the output with the expression **NOT adelay**. Appropriate delay is introduced through the use of the **f_delay** function as discussed in section 7.6. This function utilizes the output rise and fall delays to choose the proper delay based on the new value.

3.1.6 3-Input Positive-And Gate

This and gate device performs the boolean functions $y = a \cap b \cap c$ or $y = \overline{a} \cup \overline{b} \cup \overline{c}$ in positive logic and corresponds to the Texas Instruments 11 series TTL parts [TI86]. Figure 3.6 shows the symbolic representation of

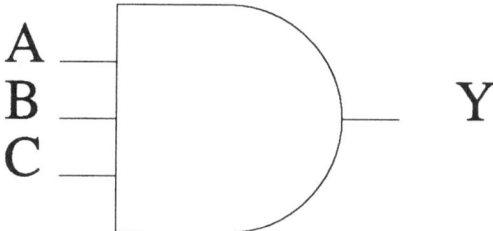

Figure 3.6: Logic Diagram 3-Input Positive-And

this device. The function table for the device is shown below

3.1. SIMPLE GATES

Function Table

inputs			output
a	b	c	y
H	H	H	H
L	X	X	L
X	L	X	L
X	X	L	L

The following shows the full model for an and gate with the characteristics
- Positive logic
- 3 inputs

3-Input Positive-And Gate

```
USE std.std_logic.ALL;
USE std.std_ttl.ALL;
ENTITY pos3and IS
    GENERIC (a_i01,a_i10,
             b_i01,b_i10,
             c_i01,c_i10,
             y_o01,y_o10,
             a_min,b_min,c_min : TIME := 2ns);
    PORT    (a,b,c: IN t_wlogic; y: OUT t_wlogic);
END pos3and;

ARCHITECTURE full OF pos3and IS
    SIGNAL adelay, bdelay, cdelay : t_wlogic;
BEGIN
    -- check for spikes
    PROCESS (a)
        VARIABLE alastev : TIME := 0ns;
        BEGIN
        ASSERT (NOW = 0ns) OR ((NOW - alastev) >= a_min)
            REPORT "Spike detected" SEVERITY warning;
        alastev := NOW;
        END PROCESS;
    PROCESS (b)
        VARIABLE blastev : TIME := 0ns;
        BEGIN
        ASSERT (NOW = 0ns) OR ((NOW - blastev) >= b_min)
            REPORT "Spike detected" SEVERITY warning;
        blastev := NOW;
        END PROCESS;
    PROCESS (c)
        VARIABLE clastev : TIME := 0ns;
        BEGIN
        ASSERT (NOW = 0ns) OR ((NOW - clastev) >= c_min)
```

```
                REPORT "Spike detected" SEVERITY warning;
        clastev := NOW;
        END PROCESS;

    -- input delay processing
    adelay <= a AFTER f_delay(a,a_i01,a_i10);
    bdelay <= b AFTER f_delay(b,b_i01,b_i10);
    cdelay <= c AFTER f_delay(c,c_i01,c_i10);

    -- gate function
    y <= adelay AND bdelay AND cdelay
         AFTER f_delay(adelay AND bdelay AND cdelay, y_o01, y_o10);
END full;
```

The ports for this model are

- a - data input
- b - data input
- c - data input
- y - data output

Generic parameters to this model are summarized here

- a_i01, a_i10 - low to high and high to low a input port delays
- b_i01, b_i10 - low to high and high to low b input port delays
- c_i01, c_i10 - low to high and high to low c input port delays
- y_o01, y_o10 - low to high and high to low y output port delays
- Minimum a pulse width a_min
- Minimum b pulse width b_min
- Minimum c pulse width c_min

This model features three error checks

- a data spike detection
- b data spike detection
- c data spike detection

as shown in the error checking section of the architecture. Each of the error checks utilizes a separate process statement. Each of the spike detection processes uses a local variable to save the previous event time for the signal being checked. This is somewhat more efficient than using the delayed attribute since it does not require the simulator to create an additional signal.

The input delay processing section of the model incorporates the appropriate delay for each of the input ports. An internal signal for each input

3.1. SIMPLE GATES

port is declared. The main device function process utilizes the delayed signals rather than the primary inputs.

The gate operation is handled by a single concurrent signal assignment statement which calculates the output with the expression **adelay AND bdelay AND cdelay**. Appropriate delay is introduced through the use of the **f_delay** function as discussed in section 7.6. This function utilizes the output rise and fall delays to choose the proper delay based on the new value.

3.1.7 3-Input Positive-Nand Gate

This nand gate device performs the boolean functions $y = \overline{a \cap b \cap c}$ or $y = \overline{a} \cup \overline{b} \cup \overline{c}$ in positive logic and corresponds to the Texas Instruments 12 series TTL parts [TI86]. Figure 3.7 shows the symbolic representation of

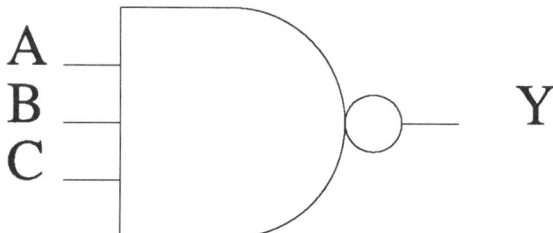

Figure 3.7: Logic Diagram 3-Input Positive-Nand

this device. The function table for the device is shown below

Function Table

inputs			output
a	b	c	y
H	H	H	L
L	X	X	H
X	L	X	H
X	X	L	H

The following shows the full model for a nand gate with the characteristics
- Positive logic
- 3 inputs

3-Input Positive-Nand Gate

```
USE std.std_logic.ALL;
USE std.std_ttl.ALL;
ENTITY pos3nand IS
    GENERIC (a_i01,a_i10,
             b_i01,b_i10,
             c_i01,c_i10,
             y_o01,y_o10,
             a_min,b_min,c_min : TIME := 2ns);
    PORT    (a,b,c: IN t_wlogic; y: OUT t_wlogic);
END pos3nand;

ARCHITECTURE full OF pos3nand IS
    SIGNAL adelay, bdelay, cdelay : t_wlogic;
BEGIN
    -- check for spikes
    PROCESS (a)
        VARIABLE alastev : TIME := 0ns;
        BEGIN
        ASSERT (NOW = 0ns) OR ((NOW - alastev) >= a_min)
            REPORT "Spike detected" SEVERITY warning;
        alastev := NOW;
        END PROCESS;
    PROCESS (b)
        VARIABLE blastev : TIME := 0ns;
        BEGIN
        ASSERT (NOW = 0ns) OR ((NOW - blastev) >= b_min)
            REPORT "Spike detected" SEVERITY warning;
        blastev := NOW;
        END PROCESS;
    PROCESS (c)
        VARIABLE clastev : TIME := 0ns;
        BEGIN
        ASSERT (NOW = 0ns) OR ((NOW - clastev) >= c_min)
            REPORT "Spike detected" SEVERITY warning;
        clastev := NOW;
        END PROCESS;

    -- input delay processing
    adelay <= a AFTER f_delay(a,a_i01,a_i10);
    bdelay <= b AFTER f_delay(b,b_i01,b_i10);
    cdelay <= c AFTER f_delay(c,c_i01,c_i10);

    -- gate function
    y <= NOT (adelay AND bdelay AND cdelay)
        AFTER f_delay(NOT(adelay AND bdelay AND cdelay), y_o01, y_o10);
END full;
```

The ports for this model are

- a - data input

3.1. SIMPLE GATES

- b - data input
- c - data input
- y - data output

Generic parameters to this model are summarized here

- a_i01, a_i10 - low to high and high to low a input port delays
- b_i01, b_i10 - low to high and high to low b input port delays
- c_i01, c_i10 - low to high and high to low c input port delays
- y_o01, y_o10 - low to high and high to low y output port delays
- Minimum a pulse width a_min
- Minimum b pulse width b_min
- Minimum c pulse width c_min

This model features three error checks

- a data spike detection
- b data spike detection
- c data spike detection

as shown in the error checking section of the architecture. Each of the error checks utilizes a separate process statement. Each of the spike detection processes uses a local variable to save the previous event time for the signal being checked. This is somewhat more efficient than using the delayed attribute since it does not require the simulator to create an additional signal.

The input delay processing section of the model incorporates the appropriate delay for each of the input ports. An internal signal for each input port is declared. The main device function process utilizes the delayed signals rather than the primary inputs.

The gate operation is handled by a single concurrent signal assignment statement which calculates the output with the expression **adelay NAND bdelay NAND cdelay**. Appropriate delay is introduced through the use of the **f_delay** function as discussed in section 7.6. This function utilizes the output rise and fall delays to choose the proper delay based on the new value.

3.1.8 2-Input Positive-Or Gate

This or gate device performs the boolean functions $y = a \cup b$ or $y = \overline{a} \cap \overline{b}$ in positive logic and corresponds to the Texas Instruments 32 series TTL parts [TI86]. Figure 3.8 shows the symbolic representation of this device. The function table for the device is shown below

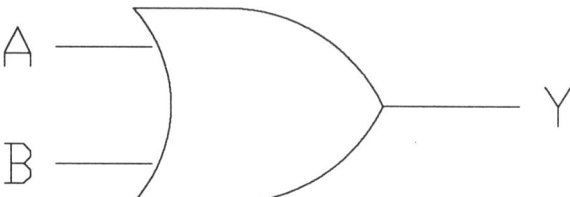

Figure 3.8: Logic Diagram 2-Input Positive-Or

Function Table

inputs		output
a	b	y
H	X	H
X	H	H
L	L	L

The following shows the full model for an or gate with the characteristics
- Positive logic
- 2 inputs

3.1. SIMPLE GATES

2-Input Positive-Or Gate

```
USE std.std_logic.ALL;
USE std.std_ttl.ALL;
ENTITY pos2or IS
    GENERIC (a_i01,a_i10,b_i01,b_i10,y_o01,y_o10,a_min,b_min : TIME := 2ns);
    PORT    (a,b: IN t_wlogic; y: OUT t_wlogic);
END pos2or;

ARCHITECTURE full OF pos2or IS
    SIGNAL adelay, bdelay : t_wlogic;
BEGIN
    -- check for spikes
    PROCESS (a)
        VARIABLE alastev : TIME := 0ns;
        BEGIN
        ASSERT (NOW = 0ns) OR ((NOW - alastev) >= a_min)
            REPORT "Spike detected" SEVERITY warning;
        alastev := NOW;
        END PROCESS;
    PROCESS (b)
        VARIABLE blastev : TIME := 0ns;
        BEGIN
        ASSERT (NOW = 0ns) OR ((NOW - blastev) >= b_min)
            REPORT "Spike detected" SEVERITY warning;
        blastev := NOW;
        END PROCESS;

    -- input delay processing
    adelay <= a AFTER f_delay(a,a_i01,a_i10);
    bdelay <= b AFTER f_delay(b,b_i01,b_i10);

    -- gate function
    y <= adelay OR bdelay
        AFTER f_delay(adelay OR bdelay, y_o01, y_o10);
END full;
```

The ports for this model are

- a - data input
- b - data input
- y - data output

Generic parameters to this model are summarized here

- a_i01, a_i10 - low to high and high to low **a** input port delays
- b_i01, b_i10 - low to high and high to low **b** input port delays
- y_o01, y_o10 - low to high and high to low **y** output port delays
- Minimum **a** pulse width **a_min**

- Minimum b pulse width **b_min**

This model features two error checks

- a data spike detection
- b data spike detection

as shown in the error checking section of the architecture. Each of the error checks utilizes a separate process statement. Each of the spike detection processes uses a local variable to save the previous event time for the signal being checked. This is somewhat more efficient than using the delayed attribute since it does not require the simulator to create an additional signal.

The input delay processing section of the model incorporates the appropriate delay for each of the input ports. An internal signal for each input port is declared. The main device function process utilizes the delayed signals rather than the primary inputs.

The gate operation is handled by a single concurrent signal assignment statement which calculates the output with the expression **adelay OR bdelay**. Appropriate delay is introduced through the use of the **f_delay** function as discussed in section 7.6. This function utilizes the output rise and fall delays to choose the proper delay based on the new value.

3.1.9 2-Input Positive-Xor Gate

This xor gate device performs the boolean functions $y = a \oplus b$ or $y = (\overline{a} \cap b) \cup (a \cap \overline{b})$ in positive logic and corresponds to the Texas Instruments 86 series TTL parts [TI86]. Figure 3.9 shows the symbolic representation

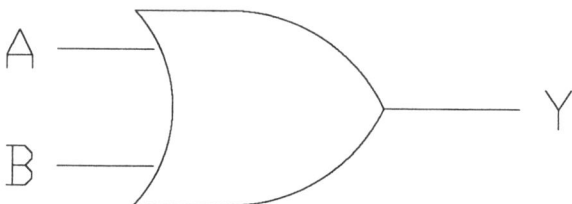

Figure 3.9: Logic Diagram 2-Input Positive-Xor

of this device. The function table for the device is shown below

3.1. SIMPLE GATES

Function Table

inputs		output
a	b	y
L	L	L
L	H	H
H	L	H
H	H	L

The following shows the full model for an xor gate.

2-Input Positive-Xor Gate

```
USE std.std_logic.ALL;
USE std.std_ttl.ALL;
ENTITY pos2xor IS
    GENERIC (a_i01,a_i10,b_i01,b_i10,y_o01,y_o10,a_min,b_min : TIME := 2ns);
    PORT    (a,b: IN t_wlogic; y: OUT t_wlogic);
END pos2xor;

ARCHITECTURE full OF pos2xor IS
    SIGNAL adelay, bdelay : t_wlogic;
BEGIN
    -- check for spikes
    PROCESS (a)
        VARIABLE alastev : TIME := 0ns;
        BEGIN
        ASSERT (NOW = 0ns) OR ((NOW - alastev) >= a_min)
            REPORT "Spike detected" SEVERITY warning;
        alastev := NOW;
        END PROCESS;
    PROCESS (b)
        VARIABLE blastev : TIME := 0ns;
        BEGIN
        ASSERT (NOW = 0ns) OR ((NOW - blastev) >= b_min)
            REPORT "Spike detected" SEVERITY warning;
        blastev := NOW;
        END PROCESS;

    -- input delay processing
    adelay <= a AFTER f_delay(a,a_i01,a_i10);
    bdelay <= b AFTER f_delay(b,b_i01,b_i10);

    -- gate function
    y <= adelay XOR bdelay
        AFTER f_delay(adelay XOR bdelay, y_o01, y_o10);
END full;
```

The device has the characteristics

- Positive logic
- 2 inputs

The ports for this model are

- **a** - data input
- **b** - data input
- **y** - data output

Generic parameters to this model are summarized here

- **a_i01, a_i10** - low to high and high to low **a** input port delays
- **b_i01, b_i10** - low to high and high to low **b** input port delays
- **y_o01, y_o10** - low to high and high to low **y** output port delays
- Minimum **a** pulse width **a_min**
- Minimum **b** pulse width **b_min**

This model features two error checks

- **a** data spike detection
- **b** data spike detection

as shown in the error checking section of the architecture. Each of the error checks utilizes a separate process statement. Each of the spike detection processes uses a local variable to save the previous event time for the signal being checked. This is somewhat more efficient than using the delayed attribute since it does not require the simulator to create an additional signal.

The input delay processing section of the model incorporates the appropriate delay for each of the input ports. An internal signal for each input port is declared. The main device function process utilizes the delayed signals rather than the primary inputs.

The gate operation is handled by a single concurrent signal assignment statement which calculates the output with the expression **adelay XOR bdelay**. Appropriate delay is introduced through the use of the **f_delay** function as discussed in section 7.6. This function utilizes the output rise and fall delays to choose the proper delay based on the new value.

3.2 Selectors/Multiplexers

The following models describe various types of multiplexer devices. These models follow a standard format and all are presented with full timing functionality. It is possible to reduce the simulation and model complexity in stages as follows

3.2. SELECTORS/MULTIPLEXERS

- Remove input delay processing

  – Remove all input timing generic parameters

  – Remove all internal delay signals

  – Remove the input delay signal assignment statements

  – Rename all references in the main behavioral process from the delayed port names to primary input port names

- Remove error checks: delete the error checking processes which occur prior to the main behavioral process

- Use simpler output delay calculations: adjust the final output assignment statements

By methodically adjusting these major components of the models it is possible to build a wide range of models which fit into the model accuracy continuum discussed in section 2.5.

The following adaptations are possible starting from these base models:

- Adjust the number of address bits; several examples are shown which range from single to 3 bit addressing. Adding additional bits is analogous to the approaches taken in these models.

3.2.1 3 to 8 Decoder/Multiplexer

This device is designed for memory-decoding or data-routing applications. The binary select inputs **a**, **b** and **c** and the three enable inputs **g1** (active-high), **g2a** and **g2b** (active-low) select one of eight input lines. This device corresponds to the Texas Instruments 138 series TTL parts [TI86]. Figure 3.10 shows the symbolic representation of this device. The function table for the device is shown below

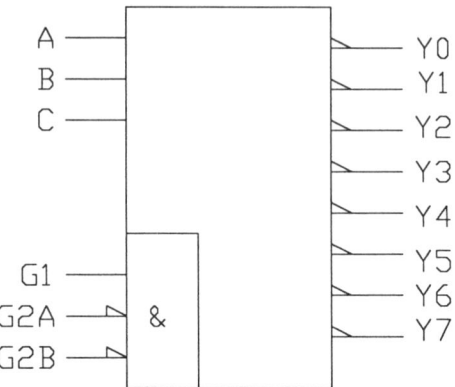

Figure 3.10: Logic Diagram 3 to 8 Decoder/Multiplexer

Function Table

enable inputs			select inputs			outputs outputs							
g1	g2a	g2b	c	b	a	y0	y1	y2	y3	y4	y5	y6	y7
X	H	X	X	X	X	H	H	H	H	H	H	H	H
X	X	H	X	X	X	H	H	H	H	H	H	H	H
L	X	X	X	X	X	H	H	H	H	H	H	H	H
H	L	L	L	L	L	L	H	H	H	H	H	H	H
H	L	L	L	L	H	H	L	H	H	H	H	H	H
H	L	L	L	H	L	H	H	L	H	H	H	H	H
H	L	L	L	H	H	H	H	H	L	H	H	H	H
H	L	L	H	L	L	H	H	H	H	L	H	H	H
H	L	L	H	L	H	H	H	H	H	H	L	H	H
H	L	L	H	H	L	H	H	H	H	H	H	L	H
H	L	L	H	H	H	H	H	H	H	H	H	H	L

The following shows the full model for a decoder with the characteristics
- 3 bit address
- 8 decode outputs
- Asynchronous preset control; two active high controls, and one active low control
- Asynchronous decode operation

3.2. SELECTORS/MULTIPLEXERS

3 to 8 Decoder/Multiplexer

```
USE std.std_logic.ALL;
USE std.std_ttl.ALL;
ENTITY decmul3to8latch IS
    GENERIC (g1_i01,g1_i10,
             g2a_i01,g2a_i10,
             g2b_i01,g2b_i10,
             a_i01,a_i10,
             b_i01,b_i10,
             c_i01,c_i10,
             y0_o01,y0_o10,
             y1_o01,y1_o10,
             y2_o01,y2_o10,
             y3_o01,y3_o10,
             y4_o01,y4_o10,
             y5_o01,y5_o10,
             y6_o01,y6_o10,
             y7_o01,y7_o10 : TIME := 2 ns);
    PORT    (g1,g2a,g2b,a,b,c: IN t_wlogic;
             y0,y1,y2,y3,y4,y5,y6,y7: OUT t_wlogic);
END decmul3to8latch;

ARCHITECTURE full OF decmul3to8latch IS
    SIGNAL g1delay,g2adelay,g2bdelay,adelay,bdelay,cdelay : t_wlogic;
BEGIN
    -- input delay processing
    g1delay <= g1 AFTER f_delay(g1,g1_i01,g1_i10);
    g2adelay <= g2a AFTER f_delay(g2a,g2a_i01,g2a_i10);
    g2bdelay <= g2b AFTER f_delay(g2b,g2b_i01,g2b_i10);
    adelay <= a AFTER f_delay(a,a_i01,a_i10);
    bdelay <= b AFTER f_delay(b,b_i01,b_i10);
    cdelay <= c AFTER f_delay(c,c_i01,c_i10);

    -- decoder/multiplexer operation
    PROCESS (g1delay,g2adelay,g2bdelay,adelay,bdelay,cdelay)
        VARIABLE baddress : t_logarray (1 TO 3);
        VARIABLE state : t_logarray (1 TO 8) := (U,U,U,U,U,U,U,U);
        VARIABLE iaddress : integer;
        VARIABLE unknown : boolean;
    BEGIN
        -- check for preset
        IF (g1delay = '0') OR (g2adelay = '1') OR (g2bdelay = '1') THEN
            FOR i IN state'RANGE LOOP state(i) := f_ttl('1'); END LOOP;

        -- check for unknown control
        ELSIF (g1delay = 'X') OR (g2adelay = 'X') OR (g2bdelay = 'X') THEN
            FOR i IN state'RANGE LOOP state(i) := f_ttl('X'); END LOOP;

        -- decode
        ELSE
            -- pickup the binary address
```

```
            baddress(1) := cdelay;
            baddress(2) := bdelay;
            baddress(3) := adelay;

            -- calculate the integer address
            f_logictoint(baddress,unknown,iaddress);

            -- watch out for unknown address
            IF unknown THEN
                FOR i IN state'RANGE LOOP state(i) := f_ttl('X'); END LOOP;
            -- decode the address
            ELSE f_intdecode(iaddress, state, ttl);
            END IF;
        END IF;

        -- assign values to output signals
        y0 <= state(1) AFTER f_delay(state(1), y0_o01, y0_o10);
        y1 <= state(2) AFTER f_delay(state(2), y1_o01, y1_o10);
        y2 <= state(3) AFTER f_delay(state(3), y2_o01, y2_o10);
        y3 <= state(4) AFTER f_delay(state(4), y3_o01, y3_o10);
        y4 <= state(5) AFTER f_delay(state(5), y4_o01, y4_o10);
        y5 <= state(6) AFTER f_delay(state(6), y5_o01, y5_o10);
        y6 <= state(7) AFTER f_delay(state(7), y6_o01, y6_o10);
        y7 <= state(8) AFTER f_delay(state(8), y7_o01, y7_o10);
    END PROCESS;
END full;
```

The ports for this model are

- **g1** - active high preset
- **g2a, g2b** - active low preset
- **a** - address input
- **b** - address input
- **c** - address input
- **y0** - decode output
- **y1** - decode output
- **y2** - decode output
- **y3** - decode output
- **y4** - decode output
- **y5** - decode output
- **y6** - decode output
- **y7** - decode output

Generic parameters to this model are summarized here

3.2. SELECTORS/MULTIPLEXERS

- **g1_i01, g1_i10** - low to high and high to low **g1** input port delays
- **g2a_i01, g2a_i10** - low to high and high to low **g2a** input port delays
- **g2b_i01, g2b_i10** - low to high and high to low **g2b** input port delays
- **a_i01, a_i10** - low to high and high to low **a** input port delays
- **b_i01, b_i10** - low to high and high to low **b** input port delays
- **c_i01, c_i10** - low to high and high to low **c** input port delays
- **y0_o01, y0_o10** - low to high and high to low **y0** output port delays
- **y1_o01, y1_o10** - low to high and high to low **y1** output port delays
- **y2_o01, y2_o10** - low to high and high to low **y2** output port delays
- **y3_o01, y3_o10** - low to high and high to low **y3** output port delays
- **y4_o01, y4_o10** - low to high and high to low **y4** output port delays
- **y5_o01, y5_o10** - low to high and high to low **y5** output port delays
- **y6_o01, y6_o10** - low to high and high to low **y6** output port delays
- **y7_o01, y7_o10** - low to high and high to low **y7** output port delays

This model performs no error checks. The input delay processing section of the model incorporates the appropriate delay for each of the input ports. An internal signal for each input port is declared. The main device function process utilizes the delayed signals rather than the primary inputs.

The decode operation is handled by a single process which is sensitive to the input delayed signals **g1delay**, **g2adelay**, **g2bdelay**, **adelay**, **bdelay** and **cdelay**. The following conditions are checked in order whenever any of the asynchronous inputs have an event

- Preset device - set all outputs high
- Check for unknown control - set all outputs to unknown
- Decode operation - set appropriate output low, all others high

Note that the order of the above checks are critical to the correct operation of the model. In particular the asynchronous preset takes precedence over the decode operation of the device.

Once the state of the device is determined (and stored in the local variable **state** the output signal assignment statements for **y0**, **y1**, **y2**, **y3**, **y4**, **y5**, **y6** and **y7** are executed. Output state changes are executed with appropriate delay introduced through the use of the **f_delay** function as discussed in section 7.6. This function utilizes the output rise and fall delays and based on the new value chooses the proper delay.

3.2.2 2 to 4 Decoder/Multiplexer

This device is designed for memory-decoding or data-routing applications. The active-low enable g input can be used as a data line in demultiplexing applications. This device corresponds to the Texas Instruments 139 series TTL parts [TI86]. Figure 3.11 shows the symbolic representation of this

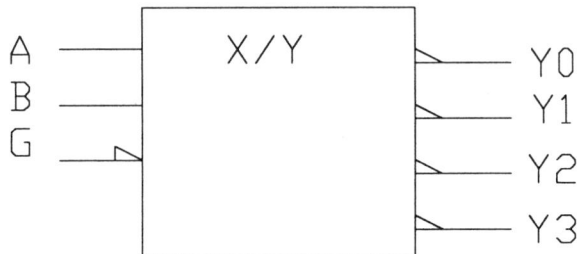

Figure 3.11: Logic Diagram 2 to 4 Decoder/Multiplexer

device. The function table for the device is shown below

Function Table

inputs		outputs			
enable	select				
g	b	a	y0	y1	y2	y3
H	X	X	H	H	H	H
L	L	L	L	H	H	H
L	L	H	H	L	H	H
L	H	L	H	H	L	H
L	H	H	H	H	H	L

The following shows the full model for a decoder with the characteristics

- 2 bit address
- 4 decode outputs
- Asynchronous preset control
- Asynchronous decode operation

3.2. SELECTORS/MULTIPLEXERS

2 to 4 Decoder/Multiplexer

```
USE std.std_logic.ALL;
USE std.std_ttl.ALL;
ENTITY decmul2to4 IS
    GENERIC (g_i01,g_i10,
             a_i01,a_i10,
             b_i01,b_i10,
             y0_o01,y0_o10,
             y1_o01,y1_o10,
             y2_o01,y2_o10,
             y3_o01,y3_o10 : TIME := 2 ns);
    PORT    (g,a,b: IN t_wlogic;
             y0,y1,y2,y3: OUT t_wlogic);
END decmul2to4;

ARCHITECTURE full OF decmul2to4 IS
    SIGNAL gdelay,adelay,bdelay : t_wlogic;
BEGIN
    -- input delay processing
    gdelay <= g AFTER f_delay(g,g_i01,g_i10);
    adelay <= a AFTER f_delay(a,a_i01,a_i10);
    bdelay <= b AFTER f_delay(b,b_i01,b_i10);

    -- decoder/multiplexer operation
    PROCESS (gdelay,adelay,bdelay)
        VARIABLE baddress : t_logarray (1 TO 2);
        VARIABLE state : t_logarray (1 TO 4) := (U,U,U,U);
        VARIABLE iaddress : integer;
        VARIABLE unknown : boolean;
    BEGIN
        -- check for preset
        IF (gdelay = '1') THEN
            FOR i IN state'RANGE LOOP state(i) := f_ttl('1'); END LOOP;

        -- check for unknown control
        ELSIF (gdelay = 'X') THEN
            FOR i IN state'RANGE LOOP state(i) := f_ttl('X'); END LOOP;

        -- decode
        ELSE
            -- pickup the binary address
            baddress(1) := bdelay;
            baddress(2) := adelay;

            -- calculate the integer address
            f_logictoint(baddress,unknown,iaddress);

            -- watch out for unknown address
            IF unknown THEN
                FOR i IN state'RANGE LOOP state(i) := f_ttl('X'); END LOOP;
            -- decode the address
```

```
            ELSE f_intdecode(iaddress, state, ttl);
            END IF;
        END IF;

        -- assign values to output signals
        y0 <= state(1) AFTER f_delay(state(1), y0_o01, y0_o10);
        y1 <= state(2) AFTER f_delay(state(2), y1_o01, y1_o10);
        y2 <= state(3) AFTER f_delay(state(3), y2_o01, y2_o10);
        y3 <= state(4) AFTER f_delay(state(4), y3_o01, y3_o10);
    END PROCESS;
END full;
```

The ports for this model are

- **g** - preset input
- **a** - address input
- **b** - address input
- **y0** - decode output
- **y1** - decode output
- **y2** - decode output
- **y3** - decode output

Generic parameters to this model are summarized here

- **g_i01, g_i10** - low to high and high to low **g** input port delays
- **a_i01, a_i10** - low to high and high to low **a** input port delays
- **b_i01, b_i10** - low to high and high to low **b** input port delays
- **y0_o01, y0_o10** - low to high and high to low **y0** output port delays
- **y1_o01, y1_o10** - low to high and high to low **y1** output port delays
- **y2_o01, y2_o10** - low to high and high to low **y2** output port delays
- **y3_o01, y3_o10** - low to high and high to low **y3** output port delays

This model performs no error checks. The input delay processing section of the model incorporates the appropriate delay for each of the input ports. An internal signal for each input port is declared. The main device function process utilizes the delayed signals rather than the primary inputs.

The decode operation is handled by a single process which is sensitive to the input delayed signals **gdelay**, **adelay** and **bdelay**. The following conditions are checked in order whenever any of the asynchronous inputs have an event

- Preset device - set all outputs high
- Check for unknown control - set all outputs to unknown

3.2. SELECTORS/MULTIPLEXERS

- Decode operation - set appropriate output low, all others high; if unknowns are seen then output all unknowns

Note that the order of the above checks are critical to the correct operation of the model. In particular the asynchronous preset takes precedence over the decode operation of the device.

Once the state of the device is determined (and stored in the local variable **state** the output signal assignment statements for **y0**, **y1**, **y2** and **y3** are executed. Output state changes are executed with appropriate delay introduced through the use of the **f_delay** function as discussed in section 7.6. This function utilizes the output rise and fall delays and based on the new value chooses the proper delay.

3.2.3 1 of 8 Selector/Multiplexer

This data selector/multiplexer provides full binary decoding to select one of eight data sources. The strobe input (**g**) must be low to enable the inputs. A high at the strobe **g** forces the **w** output high and the **y** output low. This device corresponds to the Texas Instruments 151 series TTL parts [TI86]. Figure 3.12 shows the symbolic representation of this device. The function

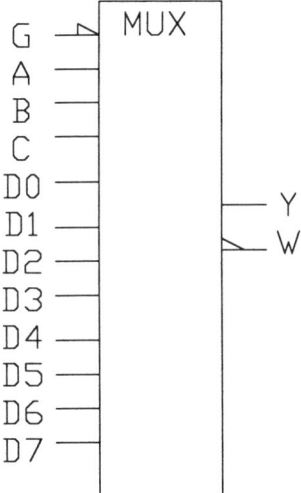

Figure 3.12: Logic Diagram 1 of 8 Selector/Multiplexer

table for the device is shown below

Function Table

inputs				outputs	
select			strobe		
c	b	a	g	y	w
X	X	X	H	L	H
L	L	L	L	d0	not d0
L	L	H	L	d1	not d1
L	H	L	L	d2	not d2
L	H	H	L	d3	not d3
H	L	L	L	d4	not d4
H	L	H	L	d5	not d5
H	H	L	L	d6	not d6
H	H	H	L	d7	not d7

The following shows the full model for a selector with the characteristics

- 3 bit address
- 8 select inputs
- Asynchronous clear control
- Asynchronous select operation
- Select output with complement

1 of 8 Selector/Multiplexer

```
USE std.std_logic.ALL;
USE std.std_ttl.ALL;
ENTITY selmul1of8 IS
    GENERIC (g_i01,g_i10,
             a_i01,a_i10,
             b_i01,b_i10,
             c_i01,c_i10,
             d0_i01,d0_i10,
             d1_i01,d1_i10,
             d2_i01,d2_i10,
             d3_i01,d3_i10,
             d4_i01,d4_i10,
             d5_i01,d5_i10,
             d6_i01,d6_i10,
             d7_i01,d7_i10,
             y_o01,y_o10,
             w_o01,w_o10 : TIME := 2 ns);
    PORT (g,a,b,c,d0,d1,d2,d3,d4,d5,d6,d7: IN t_wlogic;
          y,w: OUT t_wlogic);
END selmul1of8;
```

3.2. SELECTORS/MULTIPLEXERS

```
ARCHITECTURE full OF selmul1of8 IS
    SIGNAL gdelay,adelay,bdelay,cdelay,
           d0delay,d1delay,d2delay,d3delay,
           d4delay,d5delay,d6delay,d7delay : t_wlogic;
BEGIN
    -- input delay processing
    gdelay <= g AFTER f_delay(g,g_i01,g_i10);
    adelay <= a AFTER f_delay(a,a_i01,a_i10);
    bdelay <= b AFTER f_delay(b,b_i01,b_i10);
    cdelay <= c AFTER f_delay(c,c_i01,c_i10);

    d0delay <= d0 AFTER f_delay(d0,d0_i01,d0_i10);
    d1delay <= d1 AFTER f_delay(d1,d1_i01,d1_i10);
    d2delay <= d2 AFTER f_delay(d2,d2_i01,d2_i10);
    d3delay <= d3 AFTER f_delay(d3,d3_i01,d3_i10);
    d4delay <= d4 AFTER f_delay(d4,d4_i01,d4_i10);
    d5delay <= d5 AFTER f_delay(d5,d5_i01,d5_i10);
    d6delay <= d6 AFTER f_delay(d6,d6_i01,d6_i10);
    d7delay <= d7 AFTER f_delay(d7,d7_i01,d7_i10);

    -- select operation
    PROCESS (gdelay,adelay,bdelay,cdelay,
             d0delay,d1delay,d2delay,d3delay,
             d4delay,d5delay,d6delay,d7delay)
        VARIABLE baddress : t_logarray (1 TO 3);
        VARIABLE iaddress : integer;
        VARIABLE ystate,wstate : t_logic;
        VARIABLE unknown : boolean;
        BEGIN
        -- check for clear
        IF (gdelay = '1') THEN
            ystate := f_ttl('0');
            wstate := f_ttl('1');

        -- check for unknown control
        ELSIF (gdelay = 'X') THEN
            ystate := f_ttl('X');
            wstate := f_ttl('X');

        -- select
        ELSE
            -- pickup the binary address
            baddress(1) := cdelay;
            baddress(2) := bdelay;
            baddress(3) := adelay;

            -- calculate the integer address
            f_logictoint(baddress,unknown,iaddress);

            -- unknown address?
            IF unknown THEN
```

```
                    ystate := f_ttl('X');
                    wstate := f_ttl('X');
                ELSE
                    -- select output
                    CASE iaddress IS
                        WHEN 0 =>
                            ystate := d0delay;
                            wstate := NOT d0delay;
                        WHEN 1 =>
                            ystate := d1delay;
                            wstate := NOT d1delay;
                        WHEN 2 =>
                            ystate := d2delay;
                            wstate := NOT d2delay;
                        WHEN 3 =>
                            ystate := d3delay;
                            wstate := NOT d3delay;
                        WHEN 4 =>
                            ystate := d4delay;
                            wstate := NOT d4delay;
                        WHEN 5 =>
                            ystate := d5delay;
                            wstate := NOT d5delay;
                        WHEN 6 =>
                            ystate := d6delay;
                            wstate := NOT d6delay;
                        WHEN 7 =>
                            ystate := d7delay;
                            wstate := NOT d7delay;

                        -- watch out for unknown address
                        WHEN OTHERS =>
                            ystate := f_ttl('X');
                            wstate := f_ttl('X');
                    END CASE;
                END IF;
            END IF;

            -- assign values to output signals
            y <= ystate AFTER f_delay(ystate, y_o01, y_o10);
            w <= wstate AFTER f_delay(wstate, w_o01, w_o10);
        END PROCESS;
END full;
```

The ports for this model are

- g - strobe input
- a - address input
- b - address input
- c - address input

3.2. SELECTORS/MULTIPLEXERS

- d0 - select input
- d1 - select input
- d2 - select input
- d3 - select input
- d4 - select input
- d5 - select input
- d6 - select input
- d7 - select input
- y - select output
- w - select output complement

Generic parameters to this model are summarized here

- g_i01, g_i10 - low to high and high to low **g** input port delays
- a_i01, a_i10 - low to high and high to low **a** input port delays
- b_i01, b_i10 - low to high and high to low **b** input port delays
- c_i01, c_i10 - low to high and high to low **c** input port delays
- d0_i01, d0_i10 - low to high and high to low **d0** input port delays
- d1_i01, d1_i10 - low to high and high to low **d1** input port delays
- d2_i01, d2_i10 - low to high and high to low **d2** input port delays
- d3_i01, d3_i10 - low to high and high to low **d3** input port delays
- d4_i01, d4_i10 - low to high and high to low **d4** input port delays
- d5_i01, d5_i10 - low to high and high to low **d5** input port delays
- d6_i01, d6_i10 - low to high and high to low **d6** input port delays
- d7_i01, d7_i10 - low to high and high to low **d7** input port delays
- y_i01, y_i10 - low to high and high to low **y** output port delays
- w_i01, w_i10 - low to high and high to low **w** output port delays

This model performs no error checks. The input delay processing section of the model incorporates the appropriate delay for each of the input ports. An internal signal for each input port is declared. The main device function process utilizes the delayed signals rather than the primary inputs.

The select operation is handled by a single process which is sensitive to the input delayed signals **gdelay, adelay, bdelay, cdelay, d0delay, d1delay, d2delay, d3delay, d4delay, d5delay, d6delay** and **d7delay**. The following conditions are checked in order whenever any of the asynchronous inputs have an event

- Clear device - set output low (complement is high)
- Check for unknown control - set outputs to unknown
- Select operation - propagate selected input to output (complementary output to **w**).

Note that the order of the above checks are critical to the correct operation of the model. In particular the asynchronous clear takes precedence over the select operation of the device.

Once the state of the device is determined (and stored in the local variables **ystate** and **wstate**) the output signal assignment statements for **y** and **w** are executed. Output state changes are executed with appropriate delay introduced through the use of the **f_delay** function as discussed in section 7.6. This function utilizes the output rise and fall delays and based on the new value chooses the proper delay.

3.2.4 1 of 4 Selector/Multiplexer

This data selector/multiplexer provides full binary decoding to select one of four data sources. The strobe input (**g**) must be low to enable the inputs. A high at the strobe **g** forces the **y** output low. This device corresponds to the Texas Instruments 153 series TTL parts [TI86]. Figure 3.13 shows the

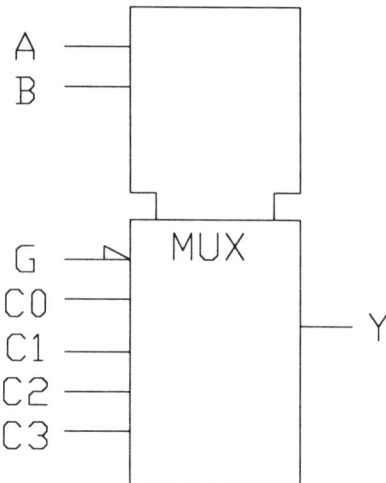

Figure 3.13: Logic Diagram 1 of 4 Selector/Multiplexer

symbolic representation of this device. The function table for the device is shown below

3.2. SELECTORS/MULTIPLEXERS

Function Table

select inputs		data				strobe	output
b	a	c0	c1	c2	c3	g	y
X	X	X	X	X	X	H	L
L	L	L	X	X	X	L	L
L	L	H	X	X	X	L	H
L	H	X	L	X	X	L	L
L	H	X	H	X	X	L	H
H	L	X	X	L	X	L	L
H	L	X	X	H	X	L	H
H	H	X	X	X	L	L	L
H	H	X	X	X	H	L	H

The following shows the full model for a selector with the characteristics

- 2 bit address
- 4 select inputs
- Asynchronous clear control
- Asynchronous select operation
- Select output

1 of 4 Selector/Multiplexer

```
USE std.std_logic.ALL;
USE std.std_ttl.ALL;
ENTITY selmul1of4 IS
    GENERIC (g_i01,g_i10,
             a_i01,a_i10,
             b_i01,b_i10,
             c0_i01,c0_i10,
             c1_i01,c1_i10,
             c2_i01,c2_i10,
             c3_i01,c3_i10,
             y_o01,y_o10 : TIME := 2 ns);
    PORT    (g,a,b,c0,c1,c2,c3: IN t_wlogic;
             y: OUT t_wlogic);
END selmul1of4;

ARCHITECTURE full OF selmul1of4 IS
    SIGNAL gdelay,adelay,bdelay,
           c0delay,c1delay,c2delay,c3delay : t_wlogic;
BEGIN
    -- input delay processing
```

```
        gdelay <= g AFTER f_delay(g,g_i01,g_i10);
        adelay <= a AFTER f_delay(a,a_i01,a_i10);
        bdelay <= b AFTER f_delay(b,b_i01,b_i10);

        c0delay <= c0 AFTER f_delay(c0,c0_i01,c0_i10);
        c1delay <= c1 AFTER f_delay(c1,c1_i01,c1_i10);
        c2delay <= c2 AFTER f_delay(c2,c2_i01,c2_i10);
        c3delay <= c3 AFTER f_delay(c3,c3_i01,c3_i10);

        -- select operation
        PROCESS (gdelay,adelay,bdelay,
                 c0delay,c1delay,c2delay,c3delay)
           VARIABLE baddress : t_logarray (1 TO 2);
           VARIABLE iaddress : integer;
           VARIABLE ystate : t_logic;
           VARIABLE unknown : boolean;
           BEGIN
           -- check for clear
           IF (gdelay = '1') THEN ystate := f_ttl('0');

           -- check for unknown control
           ELSIF (gdelay = 'X') THEN ystate := f_ttl('X');

           -- select
           ELSE
               -- pickup the binary address
               baddress(1) := bdelay;
               baddress(2) := adelay;

               -- calculate the integer address
               f_logictoint(baddress,unknown,iaddress);

               -- unknown address?
               IF unknown THEN ystate := f_ttl('X');
               ELSE
                   -- select output
                   CASE iaddress IS
                       WHEN 0 => ystate := c0delay;
                       WHEN 1 => ystate := c1delay;
                       WHEN 2 => ystate := c2delay;
                       WHEN 3 => ystate := c3delay;

                       -- watch out for unknown address
                       WHEN OTHERS => ystate := f_ttl('X');
                   END CASE;
               END IF;
           END IF;

           -- assign values to output signals
           y <= ystate AFTER f_delay(ystate, y_o01, y_o10);
        END PROCESS;
END full;
```

3.2. SELECTORS/MULTIPLEXERS

The ports for this model are
- **g** - strobe input
- **a** - address input
- **b** - address input
- **c0** - select input
- **c1** - select input
- **c2** - select input
- **c3** - select input
- **y** - select output

Generic parameters to this model are summarized here
- **g_i01, g_i10** - low to high and high to low **g** input port delays
- **a_i01, a_i10** - low to high and high to low **a** input port delays
- **b_i01, b_i10** - low to high and high to low **b** input port delays
- **c0_i01, c0_i10** - low to high and high to low **c0** input port delays
- **c1_i01, c1_i10** - low to high and high to low **c1** input port delays
- **c2_i01, c2_i10** - low to high and high to low **c2** input port delays
- **c3_i01, c3_i10** - low to high and high to low **c3** input port delays
- **y_i01, y_i10** - low to high and high to low **y** output port delays

This model performs no error checks. The input delay processing section of the model incorporates the appropriate delay for each of the input ports. An internal signal for each input port is declared. The main device function process utilizes the delayed signals rather than the primary inputs.

The select operation is handled by a single process which is sensitive to the input delayed signals **gdelay, adelay, bdelay, c0delay, c1delay, c2delay** and **c3delay**. The following conditions are checked in order whenever any of the asynchronous inputs have an event

- Clear device - set output low
- Check for unknown control - set output to unknown
- Select operation - propagate selected input to output

Note that the order of the above checks are critical to the correct operation of the model. In particular the asynchronous clear takes precedence over the select operation of the device.

Once the state of the device is determined (and stored in the local variable **ystate**) the output signal assignment statement for **y** is executed. Output state changes are executed with appropriate delay introduced through the use of the **f_delay** function as discussed in section 7.6. This function utilizes the output rise and fall delays and based on the new value chooses the proper delay.

3.2.5 1 of 2 Selector/Multiplexer

This data selector/multiplexer provides full binary decoding to select one of two data sources. The strobe input (g) must be low to enable the inputs. A high at the strobe forces the y output low. This device corresponds to the Texas Instruments 157 series TTL parts [TI86]. Figure 3.14 shows the

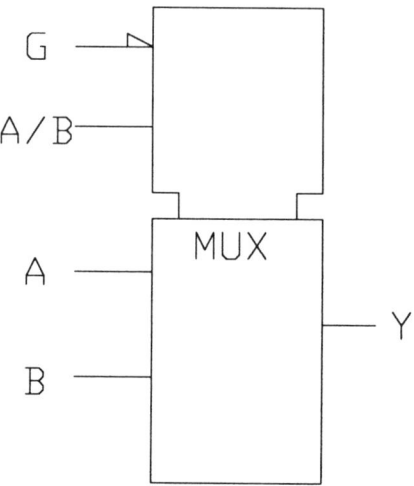

Figure 3.14: Logic Diagram 1 of 2 Selector/Multiplexer

symbolic representation of this device. The function table for the device is shown below

Function Table

inputs				output
strobe	select	data		
g	a/b	a	b	y
H	X	X	X	L
L	L	L	X	L
L	L	H	X	H
L	H	X	L	L
L	H	X	H	H

The following shows the full model for a selector.

3.2. SELECTORS/MULTIPLEXERS

1 of 2 Selector/Multiplexer

```
USE std.std_logic.ALL;
USE std.std_ttl.ALL;
ENTITY selmul1of2 IS
    GENERIC (g_i01,g_i10,
             sel_i01,sel_i10,
             a_i01,a_i10,
             b_i01,b_i10,
             y_o01,y_o10 : TIME := 2 ns);
    PORT    (g,sel,a,b: IN t_wlogic;
             y: OUT t_wlogic);
END selmul1of2;

ARCHITECTURE full OF selmul1of2 IS
    SIGNAL gdelay,seldelay, adelay,bdelay : t_wlogic;
BEGIN
    -- input delay processing
    gdelay <= g AFTER f_delay(g,g_i01,g_i10);
    seldelay <= sel AFTER f_delay(sel,sel_i01,sel_i10);
    adelay <= a AFTER f_delay(a,a_i01,a_i10);
    bdelay <= b AFTER f_delay(b,b_i01,b_i10);

    -- sel operation
    PROCESS (gdelay,seldelay,adelay,bdelay)
        VARIABLE ystate : t_logic;
        BEGIN
        -- check for clear
        IF (gdelay = '1') THEN ystate := f_ttl('0');

        -- check for unknown control
        ELSIF (gdelay = 'X') THEN ystate := f_ttl('X');

        -- select
        ELSIF (seldelay = '0') THEN ystate := adelay;
        ELSIF (seldelay = '1') THEN ystate := bdelay;

        -- select unknown
        ELSE ystate := f_ttl('X');
        END IF;

        -- assign values to output signals
        y <= ystate AFTER f_delay(ystate, y_o01, y_o10);
        END PROCESS;
END full;
```

This device has the characteristics
- 1 bit address
- 2 select inputs
- Asynchronous clear control

- Asynchronous select operation
- Select output

The ports for this model are

- **g** - strobe input
- **sel** - address input
- **a** - select input
- **b** - select input
- **y** - select output

Generic parameters to this model are summarized here

- **g_i01, g_i10** - low to high and high to low **g** input port delays
- **sel_i01, sel_i10** - low to high and high to low **sel** input port delays
- **a_i01, a_i10** - low to high and high to low **a** input port delays
- **b_i01, b_i10** - low to high and high to low **b** input port delays
- **y_i01, y_i10** - low to high and high to low **y** output port delays

This model performs no error checks. The input delay processing section of the model incorporates the appropriate delay for each of the input ports. An internal signal for each input port is declared. The main device function process utilizes the delayed signals rather than the primary inputs.

The select operation is handled by a single process which is sensitive to the input delayed signals **gdelay, seldelay, adelay** and **bdelay**. The following conditions are checked in order whenever any of the asynchronous inputs have an event

- Clear device - set output low
- Check for unknown control - set output to unknown
- Select operation - propagate selected input to output

Note that the order of the above checks are critical to the correct operation of the model. In particular the asynchronous clear takes precedence over the select operation of the device.

Once the state of the device is determined (and stored in the local variable **ystate**) the output signal assignment statement for **y** is executed. Output state changes are executed with appropriate delay introduced through the use of the **f_delay** function as discussed in section 7.6. This function utilizes the output rise and fall delays and based on the new value chooses the proper delay.

3.3 Switch Level Devices

During the original design of the VHDL language, switch level simulation was not included as a primary requirement. As a result, there is a widely held belief that VHDL cannot perform switch level modelling in which bidirectional transmission gates are simulated. Recent research into switch level algorithms and VHDL capabilities has resulted in a methodology which allows accurate switch level simulation to be performed using VHDL without extensions. This section describes the algorithm which allows switch level simulation in VHDL, and gives the source code for the basic switch devices.

The simulation algorithms for networks of bi-directional pass transistors are known under the generic name of *relaxation* algorithms. Algorithms where there is no global information, and where the code that models a given transistor can affect only data that is accessible only by transistors directly connected to the given transistor are known as *local relaxation* algorithms. Local relaxation algorithms have been used for switch-level simulation. However, it has been assumed previously that local relaxation algorithms cannot be implemented as VHDL behavioral descriptions. The work discussed in this section will show a local relaxation algorithm which has been implemented in VHDL.

Typically, switch-level semantics can be implemented more efficiently by global relaxation algorithms. A particular simulator may run a switch-level description faster because it has the semantics hardwired in its runtime kernel, but all VHDL simulators will be able to run the given switch-level description, provided that the semantics of the switch-level components is described in VHDL. Having the VHDL behavior for each switch-level component, results in code portability, and most of all, in a standard hardware description language that spans behavior, dataflow, structure, gate-level, and switch-level.

The Simulation of a network of bi-directional pass transistors is performed using a local relaxation algorithm. The core of the algorithm consists of the following steps:

1. Insert a non-zero delay (provided by the user) between the signal connected to the gate and the gate itself.

2. Whenever there is activity on any signal to which an on-transistor is connected, disconnect the on-transistor from the source and drain connections. Whenever there is activity on any signal to which an off transistor is connected, disconnect the off transistor from the source and/or drain connection(s), where the activity occurred. Note:

    - A signal that is not driven preserves its state (zero or one), but changes its strength to Z.

- Consider the disconnection of a transistor to be an activity on the signal.

3. During the delta that follows the one at step 2, drive the ports disconnected at step 2 with the resolved value of the drain and source ports if the transistor is on and with the state found after disconnecting the source or drain and Z strength if the transistor is off.

Note that the state of the gate may change only during the first delta of a time point, because there is a delay between the signal to which the gate is connected and the internal gate signal of the transistor. Therefore, a change in the state of the gate may not occur during step 3.

The VHDL code corresponding to the core algorithm is presented in the following sections. The core algorithm can be optimized by not disconnecting nor attempting to change the driving value corresponding to a signal that has the strength F. This reduces the number of unnecessary transistor disconnections.

3.3.1 Switch Modelling Utilities

The following functions provide basic capabilities which support switch level modelling in VHDL. It is possible to build all of the key switch level devices including bi-directional transmission gates with these functions.

The following shows the definition for the **f_uxfr** function which calculates the output of the drain for a transistor based on the value of the enable (**G**), source and drain.

Calculate Drain based on G, Source, Drain

```
FUNCTION f_uxfr ( G,Src,Drn  : IN t_logic ) RETURN t_logic IS
    VARIABLE t0,t1: t_logic;
    VARIABLE sOL,sOH,s1L,s1H,sOM,s1M: t_strength;
BEGIN
    CASE f_state( G ) is
        -- device is off, output is floating
        WHEN '0' => RETURN f_convz( Drn );

        -- device is on, output is source value
        WHEN '1' => RETURN Src;

        -- device status unknown
        WHEN 'X' => t0 := f_convz( Drn );
                    t1 := Src;
                    -- calculate drain interval
                    sOL := f_strengthL(t0);
                    sOH := f_strengthH(t0);
                    -- calculate source interval
```

3.3. SWITCH LEVEL DEVICES

```
                        s1L := f_strengthL(t1);
                        s1H := f_strengthH(t1);
                        -- save state?
                        IF f_state(t0) = f_state(t1) then
                            IF f_state(t0) = 'X' then
                                IF sOL < s1L THEN sOL := s1L; END IF;
                                IF sOH < s1H THEN sOH := s1H; END IF;
                                RETURN f_logicX(sOL)(sOH);
                            ELSE
                                IF sOL > s1L THEN sOL := s1L; END IF;
                                IF sOH < s1H THEN sOH := s1H; END IF;
                                IF f_state(t0) = '0' THEN
                                    RETURN f_logic0(sOL)(sOH);
                                ELSE
                                    RETURN f_logic1(sOL)(sOH);
                                END IF;
                            END IF;
                        -- different state?
                        ELSE
                            sOM := t_strength'low;
                            s1M := t_strength'low;
                            CASE f_state(t0) IS
                                WHEN '0' => sOM := sOH;
                                WHEN '1' => s1M := sOH;
                                WHEN 'X' => sOM := sOL; s1M := sOH;
                            END CASE;
                            CASE f_state(t1) IS
                                WHEN '0' => IF sOM < s1H THEN sOM := s1H; END IF;
                                WHEN '1' => IF s1M < s1H THEN s1M := s1H; END IF;
                                WHEN 'X' => IF sOM < s1L THEN sOM := s1L; END IF;
                                            IF s1M < s1H THEN s1M := s1H; END IF;
                            END CASE;
                            RETURN f_logicX(sOM)(s1M);
                        END IF;
    END CASE;
END f_uxfr;
```

The algorithm goes as follows:

- If **G** is on, then the output to drain is the source value
- If **G** is off, then the output to drain is **f_convz(drain)**
- Otherwise the output to drain is the interval between source and **f_convz(drain)**.

This function represents the behavior of a uni-directional pass transistor.
The following shows the definition of the **f_resistor** function.

Resistor Calculation

```
FUNCTION f_resistor ( s : IN t_logic; size : IN natural ) RETURN t_logic IS
```

```
        VARIABLE sl, sh : t_strength;
BEGIN
    sl := f_strengthL( s );
    sh := f_strengthH( s );
    IF t_strength'pos( sl ) > (t_strength'pos( t_strength'low ) + size) THEN
        sl := t_strength'val( t_strength'pos( sl ) - size );
    ELSE
        sl := 'U';
    END IF;
    IF t_strength'pos( sh ) > (t_strength'pos( t_strength'low ) + size) THEN
        sh := t_strength'val( t_strength'pos( sh ) - size );
    ELSE
        sh := 'U';
    END IF;
    CASE f_state(S) IS
        WHEN '0' => RETURN f_logic0( sl )( sh );
        WHEN '1' => RETURN f_logic1( sl )( sh );
        WHEN 'X' => RETURN f_logicX( sl )( sh );
    END CASE;
END f_resistor;
```

This function calculates the output of the drive output based on the current value of the source. The strength of the source is decreased by the parameter size. This function diminishes the value passed to it in strength by the amount given, but returns a value with the same state. This function is used in situations where a resistor is connected in series in a circuit and represents the drop in strength associated with current flowing through the resistor as shown in figure 3.15. This function is not used for resistors

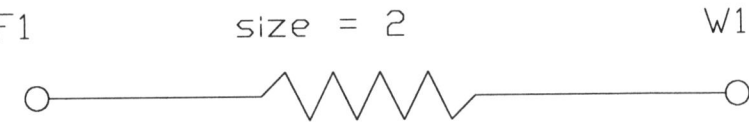

Figure 3.15: Resistor strength drop

which are tied to power or ground (a different mechanism is used in these cases).

3.3. SWITCH LEVEL DEVICES

The following shows the definition for the **f_cap** function which implements the capacitor.

Capacitor Calculation

```
FUNCTION f_cap ( s   : IN t_logic; size : IN t_strength ) RETURN t_logic IS
    VARIABLE sl, sh : t_strength;
BEGIN
    sl := f_strengthL( s );
    sh := f_strengthH( s );
    IF sl < size THEN sl := size; END IF;
    IF sh < size THEN sh := size; END IF;
    CASE f_state(S) IS
        WHEN '0' => RETURN f_logic0( sl )( sh );
        WHEN '1' => RETURN f_logic1( sl )( sh );
        WHEN 'X' => RETURN f_logicX( sl )( sh );
    END CASE;
END f_cap;
```

The algorithm is as follows:

- If the strength of the source is larger or equal to the size of the capacitor then return the source value
- Otherwise return current charged value of the capacitor

This function insures that a minimum strength level is always maintained and in essence performs a floor operation on the input value. A typical application is shown in figure 3.16. In this example, when the transistor

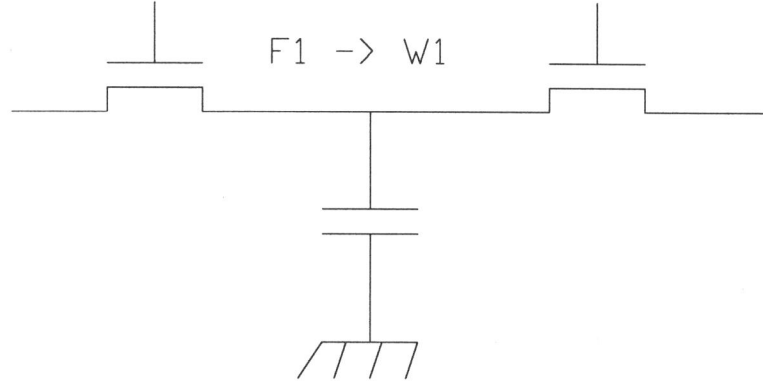

Figure 3.16: Capacitor application

providing the F1 value is disconnected, the capacitor holds the strength to a W1 where without the capacitor the value would drop to Z1.

The following shows the definition of the **f_ceil** function which calculates the value of the driving output based on the current value of the source.

Ceiling Calculation

```
FUNCTION f_ceil ( s    : IN t_logic; size : IN t_strength ) RETURN t_logic IS
    VARIABLE sl, sh : t_strength;
BEGIN
    sl := f_strengthL( s );
    sh := f_strengthH( s );
    IF sl > size THEN sl := size; END IF;
    IF sh > size THEN sh := size; END IF;
    CASE f_state(s) IS
        WHEN '0' => RETURN f_logic0( sl )( sh );
        WHEN '1' => RETURN f_logic1( sl )( sh );
        WHEN 'X' => RETURN f_logicX( sl )( sh );
    END CASE;
END f_ceil;
```

The strength of the source is decreased by **size**. This function is analogous to the **f_cap** function except that it provides a ceiling operation by forcing a value to have no more than a maximum allowed strength.

3.3.2 Bidirectional Transmission Element

The bidirectional transmission gate element is shown below

Bidirectional Transmission Element

```
ENTITY bxfr IS
    GENERIC (gdelay : time := 3 ps;
             maxstrength : t_strength := 'R');
    PORT( g: IN t_wlogic; src,drn: INOUT t_wlogic);
END bxfr;

ARCHITECTURE bxfr_behavior OF bxfr IS
        SIGNAL wakeup : boolean := false;
        SIGNAL gd : t_wlogic;
BEGIN
    -- delay element
    gd <= g AFTER gdelay;

    -- main device behavior
    PROCESS(gd,src'transaction,drn'transaction)
        VARIABLE vsrc, vdrn : t_logic;    -- computed values from outside
        VARIABLE psrc, pdrn : t_logic;    -- values driven from inside
        VARIABLE init : boolean := true;  -- powerup
    BEGIN
```

3.3. SWITCH LEVEL DEVICES

```
            -- disconnect mode
            IF NOT wakeup THEN
                -- initialize the model, once only
                IF init THEN
                    src <= D;
                    drn <= D;
                    init := false;

              -- new event on source/drain
                ELSIF (gd'EVENT OR (f_state(gd) /= '0')) THEN
                            wakeup <= true;
                            psrc := f_convz(src);
                            pdrn := f_convz(drn);
                            src <= psrc;
                            drn <= pdrn;

                -- no event on gate, and gate IS off
                ELSE
                    -- kick the src back
                    IF src'ACTIVE THEN
                        psrc := f_convz(src);
                        src <= psrc;
                    END IF;
                    -- kick the drn back
                    IF drn'ACTIVE THEN
                        pdrn := f_convz(drn);
                        drn <= pdrn;
                    END IF;
                END IF;

            -- reconnect mode
            ELSE
                CASE f_state(gd) IS
                    -- device off
                    WHEN '0' =>
                        -- f_convz computes tristate value
                        vsrc := f_convz(src);
                        vdrn := f_convz(drn);

                    -- device on
                    WHEN '1' =>
                        -- compute brf OF src and drn
                        -- f_ceil computes the resistance
                        -- f_busres computes the resolved value
                        vsrc := f_ceil(f_busres(src)(drn),maxstrength);
                        vdrn := f_ceil(f_busres(src)(drn),maxstrength);

                    -- device activity unknown
                    WHEN 'X' =>
                        -- compute range between tristate and brf OF src and drn
                        -- f_uxfr IS the function FOR unidirectional transistor
                        -- use f_uxfr to compute the range between tristate
```

```
                --      and resolved value
                vsrc := f_ceil(f_uxfr(
                    FX,f_busres(src)(drn),f_convz(src)),maxstrength);
                vdrn := f_ceil(f_uxfr(
                    FX,f_busres(src)(drn),f_convz(drn)),maxstrength);
            END CASE;

            -- update WHEN change
            IF psrc /= vsrc THEN
                psrc := vsrc;
                src <= vsrc;
            END IF;

            -- update WHEN change
            IF pdrn /= vdrn THEN
                pdrn := vdrn;
                drn <= vdrn;
            END IF;

            -- wakeup IN 1 fs, if nothing changes
            IF (psrc = vsrc) and (pdrn = vdrn) THEN
                    wakeup <= false AFTER 1 fs;
            END IF;
        END IF;
    END PROCESS;
END bxfr_behavior;
```

A bi-directional pass transistor may be on or off depending on the state of the gate (one or zero, respectively). An on-transistor drives the signals connected to the source and drain with the resolved value of all the external contributions to the source and drain. An off-transistor does not drive the source or drain, resulting in passive values on the connected signals. In the model **bxfr** shown above, a delay element is included through the concurrent signal assignment which monitors the **g** input port and assigns to the **gd** internal signal. This internal signal is used in the main device behavior process which is sensitive to changes in the source, drain and gate control.

This model operates in two modes, the *disconnect mode* and the *reconnect mode*. The former can be viewed as the normal operating mode in which normal events from other devices are accepted. The latter is associated with the local relaxation operation of the device. See section 3.3 for a detailed description of the relaxation algorithm used.

During disconnect mode the **init** variable is used to initialize the device once during circuit power. Both source and drain are set to the weakest state possible (D). When an event on the source or drain is detected and the device is on, the high-impedance value of the source and drain are assigned to the outputs and the device is scheduled for reconnect mode by placing

3.3. SWITCH LEVEL DEVICES

an event on the **wakeup** signal. If the device is off, the source and drain events are reassigned to the outputs with high-impedance strength.

In reconnect mode the following options exist:

- **device off** - tristate value is placed on both source and drain
- **device on** - bus resolution is used to resolve the new source and drain values
- **device status unknown** - a range between the tristate and bus resolved values is placed on the source and drain

The model will continue is reconnect mode as long as changes continue to ripple through the circuit. When no activity occurs on the source or drain the model will revert to disconnect mode.

3.3.3 Basic Complementary Transmission Gate

The **nfet** model below

Basic Complementary Transmission Gate

```
USE std.std_logic.ALL;
USE work.ALL;
ENTITY nfet IS
    GENERIC (gdelay : time := 3 ps; maxstrength : t_strength := 'R');
    PORT( g: IN t_wlogic; src,drn: INOUT t_wlogic);
END nfet;

ARCHITECTURE nfet_behavior OF nfet IS
    COMPONENT bxfr_type
        GENERIC (gdelay : time := 3 ps; maxstrength : t_strength := 'R');
        PORT( g: IN t_wlogic; src,drn: INOUT t_wlogic); END COMPONENT;
BEGIN
    i1: bxfr_type GENERIC MAP (gdelay, maxstrength) PORT MAP( g, src, drn );
END nfet_behavior;
```

shows the basic complementary bidirectional transmission gate.

3.3.4 Basic Transmission Gate

The **pfet** model below

Basic Transmission Gate

```
USE std.std_logic.ALL;
USE work.ALL;
ENTITY pfet IS
    GENERIC (gdelay : time := 3 ps; maxstrength : t_strength := 'R');
```

```
      PORT( g: IN t_wlogic; src,drn: INOUT t_wlogic);
END pfet;

ARCHITECTURE pfet_behavior OF pfet IS
    COMPONENT bxfr_type
        GENERIC (gdelay : time := 3 ps; maxstrength : t_strength := 'R');
        PORT( g: IN t_wlogic; src,drn: INOUT t_wlogic); END COMPONENT;
    SIGNAL tg : t_wlogic;
BEGIN
    tg <= f_logic(f_NOT(f_state(g)))('R');
    i1: bxfr_type GENERIC MAP (gdelay, maxstrength) PORT MAP( tg, src, drn );
END pfet_behavior;
```

shows the basic transmission gate.

3.4 Simple ALU's

The model described in this section performs basic arithmetic logic unit and function generation operations. This model follows a standard format and is presented with full timing functionality. It is possible to reduce the simulation and model complexity in stages as follows

- Remove input delay processing
    - Remove all input timing generic parameters
    - Remove all internal delay signals
    - Remove the input delay signal assignment statements
    - Rename all references in the main behavioral process from the delayed port names to primary input port names
- Use simpler output delay calculations: adjust the final output assignment statements

By methodically adjusting these major components of the models it is possible to build a wide range of models which fit into the model accuracy continuum discussed in section 2.5.

The following adjustments are possible to adapt this model to other applications

- Change the instruction set. By modifying the interpretation placed on the case statement which decodes the instruction the model can be adapted as appropriate. Adjustments to the number of instructions can be made by varying the number of instruction bits; these changes will affect only the number of entries in the decode case statement with the integer conversion remaining the same.
- Adjust the ripple carry outputs by changing the boolean equations towards the end of the model.

3.4. SIMPLE ALU'S

3.4.1 ALU/Function Generator

This device is an arithmetic logic unit (ALU)/function generator. The circuit performs 16 binary arithmetic operations on two 4-bit words as shown in the following table. This device corresponds to the Texas Instruments 181 series TTL parts [TI86]. Figure 3.17 shows the symbolic representation

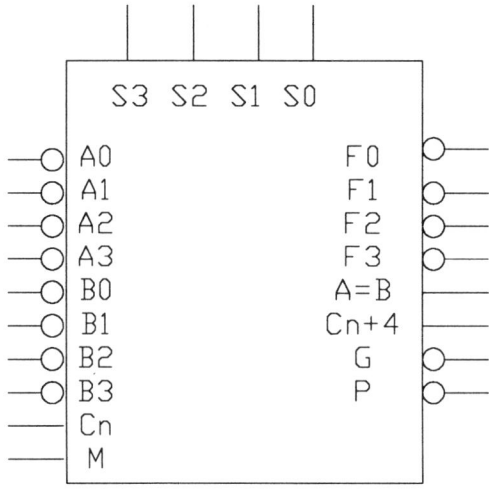

Figure 3.17: Logic Diagram ALU/Function Generator

of this device. The following shows the function table for both arithmetic and logic modes.

Function Table

selection				active-low data		
				$m = H$ logic functions	$m = L$; arithmetic ops	
					$Cn = L$ (no carry)	$Cn = H$ (carry)
s3	s2	s1	s0			
L	L	L	L	$\overline{a}$	$a - 1$	a
L	L	L	H	$\overline{a \cap b}$	$(a \cap b) - 1$	$a \cap b$
L	L	H	L	$\overline{a} \cup b$	$(a \cap \overline{b}) - 1$	$a \cap (\overline{b})$
L	L	H	H	1	-1	0
L	H	L	L	$\overline{a \cup b}$	$a + (a \cup \overline{b})$	$a + (a \cup \overline{b}) + 1$
L	H	L	H	$\overline{b}$	$(a \cap b) + (a \cap \overline{b})$	$(a \cap b) + (a \cup \overline{b}) + 1$
L	H	H	L	$a \oplus b$	$a - b - 1$	$a - b$
L	H	H	H	$a \cup \overline{b}$	$a \cup \overline{b}$	$(a \cup \overline{b}) + 1$
H	L	L	L	$\overline{a} \cap b$	$a + (a \cup b)$	$a + (a \cup b) + 1$
H	L	L	H	$a \oplus b$	$a \cup b$	$(a \cup b) + 1$
H	L	H	L	b	$(a \cap \overline{b}) + (a \cup b)$	$(a \cap \overline{b}) + (a \cup b) + 1$
H	L	H	H	$a \cup b$	$a \cup b$	$(a \cup b) + 1$
H	H	L	L	0	a	$a + 1$
H	H	L	H	$a \cap \overline{b}$	$(a \cap b) + a$	$(a \cap b) + a + 1$
H	H	H	L	$a \cap b$	$(a \cap \overline{b}) + a$	$(a \cap \overline{b}) + a + 1$
H	H	H	H	a	a	$a + 1$

These functions are selected by the four function-select lines s0, s1, s2 and s3 and include addition, subtraction, decrement, and straight transfer. When performing arithmetic manipulations, the internal carries are enabled by applying a low to the mode control input **m**. A ripple-carry input **cn** and a ripple-carry output **cnp4** are available. The following shows the full model for the alu/function generator with the characteristics

- 16 logic operations, 16 arithmetic operations
- Equality comparison
- Ripple carry input and output

ALU/Function Generator

```
USE std.std_logic.ALL;
USE std.std_ttl.ALL;
USE work.intpack.ALL;
USE work.std_ttlnew.ALL;
ENTITY alu IS
    GENERIC (s0_i01,s0_i10,
             s1_i01,s1_i10,
```

3.4. SIMPLE ALU'S

```
                    s2_i01,s2_i10,
                    s3_i01,s3_i10,

                    m_i01,m_i10,
                    cn_i01,cn_i10,

                    a0_i01,a0_i10,
                    a1_i01,a1_i10,
                    a2_i01,a2_i10,
                    a3_i01,a3_i10,

                    b0_i01,b0_i10,
                    b1_i01,b1_i10,
                    b2_i01,b2_i10,
                    b3_i01,b3_i10,

                    p_o01,p_o10,
                    g_o01,g_o10,
                    aeb_o01,aeb_o10,
                    cnp4_o01,cnp4_o10,

                    fb0_o01,fb0_o10,
                    fb1_o01,fb1_o10,
                    fb2_o01,fb2_o10,
                    fb3_o01,fb3_o10 : TIME := 2 ns);
        PORT    (s0,s1,s2,s3, m,cn, a0,a1,a2,a3, b0,b1,b2,b3 : IN t_wlogic;
                    p, g, aeb, cnp4, fb0,fb1,fb2,fb3: OUT t_wlogic);
END alu;

ARCHITECTURE full OF alu IS
    SIGNAL s0delay, s1delay, s2delay, s3delay,
            mdelay, cndelay,
            a0delay, a1delay, a2delay, a3delay,
            b0delay, b1delay, b2delay, b3delay : t_wlogic;
BEGIN
    -- input delay processing
    s0delay <= s0 AFTER f_delay(s0,s0_i01,s0_i10);
    s1delay <= s1 AFTER f_delay(s1,s1_i01,s1_i10);
    s2delay <= s2 AFTER f_delay(s2,s2_i01,s2_i10);
    s3delay <= s3 AFTER f_delay(s3,s3_i01,s3_i10);
    mdelay  <= m  AFTER f_delay(m,m_i01,m_i10);
    cndelay <= cn AFTER f_delay(cn,cn_i01,cn_i10);
    a0delay <= a0 AFTER f_delay(a0,a0_i01,a0_i10);
    a1delay <= a1 AFTER f_delay(a1,a1_i01,a1_i10);
    a2delay <= a2 AFTER f_delay(a2,a2_i01,a2_i10);
    a3delay <= a3 AFTER f_delay(a3,a3_i01,a3_i10);
    b0delay <= b0 AFTER f_delay(b0,b0_i01,b0_i10);
    b1delay <= b1 AFTER f_delay(b1,b1_i01,b1_i10);
    b2delay <= b2 AFTER f_delay(b2,b2_i01,b2_i10);
    b3delay <= b3 AFTER f_delay(b3,b3_i01,b3_i10);

    -- alu operation
```

```
PROCESS (s0delay,s1delay,s2delay,s3delay,
         mdelay,cndelay,
         a0delay,a1delay,a2delay,a3delay,
         b0delay,b1delay,b2delay,b3delay)
    VARIABLE state : t_logarray (1 TO 4);
    VARIABLE sstate : integer := 0;
    VARIABLE a : integer := 0;
    VARIABLE b : integer := 0;
    VARIABLE f : integer := 0;
    VARIABLE pwork,gwork,aebwork,cnp4work : t_logic;
    VARIABLE ab01,ab02,ab11,ab12,ab21,ab22,ab31,ab32 : t_logic;
    VARIABLE unknown : boolean := true;
    BEGIN
    -- encode the instruction
    state(4) := s0delay;
    state(3) := s1delay;
    state(2) := s2delay;
    state(1) := s3delay;
    f_logictoint(state,unknown,sstate);

    -- logic functions
    IF (mdelay = '1') AND (NOT unknown) THEN
        -- pickup a data
        state(4) := a0delay;
        state(3) := a1delay;
        state(2) := a2delay;
        state(1) := a3delay;
        f_logictoint(state,unknown,a);

        IF NOT unknown THEN
            -- pickup b data
            state(4) := b0delay;
            state(3) := b1delay;
            state(2) := b2delay;
            state(1) := b3delay;
            f_logictoint(state,unknown,b);
        END IF;

        IF NOT unknown THEN
            CASE sstate IS
                WHEN 0  => f := NOT a;
                WHEN 1  => f := NOT (a AND b);
                WHEN 2  => f := (NOT a) OR b;
                WHEN 3  => f := 1;
                WHEN 4  => f := NOT(a OR b);
                WHEN 5  => f := NOT b;
                WHEN 6  => f := NOT (a XOR b);
                WHEN 7  => f := a OR (NOT b);
                WHEN 8  => f := (NOT a) AND b;
                WHEN 9  => f := a XOR b;
                WHEN 10 => f := b;
                WHEN 11 => f := a OR b;
```

3.4. SIMPLE ALU'S

```
                WHEN 12 => f := 0;
                WHEN 13 => f := a AND (NOT b);
                WHEN 14 => f := a AND b;
                WHEN 15 => f := a;
                WHEN OTHERS => unknown := true;
            END CASE;
            f_inttologic(f,state,ttl);
        END IF;

    -- arithmetic functions
    ELSIF (mdelay = '0') AND (NOT unknown) THEN
        -- pickup a data
        state(4) := a0delay;
        state(3) := a1delay;
        state(2) := a2delay;
        state(1) := a3delay;
        f_logictoint2c(state,unknown,a);

        IF NOT unknown THEN
            -- pickup b data
            state(4) := b0delay;
            state(3) := b1delay;
            state(2) := b2delay;
            state(1) := b3delay;
            f_logictoint2c(state,unknown,b);
        END IF;

        IF NOT unknown THEN
            CASE sstate IS
                WHEN 0 =>
                    IF (cndelay = '0') THEN f := a - 1;
                    ELSE f := a;
                    END IF;
                WHEN 1 =>
                    IF (cndelay = '0') THEN f := (a AND b) - 1;
                    ELSE f := a AND b;
                    END IF;
                WHEN 2 =>
                    IF (cndelay = '0') THEN f := (a AND (NOT b)) - 1;
                    ELSE f := (a AND (NOT b));
                    END IF;
                WHEN 3 =>
                    IF (cndelay = '0') THEN f := -1;
                    ELSE f := 0;
                    END IF;
                WHEN 4 =>
                    IF (cndelay = '0') THEN f := a + (a OR (NOT b));
                    ELSE f := a + (a OR (NOT b)) + 1;
                    END IF;
                WHEN 5 =>
                    IF (cndelay = '0') THEN f:=(a AND b)+(a OR (NOT b));
                    ELSE f := (a AND b) + (a OR (NOT b)) + 1;
```

```
                    END IF;
                WHEN 6 =>
                    IF (cndelay = '0') THEN f := a - b - 1;
                    ELSE f := a - b;
                    END IF;
                WHEN 7 =>
                    IF (cndelay = '0') THEN f := a + (NOT b);
                    ELSE f := a + (NOT b) + 1;
                    END IF;
                WHEN 8 =>
                    IF (cndelay = '0') THEN f := a + (a OR b);
                    ELSE f := a + (a OR b) + 1;
                    END IF;
                WHEN 9 =>
                    IF (cndelay = '0') THEN f := a + b;
                    ELSE f := a + b + 1;
                    END IF;
                WHEN 10 =>
                    IF (cndelay = '0') THEN f:=(a AND (NOT b))+(a OR b);
                    ELSE f := (a AND (NOT b)) + (a OR b) + 1;
                    END IF;
                WHEN 11 =>
                    IF (cndelay = '0') THEN f := a OR b;
                    ELSE f := (a OR b) + 1;
                    END IF;
                WHEN 12 =>
                    IF (cndelay = '0') THEN f := a + a;
                    ELSE f := a + a + 1;
                    END IF;
                WHEN 13 =>
                    IF (cndelay = '0') THEN f := (a AND b) + a;
                    ELSE f := (a AND b) + a + 1;
                    END IF;
                WHEN 14 =>
                    IF (cndelay = '0') THEN f := (a AND (NOT b)) + a;
                    ELSE f := (a AND (NOT b)) + a + 1;
                    END IF;
                WHEN 15 =>
                    IF (cndelay = '0') THEN f := a;
                    ELSE f := a + 1;
                    END IF;
                WHEN OTHERS => unknown := true;
            END CASE;
            IF f < -15 THEN f := f + 15;
            ELSIF f > 15 THEN f := f - 15;
            END IF;
            f_inttologic2c(f,state,ttl);
        END IF;
-- unknown
ELSE unknown := true;
END IF;
```

3.4. SIMPLE ALU'S

```
   -- assign values to output signals
   IF NOT unknown THEN
      fb0 <= state(4) AFTER f_delay(state(4), fb0_o01, fb0_o10);
      fb1 <= state(3) AFTER f_delay(state(3), fb1_o01, fb1_o10);
      fb2 <= state(2) AFTER f_delay(state(2), fb2_o01, fb2_o10);
      fb3 <= state(1) AFTER f_delay(state(1), fb3_o01, fb3_o10);

      -- output A = B
      aebwork := state(1) AND state(2) AND state(3) AND state(4);
      aeb <= aebwork AFTER f_delay(aebwork, aeb_o01, aeb_o10);
   ELSE
      fb0 <= f_ttl('X') AFTER f_delay(f_ttl('X'), fb0_o01, fb0_o10);
      fb1 <= f_ttl('X') AFTER f_delay(f_ttl('X'), fb1_o01, fb1_o10);
      fb2 <= f_ttl('X') AFTER f_delay(f_ttl('X'), fb2_o01, fb2_o10);
      fb3 <= f_ttl('X') AFTER f_delay(f_ttl('X'), fb3_o01, fb3_o10);

      aeb <= f_ttl('X') AFTER f_delay(f_ttl('X'), aeb_o01, aeb_o10);
   END IF;

   -- calculate intermediate expressions
   ab01:= NOT (a0delay OR (b0delay AND s0delay) OR
            ((NOT b0delay) AND s1delay) );
   ab02:= (a0delay AND (NOT b0delay) AND s2delay) NOR
            (a0delay AND b0delay AND s3delay);

   ab11:= NOT (a1delay OR (b1delay AND s1delay) OR
            ((NOT b1delay) AND s1delay));
   ab12:= (a1delay AND (NOT b1delay) AND s2delay) NOR
            (a1delay AND b1delay AND s3delay);

   ab21:= NOT (a2delay OR (b2delay AND s2delay) OR
            ((NOT b2delay) AND s1delay));
   ab22:= (a2delay AND (NOT b2delay) AND s2delay) NOR
            (a2delay AND b2delay AND s3delay);

   ab31:= NOT (a3delay OR (b3delay AND s3delay) OR
            ((NOT b3delay) AND s1delay));
   ab32:= (a3delay AND (NOT b3delay) AND s2delay) NOR
            (a3delay AND b3delay AND s3delay);

   -- output P
   pwork := NOT (ab02 AND ab12 AND ab22 AND ab32);
   p <= pwork AFTER f_delay(pwork, p_o01, p_o10);

   -- output G
   gwork := NOT (
            ab31                                    OR
            (ab32 AND ab21)                         OR
            (ab32 AND ab22 AND ab11)                OR
            (ab32 AND ab22 AND ab12 AND ab01)
            ) ;
```

```
            g <= gwork AFTER f_delay(gwork, g_o01, g_o10);

            -- output cnp4
            cnp4work := (NOT gwork) OR ((NOT pwork) AND cndelay);
            cnp4 <= cnp4work AFTER f_delay(cnp4work, cnp4_o01, cnp4_o10);
            END PROCESS;
END full;
```

The ports for this model are

- s0 - instruction select input
- s1 - instruction select input
- s2 - instruction select input
- s3 - instruction select input
- m - arithmetic/logic mode control
- cn - carry in
- a0 - data a
- a1 - data a
- a2 - data a
- a3 - data a
- b0 - data b
- b1 - data b
- b2 - data b
- b3 - data b
- p - carry look-ahead
- g - carry look-ahead
- aeb - equality output
- cnp4 - carry output
- fb0 - data output
- fb1 - data output
- fb2 - data output
- fb3 - data output

Generic parameters to this model are summarized here

- s0_i01, s0_i10 - low to high and high to low s0 input port delays
- s1_i01, s1_i10 - low to high and high to low s1 input port delays
- s2_i01, s2_i10 - low to high and high to low s2 input port delays

3.4. SIMPLE ALU'S

- s3_i01, s3_i10 - low to high and high to low **s3** input port delays
- m_i01, m_i10 - low to high and high to low **m** input port delays
- cn_i01, cn_i10 - low to high and high to low **cn** input port delays
- a0_i01, a0_i10 - low to high and high to low **a0** input port delays
- a1_i01, a1_i10 - low to high and high to low **a1** input port delays
- a2_i01, a2_i10 - low to high and high to low **a2** input port delays
- a3_i01, a3_i10 - low to high and high to low **a3** input port delays
- b0_i01, b0_i10 - low to high and high to low **b0** input port delays
- b1_i01, b1_i10 - low to high and high to low **b1** input port delays
- b2_i01, b2_i10 - low to high and high to low **b2** input port delays
- b3_i01, b3_i10 - low to high and high to low **b3** input port delays
- p_o01, p_o10 - low to high and high to low **p** output port delays
- g_o01, g_o10 - low to high and high to low **g** output port delays
- aeb_o01, aeb_o10 - low to high and high to low **aeb** output port delays
- cnp4_o01, cnp4_o10 - low to high and high to low **cnp4** output port delays
- fb0_o01, fb0_o10 - low to high and high to low **fb0** output port delays
- fb1_o01, fb1_o10 - low to high and high to low **fb1** output port delays
- fb2_o01, fb2_o10 - low to high and high to low **fb2** output port delays
- fb3_o01, fb3_o10 - low to high and high to low **fb3** output port delays

This model has no error checks.

The input delay processing section of the model incorporates the appropriate delay for each of the input ports. An internal signal for each input port is declared. The main device function process utilizes the delayed signals rather than the primary inputs.

The alu operation is handled by a single process which is sensitive to changes in the s0delay, s1delay, s2delay, s3delay, mdelay, cndelay, a0delay, a1delay, a2delay, a3delay, b0delay, b1delay, b2delay, b3delay signals. This process performs the following operations whenever any of these signals changes values:

- Convert the incoming instruction signals to a single integer representation **sstate**.

- For the logic mode

  - Pickup the integer representation of incoming data lines (see **a** and **b** variables) using the **f_logictoint** procedure as documented in section 7.7.

  - If incoming data is unknown free, perform the appropriate logic operation and return the result in the integer variable **f**. The case statement indexes off of the **sstate** variable discussed above.

  - Convert the return value to logic values using **f_inttologic** procedure discussed in section 7.7.

- For the arithmetic mode

  - Pickup the integer representation of incoming data lines (see **a** and **b** variables) using the **f_logictoint2c** procedure as documented in section 7.7. This procedure performs conversion to twos complement representation, whereas the logic operations utilized a positive integer. This is required since the built in arithmetic operations in VHDL are utilized here, and these operations require twos complement arguments. The code shown here will operate correctly for ALUs which manipulate any number of bits up to 32 per word.

  - If incoming data is unknown free, perform the appropriate arithmetic operation and return the result in the integer variable **f**. The case statement indexes off of the **sstate** variable discussed above.

  - Convert the return value to logic values using **f_inttologic** procedure discussed in section 7.7.

- If the output values in **state** are assigned to output ports **fb0**, **fb1**, **fb2** and **fb3**. In addition, the **aeb** output is generated.

- Using several stages of intermediate results for readability, the carry look-ahead outputs **p** and **g** are generated. The output carry **cn4** is also generated.

Appropriate delay is introduced through the use of the **f_delay** function as discussed in section 7.6. This function utilizes the output rise and fall delays to choose the proper delay based on the new state value.

3.5 One Shots

The model described in this section is a one-shot device. This model follows a standard format and is presented with full timing functionality. It is possible to reduce the simulation and model complexity in stages as follows

- Remove input delay processing

  - Remove all input timing generic parameters
  - Remove all internal delay signals
  - Remove the input delay signal assignment statements
  - Rename all references in the main behavioral process from the delayed port names to primary input port names

- Use simpler output delay calculations: adjust the final output assignment statements

By methodically adjusting these major components of the models it is possible to build a wide range of models which fit into the model accuracy continuum discussed in section 2.5.

The following adjustments are possible to adapt this model to other applications

- Adjust the enable triggers, both active high and low are shown.
- Remove clear override.

3.5.1 Monostable Multivibrator

This device is a multivibrator which features a falling edge triggered input and a rising edge triggered input either of which can be used as an inhibit input. Once fired, the outputs are independent of further transitions of the a and b inputs. The output pulse can be terminated by an overriding clear. Input pulses may be of any duration relative to the output pulse. This device corresponds to the Texas Instruments 221 series TTL parts [TI86]. Figure 3.18 shows the symbolic representation of this device. The following shows the function table.

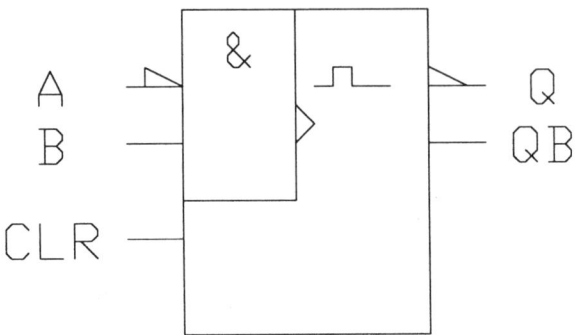

Figure 3.18: Logic Diagram One Shot

Function Table

inputs			outputs	
clear	a	b	q	qb
L	X	X	L	H
X	H	X	L	H
X	X	L	L	H
H	L	R	L/H/L pulse	H/L/H pulse
H	F	H	L/H/L pulse	H/L/H pulse
R	L	X	L/H/L pulse	H/L/H pulse
R	X	H	L/H/L pulse	H/L/H pulse

The following shows the full model for the one shot with the characteristics

- Active low and active high enables
- Clear override

One Shot

```
USE std.std_logic.ALL;
USE std.std_ttl.ALL;
ENTITY oneshot IS
    GENERIC (clear_i01,clear_i10,
             a_i01,a_i10,
             b_i01,b_i10,
             q_o01,q_o10,
             qb_o01,qb_o10 : TIME := 2 ns);
    PORT    (clear,a,b : IN t_wlogic;
             q,qb : OUT t_wlogic);
END oneshot;
```

3.5. ONE SHOTS

```
ARCHITECTURE full OF oneshot IS
    SIGNAL cleardelay,adelay,bdelay : t_wlogic;

    PROCEDURE trigger(SIGNAL q,qb : OUT t_wlogic;
                     VARIABLE q_o01,q_o10,qb_o01,qb_o10 : time) IS
        VARIABLE rise,fall : time;
        BEGIN
            -- pulse
            rise := f_delay(f_ttl('1'), q_o01, q_o10);
            fall := f_delay(f_ttl('0'), q_o01, q_o10);
            q <= f_ttl('1') AFTER rise,
                 f_ttl('0') AFTER rise+fall;

            -- complementary pulse
            rise := f_delay(f_ttl('1'), qb_o01, qb_o10);
            fall := f_delay(f_ttl('0'), qb_o01, qb_o10);
            qb <= f_ttl('0') AFTER fall,
                  f_ttl('1') AFTER fall+rise;
        END trigger;

BEGIN
    -- input delay processing
    cleardelay <= clear AFTER f_delay(clear,clear_i01,clear_i10);
    adelay <= a AFTER f_delay(a,a_i01,a_i10);
    bdelay <= b AFTER f_delay(b,b_i01,b_i10);

    -- comparator operation
    PROCESS (cleardelay,adelay,bdelay)
        VARIABLE cleared : boolean := true;
        VARIABLE conditioned : boolean := false;
        BEGIN
        -- device cleared?
        IF (cleared) AND (cleardelay='1') THEN
            IF (bdelay'EVENT) AND (adelay='0') AND (bdelay='1')THEN
                trigger(q,qb,q_o01,q_o10,qb_o01,qb_o10);
                cleared := false;
                conditioned := false;
            ELSIF (adelay'EVENT) AND (adelay='0') AND (bdelay='1')THEN
                trigger(q,qb,q_o01,q_o10,qb_o01,qb_o10);
                cleared := false;
                conditioned := false;
            ELSIF (conditioned) AND (cleardelay'EVENT) AND
                                   ((adelay='0')OR(bdelay='1'))THEN
                trigger(q,qb,q_o01,q_o10,qb_o01,qb_o10);
                cleared := false;
                conditioned := false;
            END IF;

        -- new clear?
        ELSIF cleardelay = '0' THEN
            q <= f_ttl('0') AFTER f_delay(f_ttl('0'), q_o01, q_o10);
            qb <= f_ttl('1') AFTER f_delay(f_ttl('1'), qb_o01, qb_o10);
```

```
                cleared := true;

            -- condition device?
            IF (adelay'EVENT) AND (adelay='1') THEN conditioned := true;
            ELSIF (bdelay'EVENT) AND (bdelay='0') THEN conditioned := true;
            END IF;
        END IF;

        END PROCESS;
END full;
```

The ports for this model are

- **clear** - clear override
- **a** - active low enable
- **b** - active high enable
- **q** - pulse output
- **qb** - complementary pulse output

Generic parameters to this model are summarized here

- **clear_i01, clear_i10** - low to high and high to low **clear** input port delays
- **a_i01, a_i10** - low to high and high to low **a** input port delays
- **b_i01, b_i10** - low to high and high to low **b** input port delays
- **q_o01, q_o10** - low to high and high to low **q** output port delays
- **qb_o01, qb_o10** - low to high and high to low **qb** output port delays

This model has no error checks.

The input delay processing section of the model incorporates the appropriate delay for each of the input ports. An internal signal for each input port is declared. The main device function process utilizes the delayed signals rather than the primary inputs.

The one shot operation is handled by a single process which is sensitive to changes in the **cleardelay, adelay** and **bdelay** signals. This process performs the following operations whenever any of these signals changes values:

- Device cleared (no pulse generated so far, or previous clear override seen), the trigger the device on one of the following conditions:

    - Rising transition of **bdelay**
    - Falling transition of **adelay**

3.6. COMPARATORS

- Conditioned device (**adelay** was held low while clear override was low, or **bdelay** was held high while clear override was low previously to force the conditioned mode) and a rising transition of **cleardelay** was detected.

• Device detects a clear override (**cleardelay** is low):

- Output **q** is set to low, and **qb** is set to high
- Device condition mode is checked and set as appropriate

Appropriate delay is introduced through the use of the **f_delay** function as discussed in section 7.6. This function utilizes the output rise and fall delays to choose the proper delay based on the new state value.

3.6 Comparators

The model described in this section is a comparator. This model follows a standard format and is presented with full timing functionality. It is possible to reduce the simulation and model complexity in stages as follows

• Remove input delay processing

- Remove all input timing generic parameters
- Remove all internal delay signals
- Remove the input delay signal assignment statements
- Rename all references in the main behavioral process from the delayed port names to primary input port names

• Use simpler output delay calculations: adjust the final output assignment statements

By methodically adjusting these major components of the models it is possible to build a wide range of models which fit into the model accuracy continuum discussed in section 2.5.

The following adjustments are possible to adapt this model to other applications

• Adjust the number of bits compared. The number of ports for data in can be modified affecting only the the logic to integer temporary variable **state**. The integer operations performed in the model are independent of the number of input bits up to 32 bits.

• The outputs generated; for example the less than and greater than can be removed if equality if the only output required.

3.6.1 4 Bit Magnitude Comparator

This four-bit magnitude comparator performs comparison of straight binary codes. Three comparisons about two 4-bit words **a** and **b** are made and are available at three outputs. This device corresponds to the Texas Instruments 85 series TTL parts [TI86]. Figure 3.19 shows the symbolic

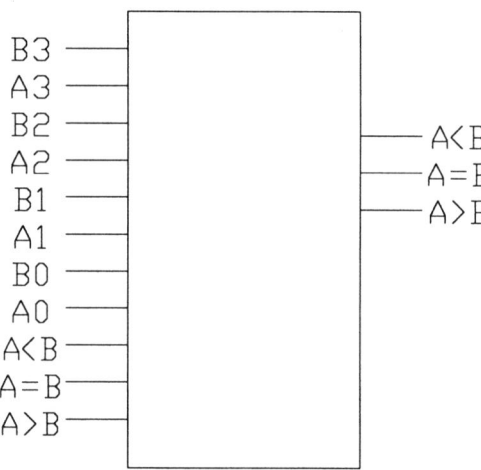

Figure 3.19: Logic Diagram 4 Bit Magnitude Comparator

representation of this device. The following shows the function table for this device.

Function Table

	comparing inputs			cascading inputs			outputs		
a3,b3	a2,b2	a1,b1	a0,b0	$a > b$	$a < b$	$a = b$	$a > b$	$a < b$	$a = b$
$a3 > b3$	X	X	X	X	X	X	H	L	L
$a3 < b3$	X	X	X	X	X	X	L	H	L
$a3 = b3$	$a2 > b2$	X	X	X	X	X	H	L	L
$a3 = b3$	$a2 < b2$	X	X	X	X	X	L	H	L
$a3 = b3$	$a2 = b2$	$a1 > b2$	X	X	X	X	H	L	L
$a3 = b3$	$a2 = b2$	$a1 < b2$	X	X	X	X	L	H	L
$a3 = b3$	$a2 = b2$	$a1 = b2$	$a0 > b0$	X	X	X	H	L	L
$a3 = b3$	$a2 = b2$	$a1 = b2$	$a0 < b0$	X	X	X	L	H	L
$a3 = b3$	$a2 = b2$	$a1 = b2$	$a0 = b0$	H	L	L	H	L	L
$a3 = b3$	$a2 = b2$	$a1 = b2$	$a0 = b0$	L	H	L	L	H	L
$a3 = b3$	$a2 = b2$	$a1 = b2$	$a0 = b0$	X	X	H	L	L	H
$a3 = b3$	$a2 = b2$	$a1 = b2$	$a0 = b0$	H	H	L	L	L	L
$a3 = b3$	$a2 = b2$	$a1 = b2$	$a0 = b0$	L	L	L	H	H	L

3.6. COMPARATORS

The following shows the full model for the comparator with the characteristics

- 4 bit words
- $a > b$, $a < b$ and $a = b$ comparison
- inputs for chaining of devices

4 Bit Magnitude Comparator

```
USE std.std_logic.ALL;
USE std.std_ttl.ALL;
ENTITY mag4comp IS
    GENERIC (a3_i01,a3_i10,
             a2_i01,a2_i10,
             a1_i01,a1_i10,
             a0_i01,a0_i10,
             b3_i01,b3_i10,
             b2_i01,b2_i10,
             b1_i01,b1_i10,
             b0_i01,b0_i10,
             aeqbin_i01,aeqbin_i10,
             agtbin_i01,agtbin_i10,
             altbin_i01,altbin_i10,
             aeqbout_o01,aeqbout_o10,
             agtbout_o01,agtbout_o10,
             altbout_o01,altbout_o10 : TIME := 2 ns);
    PORT    (a3,a2,a1,a0,b3,b2,b1,b0,aeqbin,agtbin,altbin : IN t_wlogic;
             aeqbout,agtbout,altbout : OUT t_wlogic);
END mag4comp;

ARCHITECTURE full OF mag4comp IS
    SIGNAL a3delay,a2delay,a1delay,a0delay,
           b3delay,b2delay,b1delay,b0delay,
           aeqbindelay,agtbindelay,altbindelay : t_wlogic;
BEGIN
    -- input delay processing
    a3delay <= a3 AFTER f_delay(a3,a3_i01,a3_i10);
    a2delay <= a2 AFTER f_delay(a2,a2_i01,a2_i10);
    a1delay <= a1 AFTER f_delay(a1,a1_i01,a1_i10);
    a0delay <= a0 AFTER f_delay(a0,a0_i01,a0_i10);
    b3delay <= b3 AFTER f_delay(b3,b3_i01,b3_i10);
    b2delay <= b2 AFTER f_delay(b2,b2_i01,b2_i10);
    b1delay <= b1 AFTER f_delay(b1,b1_i01,b1_i10);
    b0delay <= b0 AFTER f_delay(b0,b0_i01,b0_i10);
    aeqbindelay <= aeqbin AFTER f_delay(aeqbin,aeqbin_i01,aeqbin_i10);
    agtbindelay <= agtbin AFTER f_delay(agtbin,agtbin_i01,agtbin_i10);
    altbindelay <= altbin AFTER f_delay(altbin,altbin_i01,altbin_i10);

    -- comparator operation
    PROCESS (a3delay,a2delay,a1delay,a0delay,
```

```
            b3delay,b2delay,b1delay,b0delay,
            aeqbindelay,agtbindelay,altbindelay)
    VARIABLE astate : integer := 0;
    VARIABLE bstate : integer := 0;
    VARIABLE state : t_logarray (1 TO 4);
    VARIABLE unknown : boolean;
    BEGIN
    -- pickup a/b integer values
    state := (a3delay,a2delay,a1delay,a0delay);
    f_logictoint(state,unknown,astate);
    IF unknown THEN astate := -1; END IF;

    state := (b3delay,b2delay,b1delay,b0delay);
    f_logictoint(state,unknown,bstate);
    IF unknown THEN bstate := -1; END IF;

    -- unknown inputs?
    IF (astate < 0) OR (bstate < 0) THEN
        aeqbout <= f_ttl('X') AFTER
            f_delay(f_ttl('X'), aeqbout_o01, aeqbout_o10);
        agtbout <= f_ttl('X') AFTER
            f_delay(f_ttl('X'), agtbout_o01, agtbout_o10);
        altbout <= f_ttl('X') AFTER
            f_delay(f_ttl('X'), altbout_o01, altbout_o10);

    -- compare values
    ELSE
        -- a > b
        IF astate > bstate THEN
            aeqbout <= f_ttl('0') AFTER
                f_delay(f_ttl('0'), aeqbout_o01, aeqbout_o10);
            agtbout <= f_ttl('1') AFTER
                f_delay(f_ttl('1'), agtbout_o01, agtbout_o10);
            altbout <= f_ttl('0') AFTER
                f_delay(f_ttl('0'), altbout_o01, altbout_o10);

        -- a < b
        ELSIF astate < bstate THEN
            aeqbout <= f_ttl('0') AFTER
                f_delay(f_ttl('0'), aeqbout_o01, aeqbout_o10);
            agtbout <= f_ttl('0') AFTER
                f_delay(f_ttl('0'), agtbout_o01, agtbout_o10);
            altbout <= f_ttl('1') AFTER
                f_delay(f_ttl('1'), altbout_o01, altbout_o10);

        -- a = b
        ELSE
            IF aeqbin = '1' THEN
                aeqbout <= f_ttl('1') AFTER
                    f_delay(f_ttl('1'), aeqbout_o01, aeqbout_o10);
                agtbout <= f_ttl('0') AFTER
                    f_delay(f_ttl('0'), agtbout_o01, agtbout_o10);
```

3.6. COMPARATORS

```
                    altbout <= f_ttl('0') AFTER
                        f_delay(f_ttl('0'), altbout_o01, altbout_o10);
                ELSE
                    IF agtbin = altbin THEN
                        aeqbout <= aeqbindelay AFTER
                            f_delay(aeqbindelay, aeqbout_o01, aeqbout_o10);
                        agtbout <= NOT agtbindelay AFTER f_delay(
                            NOT agtbindelay, agtbout_o01, agtbout_o10);
                        altbout <= NOT altbindelay AFTER f_delay(
                            NOT altbindelay, altbout_o01, altbout_o10);
                    ELSE
                        aeqbout <= aeqbindelay AFTER
                            f_delay(aeqbindelay, aeqbout_o01, aeqbout_o10);
                        agtbout <= agtbindelay AFTER
                            f_delay(agtbindelay, agtbout_o01, agtbout_o10);
                        altbout <= altbindelay AFTER
                            f_delay(altbindelay, altbout_o01, altbout_o10);
                    END IF;
                END IF;
            END IF;
        END IF;
        END PROCESS;
END full;
```

The ports for this model are

- **a3** - a data
- **a2** - a data
- **a1** - a data
- **a0** - a data
- **b3** - b data
- **b2** - b data
- **b1** - b data
- **b0** - b data
- **aeqbin** - $a = b$ input
- **agtbin** - $a > b$ input
- **altbin** - $a < b$ input
- **aeqbout** - $a = b$ output
- **agtbout** - $a > b$ output
- **altbout** - $a < b$ output

Generic parameters to this model are summarized here

- **a3_i01, a3_i10** - low to high and high to low **a3** input port delays

- **a2_i01, a2_i10** - low to high and high to low **a2** input port delays
- **a1_i01, a1_i10** - low to high and high to low **a1** input port delays
- **a0_i01, a0_i10** - low to high and high to low **a0** input port delays
- **b3_i01, b3_i10** - low to high and high to low **b3** input port delays
- **b2_i01, b2_i10** - low to high and high to low **b2** input port delays
- **b1_i01, b1_i10** - low to high and high to low **b1** input port delays
- **b0_i01, b0_i10** - low to high and high to low **b0** input port delays
- **aeqbin_i01, aeqbin_i10** - low to high and high to low **aeqbin** input port delays
- **agtbin_i01, agtbin_i10** - low to high and high to low **agtbin** input port delays
- **altbin_i01, altbin_i10** - low to high and high to low **altbin** input port delays
- **aeqbout_o01, aeqbout_o10** - low to high and high to low **aeqbout** output port delays
- **agtbout_o01, agtbout_o10** - low to high and high to low **agtbout** output port delays
- **altbout_o01, altbout_o10** - low to high and high to low **altbout** output port delays

This model has no error checks.

The input delay processing section of the model incorporates the appropriate delay for each of the input ports. An internal signal for each input port is declared. The main device function process utilizes the delayed signals rather than the primary inputs.

The comparator operation is handled by a single process which is sensitive to changes in the **a3delay, a2delay, a1delay, a0delay, b3delay, b2delay, b1delay, b0delay, aeqbindelay, agtbindelay** and **altbindelay** signals. This process performs the following operations whenever any of these signals changes values:

- Converts the a and b data into integer values **astate** and **bstate** respectively.
- For unknown inputs, unknown outputs are generated.
- For known inputs, the two integer values are compared and the appropriate comparison outputs are generated. Special processing is required to handle the $a = b$ case which is dependent on the comparison inputs as shown in the model.

Appropriate delay is introduced through the use of the **f_delay** function as discussed in section 7.6. This function utilizes the output rise and fall delays to choose the proper delay based on the new value.

3.7 Parity Generators/Checkers

The model described in this section is a parity generator/checker. This model follows a standard format and is presented with full timing functionality. It is possible to reduce the simulation and model complexity in stages as follows

- Remove input delay processing
  - Remove all input timing generic parameters
  - Remove all internal delay signals
  - Remove the input delay signal assignment statements
  - Rename all references in the main behavioral process from the delayed port names to primary input port names
- Use simpler output delay calculations: adjust the final output assignment statements

By methodically adjusting these major components of the models it is possible to build a wide range of models which fit into the model accuracy continuum discussed in section 2.5.

The following adjustments are possible to adapt this model to other applications

- Adjust the parity algorithm by changing the flow of control and internal operation of the main process.
- Change the number of bits handled by adjusting the input port declarations and adding additional or less lines during high level counting.

3.7.1 9 bit Odd/Even Parity Generator/Checker

This 9 bit (8 data bits plus 1 parity bit) parity generator/checker features odd/even outputs and control inputs to facilitate operation in either odd or even parity applications. Depending on whether even or odd parity is being generated or checked, the even or odd inputs can be utilized as the parity or 9th bit input. This device corresponds to the Texas Instruments 180 series TTL parts [TI86]. Figure 3.20 shows the symbolic representation of this device. The following shows the function table for this device.

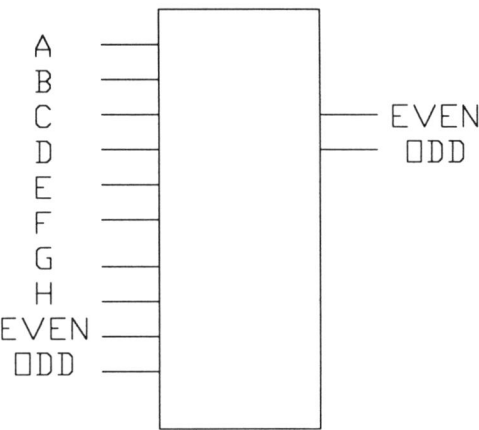

Figure 3.20: Logic Diagram 9 Bit Odd/Even Parity Generator/Checker

Function Table

inputs			outputs	
sum of H's at a thru h	even	odd	sum even	sum odd
even	H	L	H	L
odd	H	L	L	H
even	L	H	L	H
odd	L	H	H	L
X	H	H	L	L
X	L	L	H	H

The following shows the full model for the parity generator with the characteristics

- 8 data inputs, 1 parity input
- even or odd parity operation

9 bit Odd/Even Parity Generator/Checker

```
USE std.std_logic.ALL;
USE std.std_ttl.ALL;
ENTITY parity IS
    GENERIC (a_i01,a_i10,
             b_i01,b_i10,
             c_i01,c_i10,
```

3.7. PARITY GENERATORS/CHECKERS

```vhdl
                 d_i01,d_i10,
                 e_i01,e_i10,
                 f_i01,f_i10,
                 g_i01,g_i10,
                 h_i01,h_i10,
                 evenin_i01,evenin_i10,
                 oddin_i01,oddin_i10,
                 evenout_o01,evenout_o10,
                 oddout_o01,oddout_o10 : TIME := 2 ns);
    PORT  (a,b,c,d,e,f,g,h,evenin,oddin : IN t_wlogic;
                 evenout,oddout : OUT t_wlogic);
END parity;

ARCHITECTURE full OF parity IS
    SIGNAL adelay,bdelay,cdelay,ddelay,
           edelay,fdelay,gdelay,hdelay,
           evenindelay,oddindelay : t_wlogic;
BEGIN
    -- input delay processing
    adelay <= a AFTER f_delay(a,a_i01,a_i10);
    bdelay <= b AFTER f_delay(b,b_i01,b_i10);
    cdelay <= c AFTER f_delay(c,c_i01,c_i10);
    ddelay <= d AFTER f_delay(d,d_i01,d_i10);
    edelay <= e AFTER f_delay(e,e_i01,e_i10);
    fdelay <= f AFTER f_delay(f,f_i01,f_i10);
    gdelay <= g AFTER f_delay(g,g_i01,g_i10);
    hdelay <= h AFTER f_delay(h,h_i01,h_i10);
    evenindelay <= evenin AFTER f_delay(evenin,evenin_i01,evenin_i10);
    oddindelay <= oddin AFTER f_delay(oddin,oddin_i01,oddin_i10);

    -- parity generator operation
    PROCESS (adelay,bdelay,cdelay,ddelay,
             edelay,fdelay,gdelay,hdelay,
             evenindelay,oddindelay)
        VARIABLE count : integer;
        BEGIN
        -- parity error?
        IF evenindelay = oddindelay THEN
            evenout <= NOT evenindelay AFTER
                f_delay(NOT evenindelay, evenout_o01, evenout_o10);
            oddout <= NOT oddindelay AFTER
                f_delay(NOT oddindelay, oddout_o01, oddout_o10);

        -- valid parity input
        ELSE
            -- count number of 1's on input
            count := 0;
            IF adelay = '1' THEN count := count+1; END IF;
            IF bdelay = '1' THEN count := count+1; END IF;
            IF cdelay = '1' THEN count := count+1; END IF;
            IF ddelay = '1' THEN count := count+1; END IF;
            IF edelay = '1' THEN count := count+1; END IF;
```

```
            IF fdelay = '1' THEN count := count+1; END IF;
            IF gdelay = '1' THEN count := count+1; END IF;
            IF hdelay = '1' THEN count := count+1; END IF;

            -- even count?
            IF (count MOD 2) = 0 THEN
                evenout <= evenindelay AFTER
                    f_delay(evenindelay, evenout_o01, evenout_o10);
                oddout <= oddindelay AFTER
                    f_delay(oddindelay, oddout_o01, oddout_o10);

            -- odd count
            ELSE
                evenout <= NOT evenindelay AFTER
                    f_delay(NOT evenindelay, evenout_o01, evenout_o10);
                oddout <= NOT oddindelay AFTER
                    f_delay(NOT oddindelay, oddout_o01, oddout_o10);
            END IF;
        END IF;
        END PROCESS;
END full;
```

The ports for this model are

- **a** - data
- **b** - data
- **c** - data
- **d** - data
- **e** - data
- **f** - data
- **g** - data
- **h** - data
- **evenin** - even parity input
- **oddin** - odd parity input
- **evenout** - even parity output
- **oddout** - odd parity output

Generic parameters to this model are summarized here

- **a_i01, a_i10** - low to high and high to low **a** input port delays
- **b_i01, b_i10** - low to high and high to low **b** input port delays
- **c_i01, c_i10** - low to high and high to low **c** input port delays
- **d_i01, d_i10** - low to high and high to low **d** input port delays
- **e_i01, e_i10** - low to high and high to low **e** input port delays

3.7. PARITY GENERATORS/CHECKERS

- **f_i01, f_i10** - low to high and high to low **f** input port delays
- **g_i01, g_i10** - low to high and high to low **g** input port delays
- **h_i01, h_i10** - low to hihh and hihh to low **h** input port delays
- **evenin_i01, evenin_i10** - low to high and high to low **evenin** input port delays
- **oddin_i01, oddin_i10** - low to high and high to low **oddin** input port delays
- **evenout_o01, evenout_o10** - low to high and high to low **evenout** output port delays
- **oddout_o01, oddout_o10** - low to high and high to low **oddout** output port delays

This model has no error checks.

The input delay processing section of the model incorporates the appropriate delay for each of the input ports. An internal signal for each input port is declared. The main device function process utilizes the delayed signals rather than the primary inputs.

The parity generator operation is handled by a single process which is sensitive to changes in the **adelay, bdelay, cdelay, ddelay, edelay, fdelay, gdelay, hdelay, evenindelay, oddindelay** signals. This process performs the following operations whenever any of these signals changes values:

- Watch for special parity error mode, set outputs as appropriate, otherwise
- Count the number of high values on input and generate appropriate outputs for even and odd counts.

Appropriate delay is introduced through the use of the **f_delay** function as discussed in section 7.6. This function utilizes the output rise and fall delays to choose the proper delay based on the new value.

Chapter 4

Sequential Devices

4.1 Flip-Flops

The following models describe various types of flip-flop devices. These models follow a standard format and all are presented with full timing functionality. It is possible to reduce the simulation and model complexity in stages as follows

- Remove input delay processing
  - Remove all input timing generic parameters
  - Remove all internal delay signals
  - Remove the input delay signal assignment statements
  - Rename all references in the main behavioral process from the delayed port names to primary input port names
- Remove error checks: delete the error checking processes which occur prior to the main behavioral process
- Use simpler output delay calculations: adjust the final output assignment statements

By methodically adjusting these major components of the models it is possible to build a wide range of models which fit into the model accuracy continuum discussed in section 2.5.

The following adaptations are possible starting from these base models:

- Addition, deletion or adaptation of the asynchronous controls including clear, load, preset and enable inputs. A review of the various models in this chapter will give hints as to how to handle each of these features.

4.1.1 D-Type Positive-Edge Triggered Flip-Flop with Preset/Clear

This device is a D-type rising edge triggered flip-flop. A low at the **preset** or **clear** sets or resets the outputs regardless of the values of the other inputs. When **preset** or **clear** are high, data at the **d** input meeting the setup time requirements are transferred to the outputs on the rising edge of the clock pulse. Following the hold time interval, data at the **d** input may be changed without affecting the values at the outputs. This device corresponds to the Texas Instruments 74 series TTL parts [TI86]. Figure 4.1 shows the symbolic representation of this device. The following shows

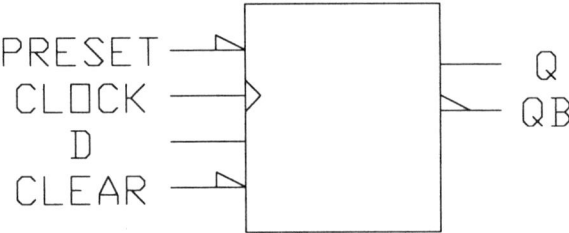

Figure 4.1: Logic Diagram D Pos Edge Flip-Flop with Preset/Clear

the function table for this device.

Function Table

inputs				outputs	
preset	clear	clock	d	q	qb
L	H	X	X	H	L
H	L	X	X	L	H
L	L	X	X	?	?
H	H	R	H	H	L
H	H	R	L	L	H
H	H	L	X	Q0	not Q0

The following shows the full model for a D-type flip-flop with characteristics

- Asynchronous preset
- Asynchronous clear
- Positive edge-triggered clocking
- Primary and complementary output

4.1. FLIP-FLOPS

D-Type Positive-Edge Triggered Flip-Flop with Preset/Clear

```
USE std.std_logic.ALL;
USE std.std_ttl.ALL;
ENTITY dff IS
    GENERIC (preset_i01,preset_i10,
             clear_i01,clear_i10,
             clock_i01,clock_i10,
             d_i01,d_i10,
             q_o01,q_o10,
             qb_o01,qb_o10,
             clock_min, preset_min, clear_min,
             d_setup,d_hold : TIME := 2 ns);
    PORT    (preset,clear,clock,d: IN t_wlogic; q,qb: OUT t_wlogic);
END dff;

ARCHITECTURE full OF dff IS
    SIGNAL presetdelay, cleardelay, clockdelay, ddelay : t_wlogic;
BEGIN
    -- check clock frequency/spike detection
    PROCESS (clock)
        VARIABLE clocklastev : TIME := 0 ns;
        BEGIN
        ASSERT (NOW = 0 ns) OR ((NOW - clocklastev) >= clock_min)
            REPORT "Spike detected on clock" SEVERITY warning;
        clocklastev := NOW;
    END PROCESS;

    -- spike detection
    PROCESS (preset)
        VARIABLE presetlastev : TIME := 0 ns;
        BEGIN
        ASSERT (NOW = 0 ns) OR ((NOW - presetlastev) >= preset_min)
            REPORT "Spike detected on preset" SEVERITY warning;
        presetlastev := NOW;
    END PROCESS;

    PROCESS (clear)
        VARIABLE clearlastev : TIME := 0 ns;
        BEGIN
        ASSERT (NOW = 0 ns) OR ((NOW - clearlastev) >= clear_min)
            REPORT "Spike detected on clear" SEVERITY warning;
        clearlastev := NOW;
    END PROCESS;

    -- check for setup/hold violations
    PROCESS (clock,d)
        VARIABLE dlastev : TIME := 0 ns;
        VARIABLE clocklastev : TIME := 0 ns;
        BEGIN
        -- Hold check
        IF d'EVENT THEN
```

```vhdl
            ASSERT (NOW = 0 ns) OR ((NOW - clocklastev) >= d_hold)
                REPORT "Hold error" SEVERITY warning;
            dlastev := NOW;
        END IF;

        -- Setup check
        IF (clock'EVENT) AND (clock = '1') THEN
            ASSERT (NOW = 0 ns) OR ((NOW - dlastev) >= d_setup)
                REPORT "Setup error" SEVERITY warning;
            clocklastev := NOW;
        END IF;
    END PROCESS;

    -- check for invalid control
    ASSERT NOT ( (preset = '0') AND (clear = '0') )
        REPORT "Preset and clear both active"
        SEVERITY warning;

    -- input delay processing
    presetdelay <= preset AFTER f_delay(preset,preset_i01,preset_i10);
    cleardelay <= clear AFTER f_delay(clear,clear_i01,clear_i10);
    clockdelay <= clock AFTER f_delay(clock,clock_i01,clock_i10);
    ddelay <= d AFTER f_delay(d,d_i01,d_i10);

    -- flip-flop operation
    PROCESS (presetdelay,cleardelay,clockdelay)
        BEGIN
        -- Check for preset
        IF (presetdelay = '0') AND (cleardelay = '1') THEN
            q <= f_ttl('1') AFTER f_delay(f_ttl('1'), q_o01, q_o10);
            qb <= f_ttl('0') AFTER f_delay(f_ttl('0'), qb_o01, qb_o10);

        -- Check for clear
        ELSIF (presetdelay = '1') AND (cleardelay = '0') THEN
            q <= f_ttl('0') AFTER f_delay(f_ttl('0'), q_o01, q_o10);
            qb <= f_ttl('1') AFTER f_delay(f_ttl('1'), qb_o01, qb_o10);

        -- Check for control error
        ELSIF (presetdelay = '0') AND (cleardelay = '0') THEN
            q <= f_ttl('X') AFTER f_delay(f_ttl('X'), q_o01, q_o10);
            qb <= f_ttl('X') AFTER f_delay(f_ttl('X'), qb_o01, qb_o10);

        -- Check for unknown controls
        ELSIF (presetdelay = 'X') OR (cleardelay = 'X') THEN
            q <= f_ttl('X') AFTER f_delay(f_ttl('X'), q_o01, q_o10);
            qb <= f_ttl('X') AFTER f_delay(f_ttl('X'), qb_o01, qb_o10);

        -- Check for unknown clock
        ELSIF (clockdelay'EVENT) AND (clockdelay = 'X') THEN
            q <= f_ttl('X') AFTER f_delay(f_ttl('X'), q_o01, q_o10);
            qb <= f_ttl('X') AFTER f_delay(f_ttl('X'), qb_o01, qb_o10);
```

4.1. FLIP-FLOPS

```
        -- Check for clocked data
        ELSIF (clockdelay'EVENT) AND (clockdelay = '1') THEN
            q <= d AFTER f_delay(d, q_o01, q_o10);
            qb <= NOT d AFTER f_delay(NOT d, qb_o01, qb_o10);
        END IF;
        END PROCESS;
END full;
```

The ports for this model are
- **preset** - preset control
- **clear** - clear control
- **clock** - clock input
- **d** - data input
- **q** - state output
- **qb** - complementary state output

Generic parameters to this model are summarized here
- **preset_i01, preset_i10** - low to high and high to low **preset** input port delays
- **clear_i01, clear_i10** - low to high and high to low **clear** input port delays
- **clock_i01, clock_i10** - low to high and high to low **clock** input port delays
- **d_i01, d_i10** - low to high and high to low **d** input port delays
- **q_o01, q_o10** - low to high and high to low **q** output port delays
- **qb_o01, qb_o10** - low to high and high to low **qb** output port delays
- Minimum **clock** pulse width **clock_min**
- Minimum **preset** pulse width **preset_min**
- Minimum **clear** pulse width **clear_min**
- Data setup **d_setup** and hold **d_hold** minimums

This model features six error checks
- Clock frequency/spike detection
- Preset spike detection
- Clear spike detection
- Setup time checking
- Hold time checking
- Invalid control: check for preset and clear both active

as shown in the error checking section of the architecture. Each of the error checks utilizes a separate process statement (the exception is the hold/setup checks which utilize a single process).

Each of the spike detection processes uses a local variable to save the previous event time for the signal being checked. This is somewhat more efficient than using the delayed attribute since it does not require the simulator to create an additional signal.

The setup/hold checking process uses two local variables **dlastev** and **clocklastev** to save the time of the last data and clock event respectively. This process assumes a positive edge triggered clock. Adjustment to the following expression is required in order to change the clock characteristics.

$$(clock'EVENT) AND (clock =' 1')$$

The input delay processing section of the model incorporates the appropriate delay for each of the input ports. An internal signal for each input port is declared. The main device function process utilizes the delayed signals rather than the primary inputs.

The flip-flop operation is handled by a single process which is sensitive to the input delayed signals **presetdelay, cleardelay** and **clockdelay**. The following conditions are checked in order whenever any of the asynchronous inputs or clock have an event:

- Preset device - set state to high
- Clear device - set state to low
- Preset and clear both active - set state to unknown
- Unknown preset/clear - set state to unknown
- Unknown clock - set state to unknown
- Clock rising edge - set state to data value

Note that the order of the above checks are critical to the correct operation of the model. In particular the asynchronous preset and clear takes precedence over the clocked operation of the device. All of the above state changes are executed with appropriate delay introduced through the use of the **f_delay** function as discussed in section 7.6. This function utilizes the output rise and fall delays and based on the new state value chooses the proper delay.

4.1.2 JK Pos-Edge Triggered Flip-Flop with Preset/Clear

This device is a JK rising edge triggered flip-flop. A low at the **preset** or **clear** inputs sets or resets the outputs regardless of the values of the other inputs. When **preset** or **clear** are high, data at the **j** and **k** inputs meeting

4.1. FLIP-FLOPS

the setup time requirements are transferred to the outputs on the rising edge of the clock pulse. Following the hold time interval, data at the **j** and **k** inputs may be changed without affecting the values at the outputs. This device corresponds to the Texas Instruments 109 series TTL parts [TI86]. Figure 4.2 shows the symbolic representation of this device. The following

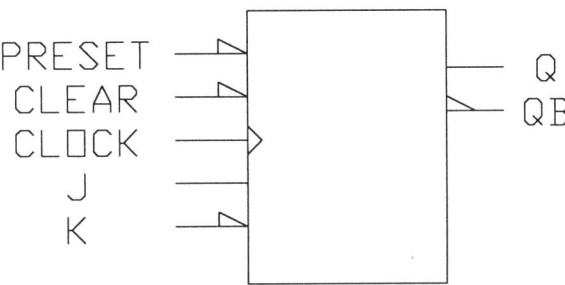

Figure 4.2: Logic Diagram JK Pos Edge Flip-Flop with Preset/Clear

shows the function table for this device.

Function Table

inputs					outputs	
preset	clear	clock	j	k	q	qb
L	H	X	X	X	H	L
H	L	X	X	X	L	H
L	L	X	X	X	?	?
H	H	R	L	L	L	H
H	H	R	H	L	toggle	
H	H	R	L	H	Q0	not Q0
H	H	R	H	H	H	L
H	H	L	X	X	Q0	not Q0

The following shows the full model for a JK flip-flop with characteristics

- Asynchronous preset
- Asynchronous clear
- Positive edge-triggered clocking
- Primary and complementary output

JK Positive-Edge Triggered Flip-Flop with Preset/Clear

```
USE std.std_logic.ALL;
USE std.std_ttl.ALL;
ENTITY jkffpospreclr IS
    GENERIC (preset_i01,preset_i10,
             clear_i01,clear_i10,
             clock_i01,clock_i10,
             j_i01,j_i10,
             k_i01,k_i10,
             q_o01,q_o10,
             qb_o01,qb_o10,
             clock_min, preset_min, clear_min,
             data_setup, data_hold : TIME := 2 ns);
    PORT    (preset,clear,clock,j,k: IN t_wlogic; q,qb: OUT t_wlogic);
END jkffpospreclr;

ARCHITECTURE full OF jkffpospreclr IS
    SIGNAL presetdelay, cleardelay, clockdelay, jdelay, kdelay : t_wlogic;
BEGIN
    -- check clock frequency/spike detection
    PROCESS (clock)
        VARIABLE clocklastev : TIME := 0 ns;
        BEGIN
        ASSERT (NOW = 0 ns) OR ((NOW - clocklastev) >= clock_min)
            REPORT "Spike detected on clock" SEVERITY warning;
        clocklastev := NOW;
    END PROCESS;

    -- spike detection
    PROCESS (preset)
        VARIABLE presetlastev : TIME := 0 ns;
        BEGIN
        ASSERT (NOW = 0 ns) OR ((NOW - presetlastev) >= preset_min)
            REPORT "Spike detected on preset" SEVERITY warning;
        presetlastev := NOW;
    END PROCESS;

    PROCESS (clear)
        VARIABLE clearlastev : TIME := 0 ns;
        BEGIN
        ASSERT (NOW = 0 ns) OR ((NOW - clearlastev) >= clear_min)
            REPORT "Spike detected on clear" SEVERITY warning;
        clearlastev := NOW;
    END PROCESS;

    -- check for setup/hold violations
    PROCESS (clock,j,k)
        VARIABLE datalastev : TIME := 0 ns;
        VARIABLE clocklastev : TIME := 0 ns;
        BEGIN
        -- hold check
        IF (j'EVENT) OR (k'EVENT) THEN
            ASSERT (NOW = 0 ns) OR ((NOW - clocklastev) >= data_hold)
```

4.1. FLIP-FLOPS

```
            REPORT "Hold error" SEVERITY warning;
        datalastev := NOW;
    END IF;

    -- setup check
    IF (clock'EVENT) AND (clock = '1') THEN
        ASSERT (NOW = 0 ns) OR ((NOW - datalastev) >= data_setup)
            REPORT "Setup error" SEVERITY warning;
        clocklastev := NOW;
    END IF;
    END PROCESS;

-- check for invalid control
ASSERT NOT ( (preset = '0') AND (clear = '0') )
    REPORT "Preset and clear both active"
    SEVERITY warning;

-- input delay processing
presetdelay <= preset AFTER f_delay(preset,preset_i01,preset_i10);
cleardelay  <= clear  AFTER f_delay(clear,clear_i01,clear_i10);
clockdelay  <= clock  AFTER f_delay(clock,clock_i01,clock_i10);
jdelay <= j AFTER f_delay(j,j_i01,j_i10);
kdelay <= k AFTER f_delay(k,k_i01,k_i10);

-- flip-flop operation
PROCESS (presetdelay,cleardelay,clockdelay)
    VARIABLE qstate, qbstate : t_wlogic := U;
    BEGIN
    -- check for preset
    IF (presetdelay = '0') AND (cleardelay = '1') THEN
        qstate  := f_ttl('1');
        qbstate := f_ttl('0');

    -- check for clear
    ELSIF (presetdelay = '1') AND (cleardelay = '0') THEN
        qstate  := f_ttl('0');
        qbstate := f_ttl('1');

    -- check for control error
    ELSIF (presetdelay = '0') AND (cleardelay = '0') THEN
        qstate  := f_ttl('X');
        qbstate := f_ttl('X');

    -- check for unknown controls
    ELSIF (presetdelay = 'X') OR (cleardelay = 'X') THEN
        qstate  := f_ttl('X');
        qbstate := f_ttl('X');

    -- check for unknown clock
    ELSIF (clockdelay'EVENT) AND (clockdelay = 'X') THEN
        qstate  := f_ttl('X');
        qbstate := f_ttl('X');
```

```
                -- check for clocked data
                ELSIF (clockdelay'EVENT) AND (clockdelay = '1') THEN
                    -- check for clear
                    IF (jdelay = '0') AND (kdelay = '0') THEN
                        qstate  := f_ttl('0');
                        qbstate := f_ttl('1');

                    -- check for toggle
                    ELSIF (jdelay = '1') AND (kdelay = '0') THEN
                        qstate  := NOT qstate;
                        qbstate := NOT qbstate;

                    -- check for preset
                    ELSIF (jdelay = '1') AND (kdelay = '1') THEN
                        qstate  := f_ttl('1');
                        qbstate := f_ttl('0');

                    -- check for unknowns
                    ELSIF (jdelay = 'X') OR (kdelay = 'X') THEN
                        qstate  := f_ttl('X');
                        qbstate := f_ttl('X');
                    END IF;
                END IF;

                -- assign values to output signals
                q  <= qstate  AFTER f_delay(qstate,  q_o01,  q_o10);
                qb <= qbstate AFTER f_delay(qbstate, qb_o01, qb_o10);
            END PROCESS;
END full;
```

The ports for this model are

- **preset** - preset control
- **clear** - clear control
- **clock** - clock input
- **j** - operation control
- **k** - operation control
- **q** - state output
- **qb** - complementary state output

Generic parameters to this model are summarized here

- **preset_i01, preset_i10** - low to high and high to low **preset** input port delays
- **clear_i01, clear_i10** - low to high and high to low **clear** input port delays

4.1. FLIP-FLOPS

- clock_i01, clock_i10 - low to high and high to low **clock** input port delays
- j_i01, j_i10 - low to high and high to low **j** input port delays
- k_i01, k_i10 - low to high and high to low **k** input port delays
- q_o01, q_o10 - low to high and high to low **q** output port delays
- qb_o01, qb_o10 - low to high and high to low **qb** output port delays
- Minimum **clock** pulse width **clock_min**
- Minimum **preset** pulse width **preset_min**
- Minimum **clear** pulse width **clear_min**
- Data setup **d_setup** and hold **d_hold** minimums

This model features six error checks

- Clock frequency/spike detection
- Preset spike detection
- Clear spike detection
- Setup time checking
- Hold time checking
- Invalid control: check for preset and clear both active

as shown in the error checking section of the architecture. Each of the error checks utilizes a separate process statement (the exception is the hold/setup checks which utilize a single process).

Each of the spike detection processes uses a local variable to save the previous event time for the signal being checked. This is somewhat more efficient than using the delayed attribute since it does not require the simulator to create an additional signal.

The setup/hold checking process uses two local variables **dlastev** and **clocklastev** to save the time of the last data and clock event respectively. This process assumes a positive edge triggered clock. Adjustment to the following expression is required in order to change the clock characteristics.

$(clock'EVENT)AND(clock ='1')$

Events from either of the data inputs **j** or **k** are treated as a single data event. A more sophisticated version of this error check (with additional generic parameters) could be crafted which tests for setup/hold errors on either of the data inputs separately through the use of two local variables, one for each data input. Most data books do not specify this level of detail in timing and therefore this approach has not been taken.

The input delay processing section of the model incorporates the appropriate delay for each of the input ports. An internal signal for each input port is declared. The main device function process utilizes the delayed signals rather than the primary inputs.

The flip-flop operation is handled by a single process which is sensitive to the input delayed signals **presetdelay**, **cleardelay** and **clockdelay**. The following conditions are checked in order whenever any of the asynchronous inputs or clock have an event

- Preset device - set state to high
- Clear device - set state to low
- Preset and clear both active - set state to unknown
- Unknown preset/clear - set state to unknown
- Unknown clock - set state to unknown
- Clock rising edge
    - Check for clear - set state to low
    - Check for toggle - negate state
    - Check for preset - set state to high
    - Check for unknown data - set state to unknown

Note that the order of the above checks are critical to the correct operation of the model. In particular the asynchronous preset and clear takes precedence over the clocked operation of the device. Once the state of the device is determined (and stored in the local variables **qstate** and **qbstate**), the output signal assignment statements for **q** and **qb** are executed. Output state changes are executed with appropriate delay introduced through the use of the **f_delay** function as discussed in section 7.6. This function utilizes the output rise and fall delays and based on the new state value chooses the proper delay.

4.1.3 JK Neg-Edge Triggered Flip-Flop with Preset/Clear

This device is a JK falling edge triggered flip-flop. A low at the **preset** or **clear** inputs sets or resets the outputs regardless of the values of the other inputs. When **preset** or **clear** are high, data at the **j** and **k** inputs meeting the setup time requirements are transferred to the outputs on the falling edge of the clock pulse. Following the hold time interval, data at the **j** and **k** inputs may be changed without affecting the values at the outputs. This device corresponds to the Texas Instruments 112 series TTL parts [TI86]. Figure 4.3 shows the symbolic representation of this device. The following shows the function table for this device.

4.1. FLIP-FLOPS

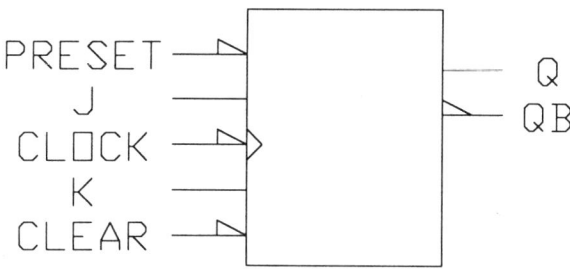

Figure 4.3: Logic Diagram JK Neg Edge Flip-Flop with Preset/Clear

Function Table

inputs					outputs	
preset	clear	clock	j	k	q	qb
L	H	X	X	X	H	L
H	L	X	X	X	L	H
L	L	X	X	X	?	?
H	H	F	L	L	Q0	not Q0
H	H	F	H	L	H	L
H	H	F	L	H	L	H
H	H	F	H	H	toggle	
H	H	H	X	X	Q0	not Q0

The following shows the full model for a JK flip-flop with characteristics

- Asynchronous preset
- Asynchronous clear
- Negative edge-triggered clocking
- Primary and complementary output

JK Negative-Edge Triggered Flip-Flop with Preset/Clear

```
USE std.std_logic.ALL;
USE std.std_ttl.ALL;
ENTITY jkffnegpreclr IS
    GENERIC (preset_i01,preset_i10,
             clear_i01,clear_i10,
             clock_i01,clock_i10,
             j_i01,j_i10,
             k_i01,k_i10,
```

```
                  q_o01,q_o10,
                  qb_o01,qb_o10,
                  clock_min, preset_min, clear_min,
                  data_setup, data_hold : TIME := 2 ns);
    PORT     (preset,clear,clock,j,k: IN t_wlogic; q,qb: OUT t_wlogic);
END jkffnegpreclr;

ARCHITECTURE full OF jkffnegpreclr IS
    SIGNAL presetdelay, cleardelay, clockdelay, jdelay, kdelay : t_wlogic;
BEGIN
    -- check clock frequency/spike detection
    PROCESS (clock)
        VARIABLE clocklastev : TIME := 0 ns;
        BEGIN
        ASSERT (NOW = 0 ns) OR ((NOW - clocklastev) >= clock_min)
            REPORT "Spike detected on clock" SEVERITY warning;
        clocklastev := NOW;
        END PROCESS;

    -- spike detection
    PROCESS (preset)
        VARIABLE presetlastev : TIME := 0 ns;
        BEGIN
        ASSERT (NOW = 0 ns) OR ((NOW - presetlastev) >= preset_min)
            REPORT "Spike detected on preset" SEVERITY warning;
        presetlastev := NOW;
        END PROCESS;

    PROCESS (clear)
        VARIABLE clearlastev : TIME := 0 ns;
        BEGIN
        ASSERT (NOW = 0 ns) OR ((NOW - clearlastev) >= clear_min)
            REPORT "Spike detected on clear" SEVERITY warning;
        clearlastev := NOW;
        END PROCESS;

    -- check for setup/hold violations
    PROCESS (clock,j,k)
        VARIABLE datalastev : TIME := 0 ns;
        VARIABLE clocklastev : TIME := 0 ns;
        BEGIN
        -- hold check
        IF (j'EVENT) OR (k'EVENT) THEN
            ASSERT (NOW = 0 ns) OR ((NOW - clocklastev) >= data_hold)
                REPORT "Hold error" SEVERITY warning;
            datalastev := NOW;
        END IF;

        -- setup check
        IF (clock'EVENT) AND (clock = '0') THEN
            ASSERT (NOW = 0 ns) OR ((NOW - datalastev) >= data_setup)
                REPORT "Setup error" SEVERITY warning;
```

4.1. FLIP-FLOPS

```
            clocklastev := NOW;
      END IF;
      END PROCESS;

   -- check for invalid control
   ASSERT NOT ( (preset = '0') AND (clear = '0') )
      REPORT "Preset and clear both active"
      SEVERITY warning;

   -- input delay processing
   presetdelay <= preset AFTER f_delay(preset,preset_i01,preset_i10);
   cleardelay <= clear AFTER f_delay(clear,clear_i01,clear_i10);
   clockdelay <= clock AFTER f_delay(clock,clock_i01,clock_i10);
   jdelay <= j AFTER f_delay(j,j_i01,j_i10);
   kdelay <= k AFTER f_delay(k,k_i01,k_i10);

   -- flip-flop operation
   PROCESS (presetdelay,cleardelay,clockdelay)
      VARIABLE qstate, qbstate : t_wlogic := U;
      BEGIN
      -- check for preset
      IF (presetdelay = '0') AND (cleardelay = '1') THEN
         qstate := f_ttl('1');
         qbstate := f_ttl('0');

      -- check for clear
      ELSIF (presetdelay = '1') AND (cleardelay = '0') THEN
         qstate := f_ttl('0');
         qbstate := f_ttl('1');

      -- check for control error
      ELSIF (presetdelay = '0') AND (cleardelay = '0') THEN
         qstate := f_ttl('X');
         qbstate := f_ttl('X');

      -- check for unknown controls
      ELSIF (presetdelay = 'X') OR (cleardelay = 'X') THEN
         qstate := f_ttl('X');
         qbstate := f_ttl('X');

      -- check for unknown clock
      ELSIF (clockdelay'EVENT) AND (clockdelay = 'X') THEN
         qstate := f_ttl('X');
         qbstate := f_ttl('X');

      -- check for clocked data
      ELSIF (clockdelay'EVENT) AND (clockdelay = '0') THEN
         -- check for clear
         IF (jdelay = '0') AND (kdelay = '1') THEN
            qstate := f_ttl('0');
            qbstate := f_ttl('1');
```

```
              -- check for toggle
              ELSIF (jdelay = '1') AND (kdelay = '1') THEN
                  qstate := NOT qstate;
                  qbstate := NOT qbstate;

              -- check for preset
              ELSIF (jdelay = '1') AND (kdelay = '0') THEN
                  qstate := f_ttl('1');
                  qbstate := f_ttl('0');

              -- check for unknowns
              ELSIF (jdelay = 'X') OR (kdelay = 'X') THEN
                  qstate := f_ttl('X');
                  qbstate := f_ttl('X');
              END IF;
          END IF;

          -- assign values to output signals
          q <= qstate AFTER f_delay(qstate, q_o01, q_o10);
          qb <= qbstate AFTER f_delay(qbstate, qb_o01, qb_o10);
          END PROCESS;
END full;
```

The ports for this model are
- **preset** - preset control
- **clear** - clear control
- **clock** - clock input
- **j** - operation control
- **k** - operation control
- **q** - state output
- **qb** - complementary state output

Generic parameters to this model are summarized here
- **preset_i01, preset_i10** - low to high and high to low **preset** input port delays
- **clear_i01, clear_i10** - low to high and high to low **clear** input port delays
- **clock_i01, clock_i10** - low to high and high to low **clock** input port delays
- **j_i01, j_i10** - low to high and high to low **j** input port delays
- **k_i01, k_i10** - low to high and high to low **k** input port delays
- **q_o01, q_o10** - low to high and high to low **q** output port delays
- **qb_o01, qb_o10** - low to high and high to low **qb** output port delays

4.1. FLIP-FLOPS

- Minimum clock pulse width **clock_min**
- Minimum preset pulse width **preset_min**
- Minimum clear pulse width **clear_min**
- Data setup **d_setup** and hold **d_hold** minimums

This model features six error checks

- Clock frequency/spike detection
- Preset spike detection
- Clear spike detection
- Setup time checking
- Hold time checking
- Invalid control: check for preset and clear both active

as shown in the error checking section of the architecture. Each of the error checks utilizes a separate process statement (the exception is the hold/setup checks which utilize a single process).

Each of the spike detection processes uses a local variable to save the previous event time for the signal being checked. This is somewhat more efficient than using the delayed attribute since it does not require the simulator to create an additional signal.

The setup/hold checking process uses two local variables **dlastev** and **clocklastev** to save the time of the last data and clock event respectively. This process assumes a negative edge triggered clock. Adjustment to the following expression is required in order to change the clock characteristics.

$(clock'EVENT) AND (clock =' 0')$

Events from either of the data inputs **j** or **k** are treated as a single data event. A more sophisticated version of this error check (with additional generic parameters) could be crafted which tests for setup/hold errors on either of the data inputs separately through the use of two local variables, one for each data input. Most data books do not specify this level of detail in timing and therefore this approach has not been taken.

The input delay processing section of the model incorporates the appropriate delay for each of the input ports. An internal signal for each input port is declared. The main device function process utilizes the delayed signals rather than the primary inputs.

The flip-flop operation is handled by a single process which is sensitive to the input delayed signals **presetdelay**, **cleardelay** and **clockdelay**. The following conditions are checked in order whenever any of the asynchronous inputs or clock have an event

- Preset device - set state to high

- Clear device - set state to low

- Preset and clear both active - set state to unknown

- Unknown preset/clear - set state to unknown

- Unknown clock - set state to unknown

- Clock falling edge

  - Check for clear - set state to low

  - Check for toggle - negate state

  - Check for preset - set state to high

  - Check for unknown data - set state to unknown

Note that the order of the above checks are critical to the correct operation of the model. In particular the asynchronous preset and clear takes precedence over the clocked operation of the device. Once the state of the device is determined (and stored in the local variables **qstate** and **qbstate**), the output signal assignment statements for **q** and **qb** are executed. Output state changes are executed with appropriate delay introduced through the use of the **f_delay** function as discussed in section 7.6. This function utilizes the output rise and fall delays and based on the new state value chooses the proper delay.

4.1.4 JK Negative-Edge Triggered Flip-Flop with Preset

This device is a JK falling edge triggered flip-flop. A low at the **preset** input sets the outputs regardless of the values of the other inputs. When **preset** is high, data at the **j** and **k** inputs meeting the setup time requirements are transferred to the outputs on the falling edge of the clock pulse. Following the hold time interval, data at the **j** and **k** inputs may be changed without affecting the values at the outputs. This device corresponds to the Texas Instruments 113 series TTL parts [TI86]. Figure 4.4 shows the symbolic representation of this device. The following shows the function table for this device.

4.1. FLIP-FLOPS

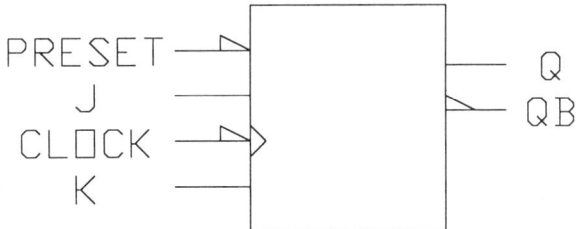

Figure 4.4: Logic Diagram JK Neg Edge Flip-Flop with Preset

Function Table

inputs				outputs	
preset	clock	j	k	q	qb
L	X	X	X	H	L
H	F	L	L	Q0	not Q0
H	F	H	L	H	L
H	F	L	H	L	H
H	F	H	H	toggle	
H	H	X	X	Q0	not Q0

The following shows the full model for a JK flip-flop with characteristics

- Asynchronous preset
- Negative edge-triggered clocking
- Primary and complementary output

JK Negative-Edge Triggered Flip-Flop with Preset

```
USE std.std_logic.ALL;
USE std.std_ttl.ALL;
ENTITY jkffnegpre IS
    GENERIC (preset_i01,preset_i10,
             clock_i01,clock_i10,
             j_i01,j_i10,
             k_i01,k_i10,
             q_o01,q_o10,
             qb_o01,qb_o10,
             clock_min, preset_min,
             data_setup, data_hold : TIME := 2 ns);
    PORT    (preset,clock,j,k: IN t_wlogic; q,qb: OUT t_wlogic);
END jkffnegpre;

ARCHITECTURE full OF jkffnegpre IS
```

CHAPTER 4. SEQUENTIAL DEVICES

```vhdl
    SIGNAL presetdelay, clockdelay, jdelay, kdelay : t_wlogic;
BEGIN
    -- check clock frequency/spike detection
    PROCESS (clock)
        VARIABLE clocklastev : TIME := 0 ns;
        BEGIN
        ASSERT (NOW = 0 ns) OR ((NOW - clocklastev) >= clock_min)
            REPORT "Spike detected on clock" SEVERITY warning;
        clocklastev := NOW;
    END PROCESS;

    -- spike detection
    PROCESS (preset)
        VARIABLE presetlastev : TIME := 0 ns;
        BEGIN
        ASSERT (NOW = 0 ns) OR ((NOW - presetlastev) >= preset_min)
            REPORT "Spike detected on preset" SEVERITY warning;
        presetlastev := NOW;
    END PROCESS;

    -- check for setup/hold violations
    PROCESS (clock,j,k)
        VARIABLE datalastev : TIME := 0 ns;
        VARIABLE clocklastev : TIME := 0 ns;
        BEGIN
        -- hold check
        IF (j'EVENT) OR (k'EVENT) THEN
            ASSERT (NOW = 0 ns) OR ((NOW - clocklastev) >= data_hold)
                REPORT "Hold error" SEVERITY warning;
            datalastev := NOW;
        END IF;

        -- setup check
        IF (clock'EVENT) AND (clock = '0') THEN
            ASSERT (NOW = 0 ns) OR ((NOW - datalastev) >= data_setup)
                REPORT "Setup error" SEVERITY warning;
            clocklastev := NOW;
        END IF;
    END PROCESS;

    -- input delay processing
    presetdelay <= preset AFTER f_delay(preset,preset_i01,preset_i10);
    clockdelay <= clock AFTER f_delay(clock,clock_i01,clock_i10);
    jdelay <= j AFTER f_delay(j,j_i01,j_i10);
    kdelay <= k AFTER f_delay(k,k_i01,k_i10);

    -- flip-flop operation
    PROCESS (presetdelay,clockdelay)
        VARIABLE qstate, qbstate : t_wlogic := U;
        BEGIN
        -- check for preset
        IF (presetdelay = '0') THEN
```

4.1. FLIP-FLOPS

```
                qstate  := f_ttl('1');
                qbstate := f_ttl('0');

            -- check for unknown controls
            ELSIF (presetdelay = 'X') THEN
                qstate  := f_ttl('X');
                qbstate := f_ttl('X');

            -- check for unknown clock
            ELSIF (clockdelay'EVENT) AND (clockdelay = 'X') THEN
                qstate  := f_ttl('X');
                qbstate := f_ttl('X');

            -- check for clocked data
            ELSIF (clockdelay'EVENT) AND (clockdelay = '0') THEN
                -- check for clear
                IF (jdelay = '0') AND (kdelay = '1') THEN
                    qstate  := f_ttl('0');
                    qbstate := f_ttl('1');

                -- check for toggle
                ELSIF (jdelay = '1') AND (kdelay = '1') THEN
                    qstate  := NOT qstate;
                    qbstate := NOT qbstate;

                -- check for preset
                ELSIF (jdelay = '1') AND (kdelay = '0') THEN
                    qstate  := f_ttl('1');
                    qbstate := f_ttl('0');

                -- check for unknowns
                ELSIF (jdelay = 'X') OR (kdelay = 'X') THEN
                    qstate  := f_ttl('X');
                    qbstate := f_ttl('X');
                END IF;
            END IF;

            -- assign values to output signals
            q  <= qstate  AFTER f_delay(qstate, q_o01, q_o10);
            qb <= qbstate AFTER f_delay(qbstate, qb_o01, qb_o10);
        END PROCESS;
END full;
```

The ports for this model are

- **preset** - preset control
- **clock** - clock input
- **j** - operation control
- **k** - operation control
- **q** - state output

- **qb** - complementary state output

Generic parameters to this model are summarized here

- **preset_i01, preset_i10** - low to high and high to low **preset** input port delays
- **clock_i01, clock_i10** - low to high and high to low **clock** input port delays
- **j_i01, j_i10** - low to high and high to low **j** input port delays
- **k_i01, k_i10** - low to high and high to low **k** input port delays
- **q_o01, q_o10** - low to high and high to low **q** output port delays
- **qb_o01, qb_o10** - low to high and high to low **qb** output port delays
- Minimum **clock** pulse width **clock_min**
- Minimum **preset** pulse width **preset_min**
- Data setup **d_setup** and hold **d_hold** minimums

This model features six error checks

- Clock frequency/spike detection
- Preset spike detection
- Setup time checking
- Hold time checking

as shown in the error checking section of the architecture. Each of the error checks utilizes a separate process statement (the exception is the hold/setup checks which utilize a single process).

Each of the spike detection processes uses a local variable to save the previous event time for the signal being checked. This is somewhat more efficient than using the delayed attribute since it does not require the simulator to create an additional signal.

The setup/hold checking process uses two local variables **dlastev** and **clocklastev** to save the time of the last data and clock event respectively. This process assumes a negative edge triggered clock. Adjustment to the following expression is required in order to change the clock characteristics.

$(clock'EVENT)AND(clock =' 0')$

Events from either of the data inputs **j** or **k** are treated as a single data event. A more sophisticated version of this error check (with additional generic parameters) could be crafted which tests for setup/hold errors on either of the data inputs separately through the use of two local variables, one for each data input. Most data books do not specify this level of detail in timing and therefore this approach has not been taken.

4.2. REGISTERS

The input delay processing section of the model incorporates the appropriate delay for each of the input ports. An internal signal for each input port is declared. The main device function process utilizes the delayed signals rather than the primary inputs.

The flip-flop operation is handled by a single process which is sensitive to the input delayed signals **presetdelay** and **clockdelay**. The following conditions are checked in order whenever any of the asynchronous inputs or clock have an event

- Preset device - set state to high
- Unknown preset - set state to unknown
- Unknown clock - set state to unknown
- Clock rising edge
  - Check for clear - set state to low
  - Check for toggle - negate state
  - Check for preset - set state to high
  - Check for unknown data - set state to unknown

Note that the order of the above checks are critical to the correct operation of the model. In particular the asynchronous preset takes precedence over the clocked operation of the device. Once the state of the device is determined (and stored in the local variables **qstate** and **qbstate**), the output signal assignment statements for **q** and **qb** are executed. Output state changes are executed with appropriate delay introduced through the use of the **f_delay** function as discussed in section 7.6. This function utilizes the output rise and fall delays and based on the new state value chooses the proper delay.

4.2 Registers

The following models describe various types of register devices. These models follow a standard format and all are presented with full timing functionality. It is possible to reduce the simulation and model complexity in stages as follows

- Remove input delay processing
  - Remove all input timing generic parameters
  - Remove all internal delay signals
  - Remove the input delay signal assignment statements

– Rename all references in the main behavioral process from the delayed port names to primary input port names

- Remove error checks: delete the error checking processes which occur prior to the main behavioral process

- Use simpler output delay calculations: adjust the final output assignment statements

By methodically adjusting these major components of the models it is possible to build a wide range of models which fit into the model accuracy continuum discussed in section 2.5.

The following adaptations are possible starting from these base models:

- Change the number of bits in register state. Adjust the generic parameter and port declarations as appropriate; adapting from a 4 bit model to an 8 bit model will require the addition of 4 more generics and 4 more ports for parallel inputs, and a similar change for parallel outputs.

- Parallel load, serial load, parallel output and serial output options can be added to the various models. By reviewing a model in this chapter which has one of these features, the capability can be added to another model.

- Asynchronous versus synchronous operation. By adjusting the sensitivity list to the main behavioral process of the model and making appropriate changes to the conditional logic of the process these adaptations can be made.

- Addition, deletion or adaptation of the asynchronous controls including clear, load, preset and enable inputs. A review of the various models in this chapter will give hints as to how to handle each of these features.

4.2.1 4-Bit Parallel-Access Shift Register

This four-bit register features parallel and serial inputs, parallel outputs, mode control, and two clock inputs. The register has two modes of operation:

- parallel load
- shift right **qa** toward **qd**

Parallel loading is accomplished by placing a high on the mode control. The data is loaded and appears at the outputs after the falling edge of the **clock2** input. During loading, the entry of serial data is inhibited. Shift right is accomplished on the falling transition of **clock1** when the mode control is low. This device corresponds to the Texas Instruments 95 series

4.2. REGISTERS

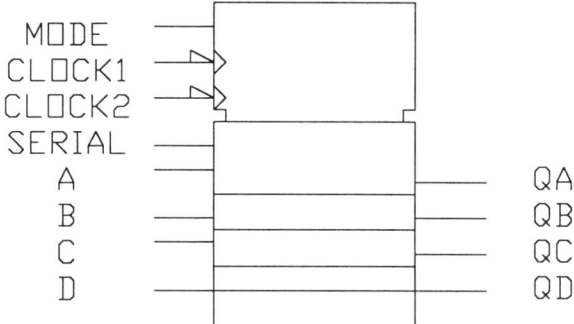

Figure 4.5: Logic Diagram 4 Bit Parallel-Access Shift Register

TTL parts [TI86]. Figure 4.5 shows the symbolic representation of this device. The following shows the function table for this device.

Function Table

mode control	clocks 2(L)	clocks 1(R)	inputs serial	parallel a	b	c	d	outputs qa	qb	qc	qd
H	H	X	X	X	X	X	X	QA0	QB0	QC0	QD0
H	F	X	X	a	b	c	d	a			
H	F	X	X	QB	QC	QD	d	QBn	QCn	QDn	d
L	L	X	X	X	X	X	X	QA0	QB0	QC0	QD0
L	X	H	H	X	X	X	X	H	QAn	QBn	QCn
L	X	L	L	X	X	X	X	L	QAn	QBn	QCn
R	L	X	X	X	X	X	X	QA0	QB0	QC0	QD0
F	L	X	X	X	X	X	X	QA0	QB0	QC0	QD0
F	L	X	X	X	X	X	X	QA0	QB0	QC0	QD0
R	H	X	X	X	X	X	X	QA0	QB0	QC0	QD0
R	H	X	X	X	X	X	X	QA0	QB0	QC0	QD0

The following shows the full model for a shift register with the characteristics

- 4 bit internal state
- Synchronous parallel data load
- Synchronous serial shift right
- Separate clocks for load and shift
- 4 bit output

4-Bit Parallel-Access Shift Register

```
USE std.std_logic.ALL;
USE std.std_ttl.ALL;
ENTITY sreg4 IS
    GENERIC (a_i01,a_i10,
             b_i01,b_i10,
             c_i01,c_i10,
             d_i01,d_i10,
             mode_i01,mode_i10,
             serial_i01,serial_i10,
             clock1_i01,clock1_i10,
             clock2_i01,clock2_i10,

             qa_o01,qa_o10,
             qb_o01,qb_o10,
             qc_o01,qc_o10,
             qd_o01,qd_o10,

             clock1_min, clock2_min,
             serial_setup,serial_hold,
             parallel_setup,parallel_hold : TIME := 2 ns);
    PORT    (a,b,c,d,mode,serial,clock1,clock2: IN t_wlogic;
             qa,qb,qc,qd: OUT t_wlogic);
END sreg4;

ARCHITECTURE full OF sreg4 IS
    SIGNAL adelay, bdelay, cdelay, ddelay,
           modedelay, serialdelay, clock1delay, clock2delay : t_wlogic;
BEGIN
    -- check clock frequency/spike detection
    PROCESS (clock1)
        VARIABLE clock1lastev : TIME := 0 ns;
        BEGIN
        ASSERT (NOW = 0 ns) OR ((NOW - clock1lastev) >= clock1_min)
            REPORT "Spike detected on clock1" SEVERITY warning;
        clock1lastev := NOW;
        END PROCESS;

    PROCESS (clock2)
        VARIABLE clock2lastev : TIME := 0 ns;
        BEGIN
        ASSERT (NOW = 0 ns) OR ((NOW - clock2lastev) >= clock2_min)
            REPORT "Spike detected on clock2" SEVERITY warning;
        clock2lastev := NOW;
        END PROCESS;

    -- check for setup/hold violations
    PROCESS (clock1,serial)
        VARIABLE seriallastev : TIME := 0 ns;
        VARIABLE clock1lastev : TIME := 0 ns;
        BEGIN
```

4.2. REGISTERS

```
        -- Hold check
        IF serial'EVENT THEN
            ASSERT (NOW = 0 ns) OR ((NOW - clock1lastev) >= serial_hold)
                REPORT "Hold error for serial input" SEVERITY warning;
            seriallastev := NOW;
        END IF;

        -- Setup check
        IF (clock1'EVENT) AND (clock1 = '1') THEN
            ASSERT (NOW = 0 ns) OR ((NOW - seriallastev) >= serial_setup)
                REPORT "Setup error for serial input" SEVERITY warning;
            clock1lastev := NOW;
        END IF;
    END PROCESS;
PROCESS (clock2,a,b,c,d)
        VARIABLE parallellastev : TIME := 0 ns;
        VARIABLE clock2lastev : TIME := 0 ns;
    BEGIN
        -- Hold check
        IF (a'EVENT OR b'EVENT OR c'EVENT OR d'EVENT) THEN
            ASSERT (NOW = 0 ns) OR ((NOW - clock2lastev) >= parallel_hold)
                REPORT "Hold error for parallel input" SEVERITY warning;
            parallellastev := NOW;
        END IF;

        -- Setup check
        IF (clock2'EVENT) AND (clock2 = '1') THEN
            ASSERT (NOW=0 ns)OR((NOW-parallellastev)>=parallel_setup)
                REPORT "Setup error for parallel input" SEVERITY warning;
            clock2lastev := NOW;
        END IF;
    END PROCESS;

    -- input delay processing
    adelay <= a AFTER f_delay(a,a_i01,a_i10);
    bdelay <= b AFTER f_delay(b,b_i01,b_i10);
    cdelay <= c AFTER f_delay(c,c_i01,c_i10);
    ddelay <= d AFTER f_delay(d,d_i01,d_i10);
    modedelay <= mode AFTER f_delay(mode,mode_i01,mode_i10);
    serialdelay <= serial AFTER f_delay(serial,serial_i01,serial_i10);
    clock1delay <= clock1 AFTER f_delay(clock1,clock1_i01,clock1_i10);
    clock2delay <= clock2 AFTER f_delay(clock2,clock2_i01,clock2_i10);

    -- register operation
    PROCESS (clock1delay,clock2delay)
        VARIABLE qastate,qbstate,qcstate,qdstate : t_wlogic := U;
    BEGIN
        -- Check for parallel load
        IF (modedelay='1')AND(clock2delay'EVENT)AND(clock2delay='0') THEN
            qastate := adelay;
            qbstate := bdelay;
```

```
            qcstate := cdelay;
            qdstate := ddelay;

        -- Serial shift right
        ELSIF(modedelay='0')AND(clock1delay'EVENT)AND(clock1delay='0')THEN
            qdstate := qcstate;
            qcstate := qbstate;
            qbstate := qastate;
            qastate := serialdelay;

        -- Check for unknown controls
        ELSIF (modedelay='X') AND ((clock1delay'EVENT) OR
                (clock2delay'EVENT)) THEN
            qastate := f_ttl('X');
            qbstate := f_ttl('X');
            qcstate := f_ttl('X');
            qdstate := f_ttl('X');

        -- Check for unknown clock
        ELSIF (modedelay = '0') AND (clock1delay'EVENT) AND
                (clock1delay = 'X') THEN
            qastate := f_ttl('X');
            qbstate := f_ttl('X');
            qcstate := f_ttl('X');
            qdstate := f_ttl('X');

        -- Check for unknown clock
        ELSIF (modedelay = '1') AND (clock2delay'EVENT) AND
                (clock2delay = 'X') THEN
            qastate := f_ttl('X');
            qbstate := f_ttl('X');
            qcstate := f_ttl('X');
            qdstate := f_ttl('X');
        END IF;

        -- Assign signal values
        qa <= qastate AFTER f_delay(qastate, qa_o01, qa_o10);
        qb <= qbstate AFTER f_delay(qbstate, qb_o01, qb_o10);
        qc <= qcstate AFTER f_delay(qcstate, qc_o01, qc_o10);
        qd <= qdstate AFTER f_delay(qdstate, qd_o01, qd_o10);
    END PROCESS;
END full;
```

The ports for this model are

- **a** - parallel data input
- **b** - parallel data input
- **c** - parallel data input
- **d** - parallel data input
- **mode** - mode control

4.2. REGISTERS

- **serial** - serial data input
- **clock1** - clock input
- **clock2** - clock input
- **qa** - parallel state output
- **qb** - parallel state output
- **qc** - parallel state output
- **qd** - parallel state output

Generic parameters to this model are summarized here

- **a_i01, a_i10** - low to high and high to low **a** input port delays
- **b_i01, b_i10** - low to high and high to low **b** input port delays
- **c_i01, c_i10** - low to high and high to low **c** input port delays
- **d_i01, d_i10** - low to high and high to low **d** input port delays
- **mode_i01, mode_i10** - low to high and high to low **mode** input port delays
- **serial_i01, serial_i10** - low to high and high to low **serial** input port delays
- **clock1_i01, clock1_i10** - low to high and high to low **clock1** input port delays
- **clock2_i01, clock2_i10** - low to high and high to low **clock2** input port delays
- **qa_o01, qa_o10** - low to high and high to low **qa** output port delays
- **qb_o01, qb_o10** - low to high and high to low **qb** output port delays
- **qc_o01, qc_o10** - low to high and high to low **qc** output port delays
- **qd_o01, qd_o10** - low to high and high to low **qd** output port delays
- Minimum **clock1** pulse width **clock1_min**
- Minimum **clock2** pulse width **clock2_min**
- Serial data setup **serial_setup** and hold **serial_hold** minimums
- Parallel data setup **parallel_setup** and hold **parallel_hold** minimums

Clock enable/disable times should be included in the output propagation times for this model.

This model features six error checks

- **clock1** frequency/spike detection
- **clock2** frequency/spike detection

- Serial data setup time checking
- Serial data hold time checking
- Parallel data setup time checking
- Parallel data hold time checking

as shown in the error checking section of the architecture. Each of the error checks utilizes a separate process statement (the exception is the hold/setup checks which utilize a single process for each clock).

Each of the spike detection processes uses a local variable to save the previous event time for the signal being checked. This is somewhat more efficient than using the delayed attribute since it does not require the simulator to create an additional signal.

The setup/hold checking processes use two local variables to save the time of the last data and clock event respectively. This process assumes a positive edge triggered clock. Adjustment to the expression

$(clock1'EVENT)AND(clock1 ='1')$

or

$(clock2'EVENT)AND(clock2 ='1')$

is required in order to change the clock characteristics. Events from the data inputs (**serial** for serial data, **a**, **b**, **c**, **d** for parallel data) are treated as a single data event. A more sophisticated version of this error check (with additional generic parameters) could be crafted which tests for setup/hold errors on any of the data inputs separately through the use of multiple local variables, one for each data input. Most data books do not specify this level of detail in timing and therefore this approach has not been taken.

The input delay processing section of the model incorporates the appropriate delay for each of the input ports. An internal signal for each input port is declared. The main device function process utilizes the delayed signals rather than the primary inputs.

The register operation is handled by a single process which is sensitive to the input delayed signals **clock1delay** and **clock2delay**. The following conditions are checked in order whenever either of the clocks have an event

- Parallel load - copy state from data inputs
- Serial shift right - shift state right, load serial input into left bit
- Unknown mode - set state to unknown
- Unknown clock1 on parallel load - set state to unknown
- Unknown clock2 on serial shift - set state to unknown

4.2. REGISTERS

Once the state of the device is determined (and stored in the local variables **qastate, qbstate, qcstate** and **qdstate**), the output signal assignment statements for **qa, qb, qc** and **qd** are executed. Output state changes are executed with appropriate delay introduced through the use of the **f_delay** function as discussed in section 7.6. This function utilizes the output rise and fall delays and based on the new state value chooses the proper delay.

4.2.2   3 to 8 Decoder/Demultiplexer with Register

This device is a three-line to eight-line decoder/demultiplexer with registers on the three address inputs. When the clock input **clock** rises, this device acts as a decoder/demultiplexer and the address present at the select inputs **a, b** and **c** is stored in the register. Further address changes are ignored until the next rising transition of **clock**. The output enable controls, **g1** and **g2**, control the state of the outputs independently of the select or **clock** inputs. All of the outputs are high unless **g1** is high and **g2** is low. This device corresponds to the Texas Instruments 131 series TTL parts [TI86]. Figure 4.6 shows the symbolic representation of this device. The following

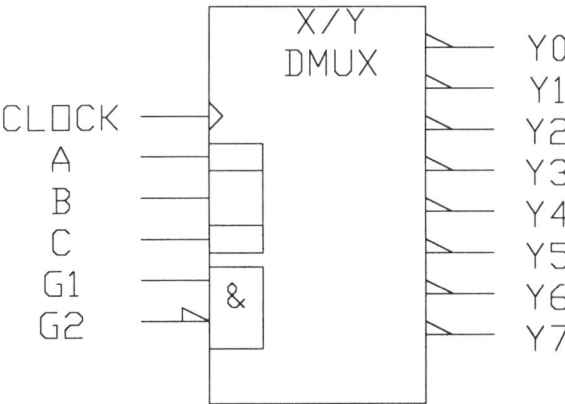

Figure 4.6: Logic Diagram 3 to 8 Decoder/Demultiplexer

shows the function table for this device.

Function Table

clk	inputs enable		inputs select			outputs							
	g1	g2	c	b	a	y0	y1	y2	y3	y4	y5	y6	y7
X	X	H	X	X	X	H	H	H	H	H	H	H	H
X	L	X	X	X	X	H	H	H	H	H	H	H	H
R	H	L	L	L	L	L	H	H	H	H	H	H	H
R	H	L	L	L	H	H	L	H	H	H	H	H	H
R	H	L	L	H	L	H	H	L	H	H	H	H	H
R	H	L	L	H	H	H	H	H	L	H	H	H	H
R	H	L	H	L	L	H	H	H	H	L	H	H	H
R	H	L	H	L	H	H	H	H	H	H	L	H	H
R	H	L	H	H	L	H	H	H	H	H	H	L	H
R	H	L	H	H	H	H	H	H	H	H	H	H	L
L or H	H	L	X	X	X	L for stored address; H for others							

The following shows the full model for a decoder with the characteristics

- 3 bit address
- 8 bit output
- Synchronous/clocked decode operation
- Asynchronous preset via dual control inputs (one active high, the other active low)

3 to 8 Decoder/Multiplexer with Register

```
USE std.std_logic.ALL;
USE std.std_ttl.ALL;
ENTITY decmul3to8reg IS
    GENERIC (clock_i01,clock_i10,
             g1_i01,g1_i10,
             g2_i01,g2_i10,
             a_i01,a_i10,
             b_i01,b_i10,
             c_i01,c_i10,
             clock_min,
             y0_o01,y0_o10,
             y1_o01,y1_o10,
             y2_o01,y2_o10,
             y3_o01,y3_o10,
             y4_o01,y4_o10,
             y5_o01,y5_o10,
             y6_o01,y6_o10,
             y7_o01,y7_o10,
```

4.2. REGISTERS

```
                data_setup, data_hold : TIME := 2 ns);
    PORT    (clock,g1,g2,a,b,c: IN t_wlogic;
             y0,y1,y2,y3,y4,y5,y6,y7: OUT t_wlogic);
END decmul3to8reg;

ARCHITECTURE full OF decmul3to8reg IS
    SIGNAL clockdelay,g1delay,g2delay,adelay,bdelay,cdelay : t_wlogic;
BEGIN
    -- check clock frequency/spike detection
    PROCESS (clock)
        VARIABLE clocklastev : TIME := 0 ns;
        BEGIN
        ASSERT (NOW = 0 ns) OR ((NOW - clocklastev) >= clock_min)
            REPORT "Spike detected on clock" SEVERITY warning;
        clocklastev := NOW;
        END PROCESS;

    -- check for setup/hold violations
    PROCESS (clock,a,b,c)
        VARIABLE datalastev : TIME := 0 ns;
        VARIABLE clocklastev : TIME := 0 ns;
        BEGIN
        -- hold check
        IF (a'EVENT) OR (b'EVENT) OR (c'EVENT) THEN
            ASSERT (NOW = 0 ns) OR ((NOW - clocklastev) >= data_hold)
                REPORT "Hold error" SEVERITY warning;
            datalastev := NOW;
        END IF;

        -- setup check
        IF (clock'EVENT) AND (clock = '1') THEN
            ASSERT (NOW = 0 ns) OR ((NOW - datalastev) >= data_setup)
                REPORT "Setup error" SEVERITY warning;
            clocklastev := NOW;
        END IF;
        END PROCESS;

    -- input delay processing
    clockdelay <= clock AFTER f_delay(clock,clock_i01,clock_i10);
    g1delay <= g1 AFTER f_delay(g1,g1_i01,g1_i10);
    g2delay <= g2 AFTER f_delay(g2,g2_i01,g2_i10);
    adelay <= a AFTER f_delay(a,a_i01,a_i10);
    bdelay <= b AFTER f_delay(b,b_i01,b_i10);
    cdelay <= c AFTER f_delay(c,c_i01,c_i10);

    -- decoder/multiplexer operation
    PROCESS (clockdelay,g1delay,g2delay)
        VARIABLE baddress : t_logarray (1 TO 3);
        VARIABLE state : t_logarray (1 TO 8);
        VARIABLE iaddress : integer;
        VARIABLE unknown : boolean;
        BEGIN
```

```
            -- check for preset
            IF (g1delay = '0') OR (g2delay = '1') THEN
                FOR i IN state'RANGE LOOP state(i) := f_ttl('1'); END LOOP;

            -- check for unknown control
            ELSIF (g1delay = 'X') OR (g2delay = 'X') THEN
                FOR i IN state'RANGE LOOP state(i) := f_ttl('X'); END LOOP;

            -- check for unknown clockdelay
            ELSIF (clockdelay'EVENT) AND (clockdelay = 'X') THEN
                FOR i IN state'RANGE LOOP state(i) := f_ttl('X'); END LOOP;

            -- clocked address
            ELSIF (clockdelay'EVENT) AND (clockdelay = '1') THEN
                -- pickup the binary address
                baddress(1) := cdelay;
                baddress(2) := bdelay;
                baddress(3) := adelay;

                -- calculate the integer address
                f_logictoint(baddress,unknown,iaddress);

                -- watch out for unknown address
                IF unknown THEN
                    FOR i IN state'RANGE LOOP state(i) := f_ttl('X'); END LOOP;
                -- decode the address
                ELSE f_intdecode(iaddress, state, ttl);
                END IF;
            END IF;

            -- assign values to output signals
            y0 <= state(1) AFTER f_delay(state(1), y0_o01, y0_o10);
            y1 <= state(2) AFTER f_delay(state(2), y1_o01, y1_o10);
            y2 <= state(3) AFTER f_delay(state(3), y2_o01, y2_o10);
            y3 <= state(4) AFTER f_delay(state(4), y3_o01, y3_o10);
            y4 <= state(5) AFTER f_delay(state(5), y4_o01, y4_o10);
            y5 <= state(6) AFTER f_delay(state(6), y5_o01, y5_o10);
            y6 <= state(7) AFTER f_delay(state(7), y6_o01, y6_o10);
            y7 <= state(8) AFTER f_delay(state(8), y7_o01, y7_o10);
        END PROCESS;
END full;
```

The ports for this model are

- **clock** - clock input
- **g1** - enable input
- **g2** - enable input
- **a** - address select
- **b** - address select

4.2. REGISTERS

- c - address select
- y0 - decode output
- y1 - decode output
- y2 - decode output
- y3 - decode output
- y4 - decode output
- y5 - decode output
- y6 - decode output
- y7 - decode output

Generic parameters to this model are summarized here

- clock_i01, clock_i10 - low to high and high to low **clock** input port delays
- g1_i01, g1_i10 - low to high and high to low **g1** input port delays
- g2_i01, g2_i10 - low to high and high to low **g2** input port delays
- a_i01, a_i10 - low to high and high to low **a** input port delays
- b_i01, b_i10 - low to high and high to low **b** input port delays
- c_i01, c_i10 - low to high and high to low **c** input port delays
- Minimum **clock** pulse width **clock_min**
- y0_o01, y0_o10 - low to high and high to low **y0** output port delays
- y1_o01, y1_o10 - low to high and high to low **y1** output port delays
- y2_o01, y2_o10 - low to high and high to low **y2** output port delays
- y3_o01, y3_o10 - low to high and high to low **y3** output port delays
- y4_o01, y4_o10 - low to high and high to low **y4** output port delays
- y5_o01, y5_o10 - low to high and high to low **y5** output port delays
- y6_o01, y6_o10 - low to high and high to low **y6** output port delays
- y7_o01, y7_o10 - low to high and high to low **y7** output port delays
- Data data setup **data_setup** and hold **data_hold** minimums

This model features three error checks

- **clock** frequency/spike detection
- Address setup time checking
- Address hold time checking

as shown in the error checking section of the architecture. Each of the error checks utilizes a separate process statement (the exception is the hold/setup check which utilizes a single process).

The spike detection processes uses a local variable to save the previous event time for the signal being checked. This is somewhat more efficient than using the delayed attribute since it does not require the simulator to create an additional signal.

The setup/hold checking process uses two local variables to save the time of the last data and clock event respectively. This process assumes a positive edge triggered clock. Adjustment to the following expression is required in order to change the clock characteristics.

$$(clock'EVENT)AND(clock ='1')$$

Events from the address inputs **a**, **b** and **c** are treated as a single data event. A more sophisticated version of this error check (with additional generic parameters) could be crafted which tests for setup/hold errors on any of the data inputs separately through the use of multiple local variables, one for each data input. Most data books do not specify this level of detail in timing and therefore this approach has not been taken.

The input delay processing section of the model incorporates the appropriate delay for each of the input ports. An internal signal for each input port is declared. The main device function process utilizes the delayed signals rather than the primary inputs.

The decoder operation is handled by a single process which is sensitive to the input delayed signals **clockdelay**, **g1delay** and **g2delay**. The following conditions are checked in order whenever the clock or asynchronous control inputs have an event

- Preset - set all outputs to high
- Unknown control - set all outputs to unknown
- Unknown clock - set all outputs to unknown
- Clocked address - calculate integer address
    - If address has an unknown - set all outputs to unknown
    - Otherwise set addressed output to low, all others to high (see the **f_intdecode** function as described in section 7.7)

Once the state of the device is determined (and stored in the local array variable **state**, the output signal assignment statements for **y0**, **y1**, **y2**, **y3**, **y4**, **y5**, **y6** and **y7** are executed. Output state changes are executed with appropriate delay introduced through the use of the **f_delay** function as discussed in section 7.6. This function utilizes the output rise and fall delays and based on the new state value chooses the proper delay.

4.2.3 3 to 8 Decoder/Demultiplexer with Latch

This device is a three-line to eight-line decoder/demultiplexer with latches on the three address inputs. When the latch-enable input **gl** is low, the device acts as a decoder/demultiplexer. When **gl** goes from low to high, the address present at the select inputs **a**, **b** and **c** is stored in the latches. Further address changes are ignored as long as **gl** remains high. The output enable controls, **g1** and **g2**, control the outputs independently of the select or latch-enable inputs. All of the outputs are forced high if **g1** is low or **g2** is high. This device corresponds to the Texas Instruments 137 series TTL parts [TI86]. Figure 4.7 shows the symbolic representation of this device.

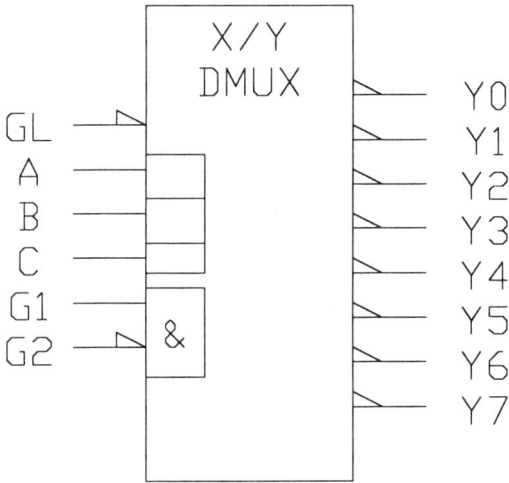

Figure 4.7: Logic Diagram 3 to 8 Decoder/Demultiplexer with Latch

The following shows the function table for this device.

Function Table

inputs						outputs							
enable			select										
gl	g1	g2	c	b	a	y0	y1	y2	y3	y4	y5	y6	y7
X	X	H	X	X	X	H	H	H	H	H	H	H	H
X	L	X	X	X	X	H	H	H	H	H	H	H	H
L	H	L	L	L	L	L	H	H	H	H	H	H	H
L	H	L	L	L	H	H	L	H	H	H	H	H	H
L	H	L	L	H	L	H	H	L	H	H	H	H	H
L	H	L	L	H	H	H	H	H	L	H	H	H	H
L	H	L	H	L	L	H	H	H	H	L	H	H	H
L	H	L	H	L	H	H	H	H	H	H	L	H	H
L	H	L	H	H	L	H	H	H	H	H	H	L	H
L	H	L	H	H	H	H	H	H	H	H	H	H	L
H	H	L	X	X	X	L for stored address; H for others							

The following shows the full model for a decoder with the characteristics

- 3 bit address
- 8 bit output
- Latched output operation
- Asynchronous preset via dual control inputs (one active high, the other active low)

3 to 8 Decoder/Multiplexer with Latch

```
USE std.std_logic.ALL;
USE std.std_ttl.ALL;
ENTITY decmul3to8latch IS
    GENERIC (gl_i01,gl_i10,
             g1_i01,g1_i10,
             g2_i01,g2_i10,
             a_i01,a_i10,
             b_i01,b_i10,
             c_i01,c_i10,
             gl_min,
             y0_o01,y0_o10,
             y1_o01,y1_o10,
             y2_o01,y2_o10,
             y3_o01,y3_o10,
             y4_o01,y4_o10,
             y5_o01,y5_o10,
             y6_o01,y6_o10,
             y7_o01,y7_o10,
```

4.2. REGISTERS

```vhdl
                    data_setup, data_hold : TIME := 2 ns);
    PORT    (gl,g1,g2,a,b,c: IN t_wlogic;
             y0,y1,y2,y3,y4,y5,y6,y7: OUT t_wlogic);
END decmul3to8latch;

ARCHITECTURE full OF decmul3to8latch IS
    SIGNAL gldelay,g1delay,g2delay,adelay,bdelay,cdelay : t_wlogic;
BEGIN
    -- latch spike detection
    PROCESS (gl)
        VARIABLE gllastev : TIME := 0 ns;
        BEGIN
        ASSERT (NOW = 0 ns) OR ((NOW - gllastev) >= gl_min)
            REPORT "Spike detected on gl" SEVERITY warning;
        gllastev := NOW;
        END PROCESS;

    -- check for setup/hold violations
    PROCESS (gl,a,b,c)
        VARIABLE datalastev : TIME := 0 ns;
        VARIABLE gllastev : TIME := 0 ns;
        BEGIN
        -- hold check
        IF (a'EVENT) OR (b'EVENT) OR (c'EVENT) THEN
            ASSERT (NOW = 0 ns) OR ((NOW - gllastev) >= data_hold)
                REPORT "Hold error" SEVERITY warning;
            datalastev := NOW;
        END IF;

        -- setup check
        IF (gl'EVENT) AND (gl = '1') THEN
            ASSERT (NOW = 0 ns) OR ((NOW - datalastev) >= data_setup)
                REPORT "Setup error" SEVERITY warning;
            gllastev := NOW;
        END IF;
        END PROCESS;

    -- input delay processing
    gldelay  <= gl AFTER f_delay(gl,gl_i01,gl_i10);
    g1delay  <= g1 AFTER f_delay(g1,g1_i01,g1_i10);
    g2delay  <= g2 AFTER f_delay(g2,g2_i01,g2_i10);
    adelay   <= a  AFTER f_delay(a,a_i01,a_i10);
    bdelay   <= b  AFTER f_delay(b,b_i01,b_i10);
    cdelay   <= c  AFTER f_delay(c,c_i01,c_i10);

    -- decoder/multiplexer operation
    PROCESS (gldelay,g1delay,g2delay,adelay,bdelay,cdelay)
        VARIABLE baddress : t_logarray (1 TO 3);
        VARIABLE state : t_logarray (1 TO 8) := (U,U,U,U,U,U,U,U);
        VARIABLE iaddress : integer;
        VARIABLE unknown : boolean;
        BEGIN
```

```
                -- check for preset
                IF (g1delay = '0') OR (g2delay = '1') THEN
                    FOR i IN state'RANGE LOOP state(i) := f_ttl('1'); END LOOP;

                -- check for unknown control
                ELSIF (g1delay = 'X') OR (g2delay = 'X') THEN
                    FOR i IN state'RANGE LOOP state(i) := f_ttl('X'); END LOOP;

                -- check for unknown gldelay
                ELSIF (gldelay = 'X') THEN
                    FOR i IN state'RANGE LOOP state(i) := f_ttl('X'); END LOOP;

                -- latch address
                ELSIF (gldelay = '0') THEN
                    -- pickup the binary address
                    baddress(1) := cdelay;
                    baddress(2) := bdelay;
                    baddress(3) := adelay;

                    -- calculate the integer address
                    f_logictoint(baddress,unknown,iaddress);

                    -- watch out for unknown address
                    IF unknown THEN
                        FOR i IN state'RANGE LOOP state(i) := f_ttl('X'); END LOOP;
                    -- decode the address
                    ELSE f_intdecode(iaddress, state, ttl);
                    END IF;
                END IF;

                -- assign values to output signals
                y0 <= state(1) AFTER f_delay(state(1), y0_o01, y0_o10);
                y1 <= state(2) AFTER f_delay(state(2), y1_o01, y1_o10);
                y2 <= state(3) AFTER f_delay(state(3), y2_o01, y2_o10);
                y3 <= state(4) AFTER f_delay(state(4), y3_o01, y3_o10);
                y4 <= state(5) AFTER f_delay(state(5), y4_o01, y4_o10);
                y5 <= state(6) AFTER f_delay(state(6), y5_o01, y5_o10);
                y6 <= state(7) AFTER f_delay(state(7), y6_o01, y6_o10);
                y7 <= state(8) AFTER f_delay(state(8), y7_o01, y7_o10);
                END PROCESS;
END full;
```

The ports for this model are

- gl - latch control
- g1 - enable input
- g2 - enable input
- a - address select
- b - address select

4.2. REGISTERS

- c - address select
- y0 - decode output
- y1 - decode output
- y2 - decode output
- y3 - decode output
- y4 - decode output
- y5 - decode output
- y6 - decode output
- y7 - decode output

Generic parameters to this model are summarized here

- gl_i01, gl_i10 - low to high and high to low **gl** input port delays
- g1_i01, g1_i10 - low to high and high to low **g1** input port delays
- g2_i01, g2_i10 - low to high and high to low **g2** input port delays
- a_i01, a_i10 - low to high and high to low **a** input port delays
- b_i01, b_i10 - low to high and high to low **b** input port delays
- c_i01, c_i10 - low to high and high to low **c** input port delays
- Minimum **gl** pulse width **gl_min**
- y0_o01, y0_o10 - low to high and high to low **y0** output port delays
- y1_o01, y1_o10 - low to high and high to low **y1** output port delays
- y2_o01, y2_o10 - low to high and high to low **y2** output port delays
- y3_o01, y3_o10 - low to high and high to low **y3** output port delays
- y4_o01, y4_o10 - low to high and high to low **y4** output port delays
- y5_o01, y5_o10 - low to high and high to low **y5** output port delays
- y6_o01, y6_o10 - low to high and high to low **y6** output port delays
- y7_o01, y7_o10 - low to high and high to low **y7** output port delays
- Data data setup **data_setup** and hold **data_hold** minimums

This model features three error checks

- Latch control **gl** frequency/spike detection
- Address setup time checking
- Address hold time checking

as shown in the error checking section of the architecture. The spike error checks utilizes a separate process statement. A second process performs the hold and setup checks.

The spike detection processes uses a local variable to save the previous event time for the signal being checked. This is somewhat more efficient than using the delayed attribute since it does not require the simulator to create an additional signal.

The setup/hold checking process uses two local variables to save the time of the last data and clock event respectively. This process assumes that the data must be settled before the latch control becomes active high. Adjustment to the following expression is required in order to change the latching control

$(gl'EVENT) AND (gl =' 1')$

characteristics. Events from the address inputs **a**, **b** and **c** are treated as a single data event. A more sophisticated version of this error check (with additional generic parameters) could be crafted which tests for setup/hold errors on any of the data inputs separately through the use of multiple local variables, one for each data input. Most data books do not specify this level of detail in timing and therefore this approach has not been taken.

The input delay processing section of the model incorporates the appropriate delay for each of the input ports. An internal signal for each input port is declared. The main device function process utilizes the delayed signals rather than the primary inputs.

The decoder operation is handled by a single process which is sensitive to the input delayed signals **gldelay, g1delay, g2delay, adelay, bdelay** and **cdelay**. Note that unlike synchronous devices, the data signals are placed in the sensitivity list along with the control signals. When the latch is not locked, each change in the address inputs must be propagated to the output. As a result, the address inputs must be placed in the sensitivity list for this mode of operation.

The following conditions are checked in order whenever the asynchronous control or address inputs have an event

- Preset - set all outputs to high
- Unknown preset control - set all outputs to unknown
- Unknown latch control - set all outputs to unknown
- Generate decoded output for current address - calculate integer address (see **iaddress**)
    - If address has an unknown - set all outputs to unknown (the **f_intdecode** function returns -1 when an unknown is detected

4.2. REGISTERS

in the address)
- Otherwise set addressed output to low, all others to high (see the f_intdecode function as described in section 7.7)

No special case is required to handle the locked latch mode since in this mode any changes in the input address are ignored by the device. Once the state of the device is determined (and stored in the local array variable **state**, the output signal assignment statements for **y0, y1, y2, y3, y4, y5, y6** and **y7** are executed. Output state changes are executed with appropriate delay introduced through the use of the **f_delay** function as discussed in section 7.6. This function utilizes the output rise and fall delays and based on the new state value chooses the proper delay.

4.2.4 8 Bit Parallel-Out Serial Shift Register

This 8-bit shift register features AND-gated serial inputs and an asynchronous clear. The gated serial inputs **a** and **b** permit complete control over incoming data. A low on either input inhibits entry of the new data and resets the first flip-flop to the low at the next clock pulse. A high input enables the other input, which will then determine the state of the first flip-flop. Data at the serial inputs may be changed while the clock is high or low, provided the minimum setup time requirements are met. Clocking occurs on the low-to-high transition of the clock input. This device corresponds to the Texas Instruments 164 series TTL parts [TI86]. Figure 4.8

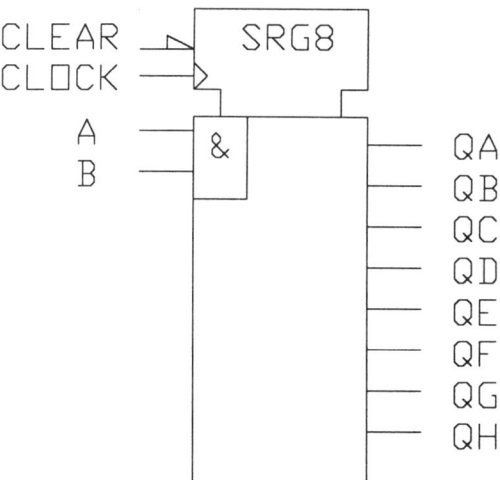

Figure 4.8: Logic Diagram 8 Bit Parallel-Out Serial Shift Register

shows the symbolic representation of this device. The following shows the function table for this device.

Function Table

inputs				outputs							
clear	clock	a	b	QA	QB	QC	QD	QE	QF	QG	QH
L	X	X	X	L	L	L	L	L	L	L	L
H	L	X	X	QA0	QB0	QC0	QD0	QE0	QF0	QG0	QH0
H	R	H	H	H	QAn	QBn	QCn	QDn	QEn	QFn	QGn
H	R	L	X	L	QAn	QBn	QCn	QDn	QEn	QFn	QGn
H	R	X	L	L	QAn	QBn	QCn	QDn	QEn	QFn	QGn

The following shows the full model for a shift register with the characteristics

- 8 bit parallel output
- Synchronous serial shift right with dual data inputs (both data inputs must be high for shift in left of high, otherwise low value is shifted in left)
- Asynchronous clear

8 Bit Parallel-Out Serial Shift Register

```
USE std.std_logic.ALL;
USE std.std_ttl.ALL;
ENTITY sreg8parser IS
    GENERIC (a_i01,a_i10,
             b_i01,b_i10,
             clock_i01,clock_i10,
             clear_i01,clear_i10,
             qa_o01,qa_o10,
             qb_o01,qb_o10,
             qc_o01,qc_o10,
             qd_o01,qd_o10,
             qe_o01,qe_o10,
             qf_o01,qf_o10,
             qg_o01,qg_o10,
             qh_o01,qh_o10,
             clock_min,
             data_setup,data_hold : TIME := 2 ns);
    PORT    (a,b,clock,clear: IN t_wlogic;
             qa,qb,qc,qd,qe,qf,qg,qh: OUT t_wlogic);
END sreg8parser;

ARCHITECTURE full OF sreg8parser IS
    SIGNAL adelay, bdelay,
           clockdelay, cleardelay : t_wlogic;
BEGIN
    -- check clock frequency/spike detection
    PROCESS (clock)
```

4.2. REGISTERS

```
        VARIABLE clocklastev : TIME := 0 ns;
        BEGIN
        ASSERT (NOW = 0 ns) OR ((NOW - clocklastev) >= clock_min)
            REPORT "Spike detected on clock" SEVERITY warning;
        clocklastev := NOW;
        END PROCESS;

    -- check for setup/hold violations
    PROCESS (clock,a,b)
        VARIABLE datalastev : TIME := 0 ns;
        VARIABLE clocklastev : TIME := 0 ns;
        BEGIN
        -- Hold check
        IF (a'EVENT) OR (b'EVENT) THEN
            ASSERT (NOW = 0 ns) OR ((NOW - clocklastev) >= data_hold)
                REPORT "Hold error for data input" SEVERITY warning;
            datalastev := NOW;
        END IF;

        -- Setup check
        IF (clock'EVENT) AND (clock = '1') THEN
            ASSERT (NOW = 0 ns) OR ((NOW - datalastev) >= data_setup)
                REPORT "Setup error for data input" SEVERITY warning;
            clocklastev := NOW;
        END IF;
        END PROCESS;

    -- input delay processing
    adelay <= a AFTER f_delay(a,a_i01,a_i10);
    bdelay <= b AFTER f_delay(b,b_i01,b_i10);
    cleardelay <= clear AFTER f_delay(clear,clear_i01,clear_i10);
    clockdelay <= clock AFTER f_delay(clock,clock_i01,clock_i10);

    -- register operation
    PROCESS (cleardelay,clockdelay)
        VARIABLE bstate : t_logarray (1 TO 8) := (U,U,U,U,U,U,U,U);
        BEGIN
        -- Check for clear
        IF (cleardelay = '0') THEN
            FOR i IN bstate'RANGE LOOP bstate(i) := f_ttl('0'); END LOOP;

        -- Check for unknown controls
        ELSIF (cleardelay = 'X') THEN
            FOR i IN bstate'RANGE LOOP bstate(i) := f_ttl('X'); END LOOP;

        -- Check for unknown clock
        ELSIF (cleardelay = '1') AND (clockdelay'EVENT) AND
              (clockdelay = 'X') THEN
            FOR i IN bstate'RANGE LOOP bstate(i) := f_ttl('X'); END LOOP;

        -- Check for high shift
        ELSIF (cleardelay = '1') AND (clockdelay'EVENT) AND
```

```
            (clockdelay = '1') AND (adelay = '1') AND (bdelay = '1') THEN
            FOR i IN bstate'HIGH DOWNTO bstate'LOW+1 LOOP
                bstate(i) := bstate(i-1); END LOOP;
            bstate(bstate'LOW) := f_ttl('1');

        -- Check for low shift
        ELSIF (cleardelay = '1') AND (clockdelay'EVENT) AND
            (clockdelay = '1') AND ((adelay = '0') OR (bdelay = '0')) THEN
            FOR i IN bstate'HIGH DOWNTO bstate'LOW+1 LOOP
                bstate(i) := bstate(i-1); END LOOP;
            bstate(bstate'LOW) := f_ttl('0');
        END IF;

        -- Assign signal values
        qa <= bstate(1) AFTER f_delay(bstate(1), qa_o01, qa_o10);
        qb <= bstate(2) AFTER f_delay(bstate(2), qb_o01, qb_o10);
        qc <= bstate(3) AFTER f_delay(bstate(3), qc_o01, qc_o10);
        qd <= bstate(4) AFTER f_delay(bstate(4), qd_o01, qd_o10);
        qe <= bstate(5) AFTER f_delay(bstate(5), qe_o01, qe_o10);
        qf <= bstate(6) AFTER f_delay(bstate(6), qf_o01, qf_o10);
        qg <= bstate(7) AFTER f_delay(bstate(7), qg_o01, qg_o10);
        qh <= bstate(8) AFTER f_delay(bstate(8), qh_o01, qh_o10);
    END PROCESS;
END full;
```

The ports for this model are

- **a** - address select
- **b** - address select
- **clock** - clock input
- **clear** - clear input
- **qa** - parallel state output
- **qb** - parallel state output
- **qc** - parallel state output
- **qd** - parallel state output
- **qe** - parallel state output
- **qf** - parallel state output
- **qg** - parallel state output
- **qh** - parallel state output

Generic parameters to this model are summarized here

- **a_i01, a_i10** - low to high and high to low **a** input port delays
- **b_i01, b_i10** - low to high and high to low **b** input port delays

4.2. REGISTERS

- clock_i01, clock_i10 - low to high and high to low **clock** input port delays
- clear_i01, clear_i10 - low to high and high to low **clear** input port delays
- Minimum **clock** pulse width **clock_min**
- qa_o01, qa_o10 - low to high and high to low **qa** output port delays
- qb_o01, qb_o10 - low to high and high to low **qb** output port delays
- qc_o01, qc_o10 - low to high and high to low **qc** output port delays
- qd_o01, qd_o10 - low to high and high to low **qd** output port delays
- qe_o01, qe_o10 - low to high and high to low **qe** output port delays
- qf_o01, qf_o10 - low to high and high to low **qf** output port delays
- qg_o01, qg_o10 - low to high and high to low **qg** output port delays
- qh_o01, qh_o10 - low to high and high to low **qh** output port delays
- Data data setup **data_setup** and hold **data_hold** minimums

This model features three error checks

- **clock** frequency/spike detection
- Serial data setup time checking
- Serial data hold time checking

as shown in the error checking section of the architecture. The spike error checks utilizes a separate process statement. A second process performs the hold and setup checks.

The spike detection processes uses a local variable to save the previous event time for the signal being checked. This is somewhat more efficient than using the delayed attribute since it does not require the simulator to create an additional signal.

The setup/hold checking process uses two local variables to save the time of the last data and clock event respectively. This process assumes that the data must be settled before the latch control becomes active high. Adjustment to the following expression is required in order to change the latching control characteristics.

$(clock'EVENT) AND (clock =' 1')$

Events from the serial data inputs **a** and **b** are treated as a single data event. A more sophisticated version of this error check (with additional generic parameters) could be crafted which tests for setup/hold errors on any of the data inputs separately through the use of multiple local variables, one for each data input. Most data books do not specify this level of detail in timing and therefore this approach has not been taken.

The input delay processing section of the model incorporates the appropriate delay for each of the input ports. An internal signal for each input port is declared. The main device function process utilizes the delayed signals rather than the primary inputs.

The register operation is handled by a single process which is sensitive to the input delayed signals **clock, a** and **b**. The following conditions are checked in order whenever the asynchronous control or address inputs have an event

- Clear - set state to all low
- Unknown clear control - set state to all unknown
- Unknown clock - set state to all unknown
- Clocked shift with high input - shift right with left input of high
- Clocked shift with low input - shift right with left input of low

Once the state of the device is determined (and stored in the local array variable **bstate**, the output signal assignment statements for **qa, qb, qc, qd, qe, qf, qg** and **qh** are executed. Output state changes are executed with appropriate delay introduced through the use of the **f_delay** function as discussed in section 7.6. This function utilizes the output rise and fall delays and based on the new state value chooses the proper delay.

4.2.5 Parallel Load 8 Bit Shift Register

This device is an 8-bit serial shift register that, when clocked, shifts the data toward serial output **qh**. Parallel-in access to each stage is provided by eight direct data inputs that are enabled by a low at the **shld** input. This device features a clock inhibit function and a complemented serial output **qh**.

Clocking is accomplished by a rising transition of the **clk** input while **shld** is held high and **clkinh** is held low. The functions of the **clk** and **clkinh** (clock inhibit) inputs are interchangeable. Since a low **clk** input and a rising transition of **clkinh** will accomplish clocking, **clkinh** should be changed to the high only while the **clk** input is high. Parallel loading is inhibited when **shld** is held high. The parallel inputs to the register are enabled while **shld** is low independently of the values of **clk, clkinh** or **ser** inputs. This device corresponds to the Texas Instruments 165 series TTL parts [TI86]. Figure 4.9 shows the symbolic representation of this device. The following shows the function table for this device.

4.2. REGISTERS

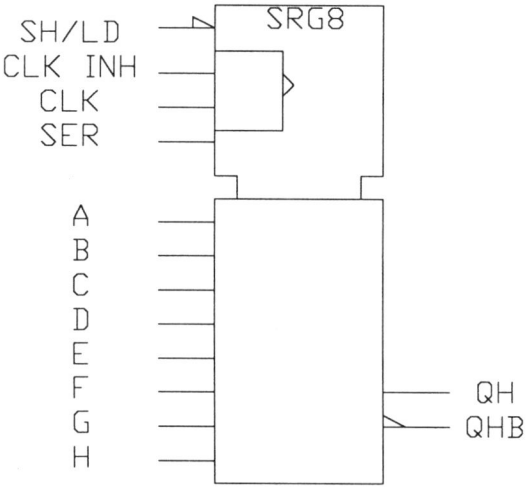

Figure 4.9: Logic Diagram 8 Bit Parallel Load Shift Register

Function Table

inputs			function
shld	clk	clkinh	
L	X	X	parallel load
H	H	X	no change
H	X	H	no change
H	L	R	shift
H	R	L	shift

The following shows the full model for a shift register with the characteristics

- 8 bit
- Serial output with complement
- Asynchronous parallel load
- Clocked shift, with serial data in
- Asynchronous shift enable control

Parallel Load 8 Bit Shift Register

```
USE std.std_logic.ALL;
USE std.std_ttl.ALL;
```

```
ENTITY sreg8parload IS
    GENERIC (clk_i01,clk_i10,
             clkinh_i01,clkinh_i10,
             shld_i01,shld_i10,
             ser_i01,ser_i10,
             a_i01,a_i10,
             b_i01,b_i10,
             c_i01,c_i10,
             d_i01,d_i10,
             e_i01,e_i10,
             f_i01,f_i10,
             g_i01,g_i10,
             h_i01,h_i10,
             q_o01,q_o10,
             qb_o01,qb_o10,
             clk_min,
             data_setup,data_hold  : TIME := 2 ns);
    PORT    (clk,clkinh,shld,ser,a,b,c,d,e,f,g,h: IN t_wlogic;
             q,qb: OUT t_wlogic);
END sreg8parload;

ARCHITECTURE full OF sreg8parload IS
    SIGNAL clkdelay,clkinhdelay,shlddelay,serdelay,
           adelay,bdelay,cdelay,ddelay,
           edelay,fdelay,gdelay,hdelay : t_wlogic;
BEGIN
    -- check clk frequency/spike detection
    PROCESS (clk)
        VARIABLE clklastev : TIME := 0 ns;
        BEGIN
        ASSERT (NOW = 0 ns) OR ((NOW - clklastev) >= clk_min)
            REPORT "Spike detected on clk" SEVERITY warning;
        clklastev := NOW;
    END PROCESS;

    -- check for setup/hold violations
    PROCESS (clk,shld,a,b,c,d,e,f,g,h)
        VARIABLE datalastev : TIME := 0 ns;
        VARIABLE clklastev  : TIME := 0 ns;
        BEGIN
        -- Hold check
        IF (a'EVENT) OR (b'EVENT) OR (c'EVENT) OR (d'EVENT) OR
           (e'EVENT) OR (f'EVENT) OR (g'EVENT) OR (h'EVENT) OR
           (shld'EVENT) THEN
               ASSERT (NOW = 0 ns) OR ((NOW - clklastev) >= data_hold)
                   REPORT "Hold error for data input" SEVERITY warning;
               datalastev := NOW;
        END IF;

        -- Setup check
        IF (clk'EVENT) AND (clk = '1') THEN
            ASSERT (NOW = 0 ns) OR ((NOW - datalastev) >= data_setup)
```

4.2. REGISTERS

```
                REPORT "Setup error for data input" SEVERITY warning;
        clklastev := NOW;
    END IF;
    END PROCESS;

-- input delay processing
adelay <= a AFTER f_delay(a,a_i01,a_i10);
bdelay <= b AFTER f_delay(b,b_i01,b_i10);
cdelay <= c AFTER f_delay(c,c_i01,c_i10);
ddelay <= d AFTER f_delay(d,d_i01,d_i10);
edelay <= e AFTER f_delay(e,e_i01,e_i10);
fdelay <= f AFTER f_delay(f,f_i01,f_i10);
gdelay <= g AFTER f_delay(g,g_i01,g_i10);
hdelay <= h AFTER f_delay(h,h_i01,h_i10);
clkdelay <= clk AFTER f_delay(clk,clk_i01,clk_i10);
clkinhdelay <= clkinh AFTER f_delay(clkinh,clkinh_i01,clkinh_i10);
shlddelay <= shld AFTER f_delay(shld,shld_i01,shld_i10);
serdelay <= ser AFTER f_delay(ser,ser_i01,ser_i10);

-- register operation
PROCESS (clkdelay,shlddelay,clkinhdelay)
    VARIABLE bstate : t_logarray (1 TO 8) := (U,U,U,U,U,U,U,U);
    BEGIN
    -- Check for parallel load
    IF (shlddelay'EVENT) AND (shlddelay = '0') THEN
        bstate(1) := adelay;
        bstate(2) := bdelay;
        bstate(3) := cdelay;
        bstate(4) := ddelay;
        bstate(5) := edelay;
        bstate(6) := fdelay;
        bstate(7) := gdelay;
        bstate(8) := hdelay;

    -- Check for unknown controls
    ELSIF (shlddelay'EVENT) AND (shlddelay = 'X') THEN
        FOR i IN bstate'RANGE LOOP bstate(i) := f_ttl('X'); END LOOP;

    -- Check for clocked shift
    ELSIF (clkinhdelay = '0') AND (clkdelay'EVENT) AND
            (clkdelay = '1') THEN
        FOR i IN bstate'HIGH DOWNTO bstate'LOW+1 LOOP
            bstate(i) := bstate(i-1); END LOOP;
        bstate(bstate'LOW) := serdelay;

    -- Check for enabled shift
    ELSIF (clkdelay = '0') AND (clkinhdelay'EVENT) AND
            (clkinhdelay = '1') THEN
        FOR i IN bstate'HIGH DOWNTO bstate'LOW+1 LOOP
            bstate(i) := bstate(i-1); END LOOP;
        bstate(bstate'LOW) := serdelay;
```

```
            -- Check for unknown clock
            ELSIF (clkinhdelay = '0') AND (clkdelay'EVENT) AND
                  (clkdelay = 'X') THEN
               FOR i IN bstate'RANGE LOOP bstate(i) := f_ttl('X'); END LOOP;

            -- Check for unknown enable
            ELSIF (clkdelay = '0') AND (clkinhdelay'EVENT) AND
                  (clkinhdelay = 'X') THEN
               FOR i IN bstate'RANGE LOOP bstate(i) := f_ttl('X'); END LOOP;
            END IF;

            -- Assign signal values
            q <= bstate(bstate'HIGH) AFTER
                f_delay(bstate(bstate'HIGH), q_o01, q_o10);
            qb <= NOT bstate(bstate'HIGH) AFTER
                f_delay(NOT bstate(bstate'HIGH), qb_o01, qb_o10);
         END PROCESS;
END full;
```

The ports for this model are

- **clk** - clock input
- **clkinh** - inhibit control
- **shld** - shift/load control
- **ser** - serial input
- **a** - parallel data input
- **b** - parallel data input
- **c** - parallel data input
- **d** - parallel data input
- **e** - parallel data input
- **f** - parallel data input
- **g** - parallel data input
- **h** - parallel data input
- **q** - serial output
- **qb** - complementary serial output

Generic parameters to this model are summarized here

- **clk_i01, clk_i10** - low to high and high to low **clk** input port delays
- **clkinh_i01, clkinh_i10** - low to high and high to low **clkinh** input port delays
- **shld_i01, shld_i10** - low to high and high to low **shld** input port delays

4.2. REGISTERS

- ser_i01, ser_i10 - low to high and high to low **ser** input port delays
- a_i01, a_i10 - low to high and high to low **a** input port delays
- b_i01, b_i10 - low to high and high to low **b** input port delays
- c_i01, c_i10 - low to high and high to low **c** input port delays
- d_i01, d_i10 - low to high and high to low **d** input port delays
- e_i01, e_i10 - low to high and high to low **e** input port delays
- f_i01, f_i10 - low to high and high to low **f** input port delays
- g_i01, g_i10 - low to high and high to low **g** input port delays
- h_i01, h_i10 - low to high and high to low **h** input port delays
- q_o01, q_o10 - low to high and high to low **q** output port delays
- qb_o01, qb_o10 - low to high and high to low **qb** output port delays
- Minimum clk pulse width **clk_min**
- Data data setup **data_setup** and hold **data_hold** minimums

This model features three error checks

- clk frequency/spike detection
- Serial data setup time checking
- Serial data hold time checking

as shown in the error checking section of the architecture. The spike error checks utilizes a separate process statement. A second process performs the hold and setup checks.

The spike detection processes uses a local variable to save the previous event time for the signal being checked. This is somewhat more efficient than using the delayed attribute since it does not require the simulator to create an additional signal.

The setup/hold checking process uses two local variables to save the time of the last data and clock event respectively. This process assumes that the data must be settled before the latch control becomes active high. Adjustment to the following expression is required in order to change the latching control characteristics.

$(clk'EVENT) AND (clk =' 1')$

Events from the parallel data inputs **shld**, **a**, **b**, **c**, **d**, **e**, **f**, **g** and **h** are treated as a single data event. A more sophisticated version of this error check (with additional generic parameters) could be crafted which tests for setup/hold errors on any of the data inputs separately through the use of multiple local variables, one for each data input. Most data books do not specify this level of detail in timing and therefore this approach has not been taken.

The input delay processing section of the model incorporates the appropriate delay for each of the input ports. An internal signal for each input port is declared. The main device function process utilizes the delayed signals rather than the primary inputs.

The register operation is handled by a single process which is sensitive to the input delayed signals **clkdelay**, **shlddelay** and **clkinhdelay**. The following conditions are checked in order whenever the asynchronous control or address inputs have an event

- Asynchronous parallel load
- Unknown control inputs
- Clocked shift
- Asynchronous enabled shift
- Unknown clock
- Unknown shift enable

Once the state of the device is determined (and stored in the local array variable **bstate**, the output signal assignment statements for **q** and **qb** are executed. Output state changes are executed with appropriate delay introduced through the use of the **f_delay** function as discussed in section 7.6. This function utilizes the output rise and fall delays and based on the new state value chooses the proper delay.

4.2.6 Parallel Load 8 Bit Shift Register with Clear

This register features parallel or serial input, and serial output. Inputs are clocked with an overriding clear. The parallel-in or serial-in modes are established by the **shld** input. When high, this input enables the serial data input and couples the eight flip-flops for serial shifting with each clock pulse. When low, the parallel data inputs are enabled and synchronous loading occurs on the next clock pulse. During parallel loading, serial data flow is inhibited. Clocking is accomplished on the rising edge of the clock pulse through a two-input positive NOR gate permitting one input to be used as a clock-enable or clock-inhibit function. Holding either of the clock inputs high inhibits clocking; holding either low enables the other clock input. The clock inhibit should be changed to high only when the clock input is high. This device corresponds to the Texas Instruments 166 series TTL parts [TI86]. Figure 4.10 shows the symbolic representation of this device. The following shows the function table for this device.

4.2. REGISTERS

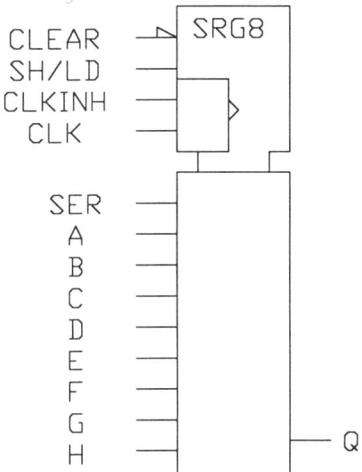

Figure 4.10: Logic Diagram 8 Bit Parallel Load Shift Register with Clear

Function Table

inputs					output
clear	shld	clkinh	clk	ser	q
L	X	X	X	X	L
H	X	L	L	X	QH0
H	L	L	R	X	h
H	H	L	R	H	QGn
H	H	L	R	L	QGn
H	X	H	R	X	QH0

The following shows the full model for a shift register with the characteristics

- 8 bit
- Serial output
- Asynchronous clear
- Asynchronous parallel load
- Clocked shift, with serial data in
- Asynchronous shift enable control

Parallel Load 8 Bit Shift Register with Clear

```vhdl
USE std.std_logic.ALL;
USE std.std_ttl.ALL;
ENTITY sreg8parlwclr IS
    GENERIC (clk_i01,clk_i10,
             clkinh_i01,clkinh_i10,
             shld_i01,shld_i10,
             ser_i01,ser_i10,
             clr_i01,clr_i10,
             a_i01,a_i10,
             b_i01,b_i10,
             c_i01,c_i10,
             d_i01,d_i10,
             e_i01,e_i10,
             f_i01,f_i10,
             g_i01,g_i10,
             h_i01,h_i10,
             q_o01,q_o10,
             clk_min, clr_min,
             data_setup,data_hold  : TIME := 2 ns);
    PORT    (clk,clkinh,shld,ser,clr,a,b,c,d,e,f,g,h: IN t_wlogic;
             q: OUT t_wlogic);
END sreg8parlwclr;

ARCHITECTURE full OF sreg8parlwclr IS
    SIGNAL clkdelay,clkinhdelay,shlddelay,serdelay,clrdelay,
           adelay,bdelay,cdelay,ddelay,
           edelay,fdelay,gdelay,hdelay : t_wlogic;
BEGIN
    -- check clk frequency/spike detection
    PROCESS (clk)
        VARIABLE clklastev : TIME := 0 ns;
        BEGIN
        ASSERT (NOW = 0 ns) OR ((NOW - clklastev) >= clk_min)
            REPORT "Spike detected on clk" SEVERITY warning;
        clklastev := NOW;
        END PROCESS;

    PROCESS (clr)
        VARIABLE clrlastev : TIME := 0 ns;
        BEGIN
        ASSERT (NOW = 0 ns) OR ((NOW - clrlastev) >= clr_min)
            REPORT "Spike detected on clr" SEVERITY warning;
        clrlastev := NOW;
        END PROCESS;

    -- check for setup/hold violations
    PROCESS (clk,shld,a,b,c,d,e,f,g,h)
        VARIABLE datalastev : TIME := 0 ns;
        VARIABLE clklastev : TIME := 0 ns;
        BEGIN
```

4.2. REGISTERS

```
        -- Hold check
        IF (a'EVENT) OR (b'EVENT) OR (c'EVENT) OR (d'EVENT) OR
           (e'EVENT) OR (f'EVENT) OR (g'EVENT) OR (h'EVENT) OR
           (shld'EVENT) THEN
             ASSERT (NOW = 0 ns) OR ((NOW - clklastev) >= data_hold)
                 REPORT "Hold error for data input" SEVERITY warning;
             datalastev := NOW;
        END IF;

        -- Setup check
        IF (clk'EVENT) AND (clk = '1') THEN
             ASSERT (NOW = 0 ns) OR ((NOW - datalastev) >= data_setup)
                 REPORT "Setup error for data input" SEVERITY warning;
             clklastev := NOW;
        END IF;
        END PROCESS;

-- input delay processing
adelay    <= a    AFTER f_delay(a,a_i01,a_i10);
bdelay    <= b    AFTER f_delay(b,b_i01,b_i10);
cdelay    <= c    AFTER f_delay(c,c_i01,c_i10);
ddelay    <= d    AFTER f_delay(d,d_i01,d_i10);
edelay    <= e    AFTER f_delay(e,e_i01,e_i10);
fdelay    <= f    AFTER f_delay(f,f_i01,f_i10);
gdelay    <= g    AFTER f_delay(g,g_i01,g_i10);
hdelay    <= h    AFTER f_delay(h,h_i01,h_i10);
clkdelay  <= clk  AFTER f_delay(clk,clk_i01,clk_i10);
clkinhdelay <= clkinh AFTER f_delay(clkinh,clkinh_i01,clkinh_i10);
shlddelay <= shld AFTER f_delay(shld,shld_i01,shld_i10);
serdelay  <= ser  AFTER f_delay(ser,ser_i01,ser_i10);
clrdelay  <= clr  AFTER f_delay(clr,clr_i01,clr_i10);

-- register operation
PROCESS (clkdelay,shlddelay,clkinhdelay,clrdelay)
    VARIABLE bstate : t_logarray (1 TO 8) := (U,U,U,U,U,U,U,U);
    BEGIN
    -- Check for clear
    IF (clr = '0') THEN
        FOR i IN bstate'RANGE LOOP bstate(i) := f_ttl('0'); END LOOP;

    -- Unknown clear
    ELSIF (clr = 'X') THEN
        FOR i IN bstate'RANGE LOOP bstate(i) := f_ttl('X'); END LOOP;

    -- Check for parallel load
    ELSIF (shlddelay'EVENT) AND (shlddelay = '0') THEN
        bstate(1) := adelay;
        bstate(2) := bdelay;
        bstate(3) := cdelay;
        bstate(4) := ddelay;
        bstate(5) := edelay;
        bstate(6) := fdelay;
```

```
            bstate(7) := gdelay;
            bstate(8) := hdelay;

        -- Check for unknown controls
        ELSIF (shlddelay'EVENT) AND (shlddelay = 'X') THEN
            FOR i IN bstate'RANGE LOOP bstate(i) := f_ttl('X'); END LOOP;

        -- Check for clocked shift
        ELSIF (clkinhdelay = '0') AND (clkdelay'EVENT) AND
              (clkdelay = '1') THEN
            FOR i IN bstate'HIGH DOWNTO bstate'LOW+1 LOOP
                bstate(i) := bstate(i-1); END LOOP;
            bstate(bstate'LOW) := serdelay;

        -- Check for enabled shift
        ELSIF (clkdelay = '0') AND (clkinhdelay'EVENT) AND
              (clkinhdelay = '1') THEN
            FOR i IN bstate'HIGH DOWNTO bstate'LOW+1 LOOP
                bstate(i) := bstate(i-1); END LOOP;
            bstate(bstate'LOW) := serdelay;

        -- Check for unknown clock
        ELSIF (clkinhdelay = '0') AND (clkdelay'EVENT) AND
              (clkdelay = 'X') THEN
            FOR i IN bstate'RANGE LOOP bstate(i) := f_ttl('X'); END LOOP;

        -- Check for unknown enable
        ELSIF (clkdelay = '0') AND (clkinhdelay'EVENT) AND
              (clkinhdelay = 'X') THEN
            FOR i IN bstate'RANGE LOOP bstate(i) := f_ttl('X'); END LOOP;
        END IF;

        -- Assign signal values
        q <= bstate(bstate'HIGH) AFTER
            f_delay(bstate(bstate'HIGH), q_o01, q_o10);
        END PROCESS;
END full;
```

The ports for this model are

- **clk** - clock input
- **clkinh** - inhibit control
- **shld** - shift/load control
- **ser** - serial input
- **clr** - clear input
- **a** - parallel data input
- **b** - parallel data input
- **c** - parallel data input

4.2. REGISTERS

- **d** - parallel data input
- **e** - parallel data input
- **f** - parallel data input
- **g** - parallel data input
- **h** - parallel data input
- **q** - serial output

Generic parameters to this model are summarized here

- **clk_i01, clk_i10** - low to high and high to low **clk** input port delays
- **clkinh_i01, clkinh_i10** - low to high and high to low **clkinh** input port delays
- **shld_i01, shld_i10** - low to high and high to low **shld** input port delays
- **ser_i01, ser_i10** - low to high and high to low **ser** input port delays
- **clr_i01, clr_i10** - low to high and high to low **clr** input port delays
- **a_i01, a_i10** - low to high and high to low **a** input port delays
- **b_i01, b_i10** - low to high and high to low **b** input port delays
- **c_i01, c_i10** - low to high and high to low **c** input port delays
- **d_i01, d_i10** - low to high and high to low **d** input port delays
- **e_i01, e_i10** - low to high and high to low **e** input port delays
- **f_i01, f_i10** - low to high and high to low **f** input port delays
- **g_i01, g_i10** - low to high and high to low **g** input port delays
- **h_i01, h_i10** - low to high and high to low **h** input port delays
- **q_o01, q_o10** - low to high and high to low **q** output port delays
- Minimum **clk** pulse width **clk_min**
- Data data setup **data_setup** and hold **data_hold** minimums

This model features four error checks

- **clk** frequency/spike detection
- **clr** spike detection
- Serial data setup time checking
- Serial data hold time checking

as shown in the error checking section of the architecture. The spike error checks each utilize a separate process statement. An additional single process performs the hold and setup checks.

The spike detection processes use a local variable to save the previous event time for the signal being checked. This is somewhat more efficient than using the delayed attribute since it does not require the simulator to create an additional signal.

The setup/hold checking process uses two local variables to save the time of the last data and clock event respectively. This process assumes that the data must be settled before the latch control becomes active high. Adjustment to the following expression is required in order to change the latching control characteristics.

$$(clk'EVENT) AND (clk =' 1')$$

Events from the parallel data inputs **shld, a, b, c, d, e, f, g** and **h** are treated as a single data event. A more sophisticated version of this error check (with additional generic parameters) could be crafted which tests for setup/hold errors on any of the data inputs separately through the use of multiple local variables, one for each data input. Most data books do not specify this level of detail in timing and therefore this approach has not been taken.

The input delay processing section of the model incorporates the appropriate delay for each of the input ports. An internal signal for each input port is declared. The main device function process utilizes the delayed signals rather than the primary inputs.

The register operation is handled by a single process which is sensitive to the input delayed signals **clkdelay, shlddelay, clkinhdelay** and **clrdelay**. The following conditions are checked in order whenever the asynchronous control or address inputs have an event

- Asynchronous clear
- Unknown clear control
- Asynchronous parallel load
- Unknown load control
- Clocked shift
- Asynchronous enabled shift
- Unknown clock
- Unknown shift enable

Once the state of the device is determined (and stored in the local array variable **bstate**, the output signal assignment statement for **q** is executed. Output state changes are executed with appropriate delay introduced through

4.3. COUNTERS 195

the use of the **f_delay** function as discussed in section 7.6. This function utilizes the output rise and fall delays and based on the new state value chooses the proper delay.

4.3 Counters

The following models describe various types of counters. These models follow a standard format and all are presented with full timing functionality. It is possible to reduce the simulation and model complexity in stages as follows

- Remove input delay processing
  - Remove all input timing generic parameters
  - Remove all internal delay signals
  - Remove the input delay signal assignment statements
  - Rename all references in the main behavioral process from the delayed port names to primary input port names
- Remove error checks: delete the error checking processes which occur prior to the main behavioral process
- Use simpler output delay calculations: adjust the final output assignment statements

By methodically adjusting these major components of the models it is possible to build a wide range of models which fit into the model accuracy continuum discussed in section 2.5.

The following adaptations are possible starting from these base models:

- Change the number of bits in counter state. Adjust the generic parameter and port declarations as appropriate; adapting from a 4 bit model to an 8 bit model will require the addition of 4 more generics and 4 more ports for parallel inputs, and a similar change for parallel outputs.
- Changing between decade and binary counters. Comparison of the models shown below and examining the differences between the binary and decade counters will give insight into this adaptation. The primary difference occurs during the count up and count down operation. Decade counters utilize a different MOD argument from binary counters. Clearly the counter out will also operate differently.
- Addition, deletion or adaptation of the asynchronous controls including clear, load, preset and enable inputs. A review of the various models in this chapter will give hints as to how to handle each of these features.

4.3.1 Synchronous 4 Bit Decade Counter with Asynchronous Clear

This device is a synchronous presettable decade counter with 4 bits. Synchronous operation is provided by having all flip-flops clocked simultaneously so that the outputs change coincident with each other when so instructed by the count-enable inputs. A clock input triggers the four flip-flops on the rising edge of the clock. The counters can be preset to any number between 0 and 9. Presetting is synchronous; setting up a low at the load input disables the counter and causes the outputs to agree with the data inputs after the next clock pulse regardless of the values of the enable inputs. The clear function for this device is asynchronous and a low at the **clear** input sets all four of the outputs low regardless of the values of the **clk**, **load** or **enable** inputs. Changes at control inputs **enp**, **ent** and **load** that will modify the operating mode have no effect on the contents of the counter until clocking occurs. The function of the counter (whether enabled, disabled, loading, or counting) will be determined solely by the conditions meeting the stable setup and hold times. This device corresponds to the Texas Instruments 160 series TTL parts [TI86]. Figure 4.11 shows the symbolic representation of this device. The following shows

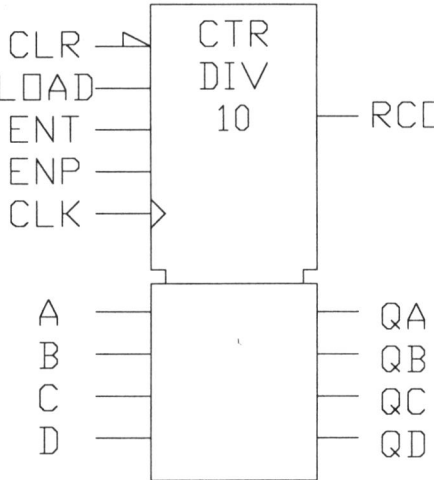

Figure 4.11: Logic Diagram 4 Bit Decade Counter with Async Clear

the full model for a counter with the characteristics
- 4 bit
- Decade count

4.3. COUNTERS

- Asynchronous clear
- Synchronous/clocked parallel load
- Synchronous/clocked count
- Dual count enable control
- Parallel output
- Carry output

Synchronous 4 Bit Decade Counter with Asynchronous Clear

```
USE std.std_logic.ALL;
USE std.std_ttl.ALL;
ENTITY syn4decade IS
    GENERIC (clr_i01,clr_i10,
             load_i01,load_i10,
             ent_i01,ent_i10,
             enp_i01,enp_i10,
             clk_i01,clk_i10,
             a_i01,a_i10,
             b_i01,b_i10,
             c_i01,c_i10,
             d_i01,d_i10,
             clk_min, clr_min, data_hold, data_setup,
             rco_o01,rco_o10,
             qa_o01,qa_o10,
             qb_o01,qb_o10,
             qc_o01,qc_o10,
             qd_o01,qd_o10 : TIME := 2 ns);
    PORT    (clr,load,ent,enp,clk,a,b,c,d: IN t_wlogic;
             rco, qa,qb,qc,qd: OUT t_wlogic);
END syn4decade;

ARCHITECTURE full OF syn4decade IS
    SIGNAL clrdelay,loaddelay,entdelay,enpdelay,clkdelay,
           adelay,bdelay,cdelay,ddelay : t_wlogic;
BEGIN
    -- check clk frequency/spike detection
    PROCESS (clk)
        VARIABLE clklastev : TIME := 0 ns;
        BEGIN
        ASSERT (NOW = 0 ns) OR ((NOW - clklastev) >= clk_min)
            REPORT "Spike detected on clock" SEVERITY warning;
        clklastev := NOW;
        END PROCESS;

    -- spike detection
    PROCESS (clr)
        VARIABLE clrlastev : TIME := 0 ns;
        BEGIN
```

```
      ASSERT (NOW = 0 ns) OR ((NOW - clrlastev) >= clr_min)
         REPORT "Spike detected on clr" SEVERITY warning;
      clrlastev := NOW;
      END PROCESS;

   -- check for setup/hold violations
   PROCESS (clk,load,ent,enp,a,b,c,d)
      VARIABLE datalastev : TIME := 0 ns;
      VARIABLE clklastrise : TIME := 0 ns;
      BEGIN
      -- hold check
      IF (load'EVENT) OR (ent'EVENT) OR (enp'EVENT) OR
         (a'EVENT) OR (b'EVENT) OR (c'EVENT) OR (d'EVENT) THEN
         ASSERT (NOW = 0 ns) OR ((NOW - clklastrise) >= data_hold)
            REPORT "Hold error" SEVERITY warning;
         datalastev := NOW;
      END IF;

      -- setup check
      IF (clk'EVENT) AND (clk = '1') THEN
         ASSERT (NOW = 0 ns) OR ((NOW - datalastev) >= data_setup)
            REPORT "Setup error" SEVERITY warning;
         clklastrise := NOW;
      END IF;
      END PROCESS;

   -- input delay processing
   clrdelay <= clr AFTER f_delay(clr,clr_i01,clr_i10);
   loaddelay <= load AFTER f_delay(load,load_i01,load_i10);
   entdelay <= ent AFTER f_delay(ent,ent_i01,ent_i10);
   enpdelay <= enp AFTER f_delay(enp,enp_i01,enp_i10);
   clkdelay <= clk AFTER f_delay(clk,clk_i01,clk_i10);
   adelay <= a AFTER f_delay(a,a_i01,a_i10);
   bdelay <= b AFTER f_delay(b,b_i01,b_i10);
   cdelay <= c AFTER f_delay(c,c_i01,c_i10);
   ddelay <= d AFTER f_delay(d,d_i01,d_i10);

   -- counter operation
   PROCESS (clkdelay,clrdelay)
      VARIABLE istate : integer := -1;
      VARIABLE bstate : t_logarray (1 TO 4);
      VARIABLE unknown : boolean := true;
      BEGIN
      -- check for clear
      IF (clrdelay'EVENT) AND (clrdelay = '0') THEN
         istate := 0;
         unknown := false;

      -- check for unknown control
      ELSIF (clrdelay'EVENT) AND (clrdelay = 'X') THEN unknown := true;
      ELSIF (clkdelay'EVENT) AND (clkdelay = '1') AND
            (loaddelay = 'X') THEN unknown := true;
```

4.3. COUNTERS

```
        ELSIF (clkdelay'EVENT) AND (clkdelay = '1') AND
            ((entdelay = 'X') OR (enpdelay = 'X')) THEN unknown := true;
        -- check for clocked load
        ELSIF (clkdelay'EVENT) AND (clkdelay='1') AND (loaddelay='0') THEN
            bstate(1) := ddelay;
            bstate(2) := cdelay;
            bstate(3) := bdelay;
            bstate(4) := adelay;
            f_logictoint(bstate,unknown,istate);
            IF NOT unknown THEN istate := istate MOD 10; END IF;
        -- check for clocked count
        ELSIF (clkdelay'EVENT) AND (clkdelay = '1') AND (entdelay ='1') AND
              (enpdelay = '1') AND (istate >= 0) THEN
            istate := (istate + 1) MOD 10;
        END IF;

        -- unknown state
        IF unknown THEN
            qa <= f_ttl('X') AFTER f_delay(f_ttl('X'), qa_o01, qa_o10);
            qb <= f_ttl('X') AFTER f_delay(f_ttl('X'), qb_o01, qb_o10);
            qc <= f_ttl('X') AFTER f_delay(f_ttl('X'), qc_o01, qc_o10);
            qd <= f_ttl('X') AFTER f_delay(f_ttl('X'), qd_o01, qd_o10);
            rco <= f_ttl('X') AFTER f_delay(f_ttl('X'), rco_o01, rco_o10);
        -- assign outputs
        ELSE
            f_inttologic(istate,bstate,ttl);
            qa <= bstate(4) AFTER f_delay(bstate(1), qa_o01, qa_o10);
            qb <= bstate(3) AFTER f_delay(bstate(2), qb_o01, qb_o10);
            qc <= bstate(2) AFTER f_delay(bstate(3), qc_o01, qc_o10);
            qd <= bstate(1) AFTER f_delay(bstate(4), qd_o01, qd_o10);
            IF istate = 9 THEN
                rco <= f_ttl('1')AFTER f_delay(f_ttl('1'),rco_o01,rco_o10);
            ELSE
                rco <= f_ttl('0')AFTER f_delay(f_ttl('0'),rco_o01,rco_o10);
            END IF;
        END IF;
        END PROCESS;
END full;
```

The ports for this model are

- **clr** - clear input
- **load** - load control
- **ent** - count enable
- **enp** - count enable
- **clk** - clock input
- **a** - parallel data input

- **b** - parallel data input
- **c** - parallel data input
- **d** - parallel data input
- **qa** - parallel data output
- **qb** - parallel data output
- **qc** - parallel data output
- **qd** - parallel data output
- **rco** - ripple carry output

Generic parameters to this model are summarized here

- **clr_i01, clr_i10** - low to high and high to low **clr** input port delays
- **load_i01, load_i10** - low to high and high to low **load** input port delays
- **ent_i01, ent_i10** - low to high and high to low **ent** input port delays
- **enp_i01, enp_i10** - low to high and high to low **enp** input port delays
- **clk_i01, clk_i10** - low to high and high to low **clk** input port delays
- **a_i01, a_i10** - low to high and high to low **a** input port delays
- **b_i01, b_i10** - low to high and high to low **b** input port delays
- **c_i01, c_i10** - low to high and high to low **c** input port delays
- **d_i01, d_i10** - low to high and high to low **d** input port delays
- Minimum **clk** pulse width **clk_min**
- Minimum **clr** pulse width **clr_min**
- Data data setup **data_setup** and hold **data_hold** minimums
- **rco_o01, rco_o10** - low to high and high to low **rco** output port delays
- **qa_o01, qa_o10** - low to high and high to low **qa** output port delays
- **qb_o01, qb_o10** - low to high and high to low **qb** output port delays
- **qc_o01, qc_o10** - low to high and high to low **qc** output port delays
- **qd_o01, qd_o10** - low to high and high to low **qd** output port delays

This model features four error checks

- **clk** frequency/spike detection
- **clr** spike detection
- Serial data setup time checking
- Serial data hold time checking

4.3. COUNTERS

as shown in the error checking section of the architecture. The spike error checks each utilize a separate process statement. An additional single process performs the hold and setup checks.

The spike detection processes use a local variable to save the previous event time for the signal being checked. This is somewhat more efficient than using the delayed attribute since it does not require the simulator to create an additional signal.

The setup/hold checking process uses two local variables to save the time of the last data and clock event respectively. This process assumes that the data must be settled before the latch control becomes active high. Adjustment to the following expression is required in order to change the latching control characteristics.

$$(clk'EVENT)AND(clk ='1')$$

Events from the data and control inputs **load, ent, enp, a, b, c** and **d** are treated as a single data event. A more sophisticated version of this error check (with additional generic parameters) could be crafted which tests for setup/hold errors on any of the data inputs separately through the use of multiple local variables, one for each data input. Most data books do not specify this level of detail in timing and therefore this approach has not been taken.

The input delay processing section of the model incorporates the appropriate delay for each of the input ports. An internal signal for each input port is declared. The main device function process utilizes the delayed signals rather than the primary inputs.

The counter operation is handled by a single process which is sensitive to the input delayed signals **clkdelay** and **clrdelay**. The following conditions are checked in order whenever the asynchronous control or address inputs have an event

- Asynchronous clear
- Unknown control or clock
- Clocked parallel load
- Clocked count

Once the state of the device is determined (and stored in the local array variable **bstate**, the output signal assignment statements for **qa, qb, qc** and **qd** are executed. The carry out assignment to **rco** is also executed. Output state changes are executed with appropriate delay introduced through the use of the **f_delay** function as discussed in section 7.6. This function utilizes the output rise and fall delays and based on the new state value chooses the proper delay.

4.3.2 Synchronous 4 Bit Binary Counter with Asynchronous Clear

This device is a synchronous presettable binary counter with 4 bits. Synchronous operation is provided by having all outputs clocked (rising edge) simultaneously. The outputs preset to any number between 0 and 15. Presetting is synchronous, setting a low at the input disables the counter and causes the outputs to match the data inputs after the next clock pulse regardless of the values of the enable inputs.

The clear function for this device is asynchronous and a low at the **clear** input sets all outputs low regardless of the values of the **clk**, **load** or **enable** inputs.

Both count enable inputs **enp** and **ent** must be low to count. Input **ent** enables the carry output. The ripple carry output **rco** will produce a low pulse while the count is zero (all inputs low) counting down or maximum (15) counting up. Changes in **enp** and **ent** are allowed regardless of the value of the clock input. Changes at control inputs **enp**, **ent** and **load** that will modify the operating mode have no effect on the contents of the counter until clocking occurs. The function of the counter (whether enabled, disabled, loading, or counting) will be determined solely by the conditions meeting the stable setup and hold times. This device corresponds to the Texas Instruments 161 series TTL parts [TI86]. Figure 4.12 shows the

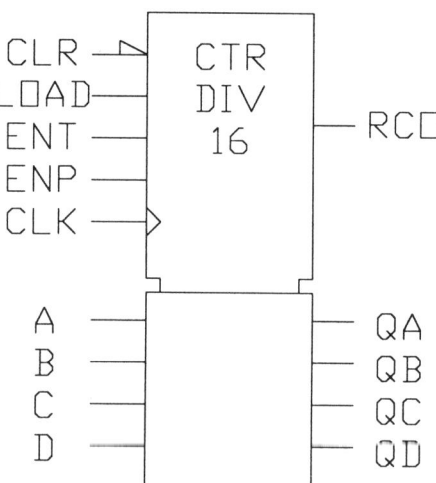

Figure 4.12: Logic Diagram 4 Bit Binary Counter with Async Clear

symbolic representation of this device. The following shows the full model

4.3. COUNTERS

for a counter with the characteristics
- 4 bit
- Binary count
- Asynchronous clear
- Synchronous/clocked parallel load
- Synchronous/clocked count
- Dual count enable control
- Parallel output
- Carry output

Synchronous 4 Bit Binary Counter with Asynchronous Clear

```
USE std.std_logic.ALL;
USE std.std_ttl.ALL;
ENTITY syn4bin IS
    GENERIC (clr_i01,clr_i10,
             load_i01,load_i10,
             ent_i01,ent_i10,
             enp_i01,enp_i10,
             clk_i01,clk_i10,
             a_i01,a_i10,
             b_i01,b_i10,
             c_i01,c_i10,
             d_i01,d_i10,
             clk_min, clr_min, data_hold, data_setup,
             rco_o01,rco_o10,
             qa_o01,qa_o10,
             qb_o01,qb_o10,
             qc_o01,qc_o10,
             qd_o01,qd_o10 : TIME := 2 ns);
    PORT    (clr,load,ent,enp,clk,a,b,c,d: IN t_wlogic;
             rco, qa,qb,qc,qd: OUT t_wlogic);
END syn4bin;

ARCHITECTURE full OF syn4bin IS
    SIGNAL clrdelay,loaddelay,entdelay,enpdelay,clkdelay,
           adelay,bdelay,cdelay,ddelay : t_wlogic;
BEGIN
    -- check clk frequency/spike detection
    PROCESS (clk)
        VARIABLE clklastev : TIME := 0 ns;
        BEGIN
        ASSERT (NOW = 0 ns) OR ((NOW - clklastev) >= clk_min)
            REPORT "Spike detected on clock" SEVERITY warning;
        clklastev := NOW;
        END PROCESS;
```

```
-- spike detection
PROCESS (clr)
   VARIABLE clrlastev : TIME := 0 ns;
   BEGIN
   ASSERT (NOW = 0 ns) OR ((NOW - clrlastev) >= clr_min)
      REPORT "Spike detected on clr" SEVERITY warning;
   clrlastev := NOW;
   END PROCESS;

-- check for setup/hold violations
PROCESS (clk,load,ent,enp,a,b,c,d)
   VARIABLE datalastev : TIME := 0 ns;
   VARIABLE clklastrise : TIME := 0 ns;
   BEGIN
   -- hold check
   IF (load'EVENT) OR (ent'EVENT) OR (enp'EVENT) OR
      (a'EVENT) OR (b'EVENT) OR (c'EVENT) OR (d'EVENT) THEN
      ASSERT (NOW = 0 ns) OR ((NOW - clklastrise) >= data_hold)
         REPORT "Hold error" SEVERITY warning;
      datalastev := NOW;
   END IF;

   -- setup check
   IF (clk'EVENT) AND (clk = '1') THEN
      ASSERT (NOW = 0 ns) OR ((NOW - datalastev) >= data_setup)
         REPORT "Setup error" SEVERITY warning;
      clklastrise := NOW;
   END IF;
   END PROCESS;

-- input delay processing
clrdelay  <= clr  AFTER f_delay(clr,clr_i01,clr_i10);
loaddelay <= load AFTER f_delay(load,load_i01,load_i10);
entdelay  <= ent  AFTER f_delay(ent,ent_i01,ent_i10);
enpdelay  <= enp  AFTER f_delay(enp,enp_i01,enp_i10);
clkdelay  <= clk  AFTER f_delay(clk,clk_i01,clk_i10);
adelay    <= a    AFTER f_delay(a,a_i01,a_i10);
bdelay    <= b    AFTER f_delay(b,b_i01,b_i10);
cdelay    <= c    AFTER f_delay(c,c_i01,c_i10);
ddelay    <= d    AFTER f_delay(d,d_i01,d_i10);

-- counter operation
PROCESS (clkdelay,clrdelay)
   VARIABLE istate : integer := 0;
   VARIABLE bstate : t_logarray (1 TO 4);
   VARIABLE unknown : boolean := true;
   BEGIN
   -- check for clear
   IF (clrdelay'EVENT) AND (clrdelay = '0') THEN
      istate := 0;
      unknown := false;
```

4.3. COUNTERS

```
            -- check for unknown control
            ELSIF (clrdelay'EVENT) AND (clrdelay = 'X') THEN unknown := true;
            ELSIF (clkdelay'EVENT) AND (clkdelay = '1') AND
                (loaddelay = 'X') THEN unknown := true;
            ELSIF (clkdelay'EVENT) AND (clkdelay = '1') AND
                ((entdelay = 'X') OR (enpdelay = 'X')) THEN unknown := true;

            -- check for clocked load
            ELSIF (clkdelay'EVENT)AND(clkdelay='1')AND(loaddelay='0') THEN
                bstate(1) := ddelay;
                bstate(2) := cdelay;
                bstate(3) := bdelay;
                bstate(4) := adelay;
                f_logictoint(bstate,unknown,istate);

            -- check for clocked count
            ELSIF (clkdelay'EVENT) AND (clkdelay = '1') AND (entdelay ='1') AND
                (enpdelay = '1') AND (istate >= 0) THEN
                    IF NOT unknown THEN istate := (istate + 1) MOD 16; END IF;
            END IF;

            -- unknown state
            IF unknown THEN
                qa <= f_ttl('X') AFTER f_delay(f_ttl('X'), qa_o01, qa_o10);
                qb <= f_ttl('X') AFTER f_delay(f_ttl('X'), qb_o01, qb_o10);
                qc <= f_ttl('X') AFTER f_delay(f_ttl('X'), qc_o01, qc_o10);
                qd <= f_ttl('X') AFTER f_delay(f_ttl('X'), qd_o01, qd_o10);
                rco <= f_ttl('X') AFTER f_delay(f_ttl('X'), rco_o01, rco_o10);
            -- assign outputs
            ELSE
                f_inttologic(istate,bstate,ttl);
                qa <= bstate(4) AFTER f_delay(bstate(1), qa_o01, qa_o10);
                qb <= bstate(3) AFTER f_delay(bstate(2), qb_o01, qb_o10);
                qc <= bstate(2) AFTER f_delay(bstate(3), qc_o01, qc_o10);
                qd <= bstate(1) AFTER f_delay(bstate(4), qd_o01, qd_o10);
                IF istate = 15 THEN
                    rco <= f_ttl('1')AFTER f_delay(f_ttl('1'),rco_o01,rco_o10);
                ELSE
                    rco <= f_ttl('0')AFTER f_delay(f_ttl('0'),rco_o01,rco_o10);
                END IF;
            END IF;
        END PROCESS;
END full;
```

The ports for this model are

- **clr** - clear input
- **load** - load control
- **ent** - count enable
- **enp** - count enable

- **clk** - clock input
- **a** - parallel data input
- **b** - parallel data input
- **c** - parallel data input
- **d** - parallel data input
- **qa** - parallel data output
- **qb** - parallel data output
- **qc** - parallel data output
- **qd** - parallel data output
- **rco** - ripple carry output

Generic parameters to this model are summarized here

- **clr_i01, clr_i10** - low to high and high to low **clr** input port delays
- **load_i01, load_i10** - low to high and high to low **load** input port delays
- **ent_i01, ent_i10** - low to high and high to low **ent** input port delays
- **enp_i01, enp_i10** - low to high and high to low **enp** input port delays
- **clk_i01, clk_i10** - low to high and high to low **clk** input port delays
- **a_i01, a_i10** - low to high and high to low **a** input port delays
- **b_i01, b_i10** - low to high and high to low **b** input port delays
- **c_i01, c_i10** - low to high and high to low **c** input port delays
- **d_i01, d_i10** - low to high and high to low **d** input port delays
- Minimum **clk** pulse width **clk_min**
- Minimum **clr** pulse width **clr_min**
- Data data setup **data_setup** and hold **data_hold** minimums
- **rco_o01, rco_o10** - low to high and high to low **rco** output port delays
- **qa_o01, qa_o10** - low to high and high to low **qa** output port delays
- **qb_o01, qb_o10** - low to high and high to low **qb** output port delays
- **qc_o01, qc_o10** - low to high and high to low **qc** output port delays
- **qd_o01, qd_o10** - low to high and high to low **qd** output port delays

This model features four error checks

- **clk** frequency/spike detection
- **clr** spike detection

4.3. COUNTERS

- Serial data setup time checking
- Serial data hold time checking

as shown in the error checking section of the architecture. The spike error checks each utilize a separate process statement. An additional single process performs the hold and setup checks.

The spike detection processes use a local variable to save the previous event time for the signal being checked. This is somewhat more efficient than using the delayed attribute since it does not require the simulator to create an additional signal.

The setup/hold checking process uses two local variables to save the time of the last data and clock event respectively. This process assumes that the data must be settled before the latch control becomes active high. Adjustment to the following expression is required in order to change the latching control characteristics.

$(clk'EVENT) AND (clk =' 1')$

Events from the data and control inputs **load, ent, enp, a, b, c** and **d** are treated as a single data event. A more sophisticated version of this error check (with additional generic parameters) could be crafted which tests for setup/hold errors on any of the data inputs separately through the use of multiple local variables, one for each data input. Most data books do not specify this level of detail in timing and therefore this approach has not been taken.

The input delay processing section of the model incorporates the appropriate delay for each of the input ports. An internal signal for each input port is declared. The main device function process utilizes the delayed signals rather than the primary inputs.

The counter operation is handled by a single process which is sensitive to the input delayed signals **clkdelay** and **clrdelay**. The following conditions are checked in order whenever the asynchronous control or address inputs have an event

- Asynchronous clear
- Unknown control or clock
- Clocked parallel load
- Clocked count

Once the state of the device is determined (and stored in the local array variable **bstate**, the output signal assignment statements for **qa, qb, qc** and **qd** are executed. The carry out assignment to **rco** is also executed. Output state changes are executed with appropriate delay introduced through the use of the **f_delay** function as discussed in section 7.6. This function utilizes the output rise and fall delays and based on the new state value chooses the proper delay.

4.3.3 Synchronous 4 Bit Decade Counter

This device is a synchronous presettable decade counter with 4 bits. Synchronous operation is provided by having all outputs clocked (rising edge) simultaneously. The outputs preset to any number between 0 and 9. Presetting is synchronous, setting a low at the input disables the counter and causes the outputs to match the data inputs after the next clock pulse regardless of the values of the enable inputs. The clear function for this device is synchronous. A low at the clear input sets all outputs low after the next clock pulse regardless of the values of the enable inputs. Both count enable inputs **enp** and **ent** must be low to count. Input **ent** enables the carry output. The ripple carry output **rco** will produce a low pulse while the count is zero (all inputs low) counting down or maximum (10) counting up. This low overflow carry pulse can be used to enable successive cascaded stages. Changes in **enp** and **ent** are allowed regardless of the value of the clock input. Changes at control inputs **enp**, **ent** and **load** that will modify the operating mode have no effect on the contents of the counter until clocking occurs. The function of the counter (whether enabled, disabled, loading, or counting) will be determined solely by the conditions meeting the stable setup and hold times. This device corresponds to the Texas Instruments 162 series TTL parts [TI86]. Figure 4.13 shows the symbolic representation of this device. The following shows the full model for a counter with the characteristics

- 4 bit
- Decade count
- Synchronous/clocked clear
- Synchronous/clocked parallel load
- Synchronous/clocked count
- Dual count enable control
- Parallel output
- Carry output

Synchronous 4 Bit Decade Counter

```
USE std.std_logic.ALL;
USE std.std_ttl.ALL;
ENTITY syn4decadesync IS
```

4.3. COUNTERS

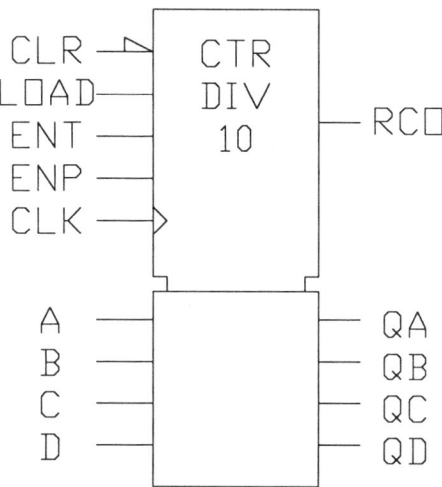

Figure 4.13: Logic Diagram 4 Bit Decade Counter

```
    GENERIC (clr_i01,clr_i10,
             load_i01,load_i10,
             ent_i01,ent_i10,
             enp_i01,enp_i10,
             clk_i01,clk_i10,
             a_i01,a_i10,
             b_i01,b_i10,
             c_i01,c_i10,
             d_i01,d_i10,
             clk_min, clr_min, data_hold, data_setup,
             rco_o01,rco_o10,
             qa_o01,qa_o10,
             qb_o01,qb_o10,
             qc_o01,qc_o10,
             qd_o01,qd_o10 : TIME := 2 ns);
    PORT    (clr,load,ent,enp,clk,a,b,c,d: IN t_wlogic;
             rco, qa,qb,qc,qd: OUT t_wlogic);
END syn4decadesync;

ARCHITECTURE full OF syn4decadesync IS
    SIGNAL clrdelay,loaddelay,entdelay,enpdelay,clkdelay,
           adelay,bdelay,cdelay,ddelay : t_wlogic;
BEGIN
    -- check clk frequency/spike detection
    PROCESS (clk)
        VARIABLE clklastev : TIME := 0 ns;
        BEGIN
        ASSERT (NOW = 0 ns) OR ((NOW - clklastev) >= clk_min)
            REPORT "Spike detected on clock" SEVERITY warning;
```

```
        clklastev := NOW;
    END PROCESS;

-- spike detection
PROCESS (clr)
    VARIABLE clrlastev : TIME := 0 ns;
    BEGIN
    ASSERT (NOW = 0 ns) OR ((NOW - clrlastev) >= clr_min)
        REPORT "Spike detected on clr" SEVERITY warning;
    clrlastev := NOW;
    END PROCESS;

-- check for setup/hold violations
PROCESS (clk,load,ent,enp,a,b,c,d)
    VARIABLE datalastev : TIME := 0 ns;
    VARIABLE clklastrise : TIME := 0 ns;
    BEGIN
    -- hold check
    IF (load'EVENT) OR (ent'EVENT) OR (enp'EVENT) OR
       (a'EVENT) OR (b'EVENT) OR (c'EVENT) OR (d'EVENT) THEN
        ASSERT (NOW = 0 ns) OR ((NOW - clklastrise) >= data_hold)
            REPORT "Hold error" SEVERITY warning;
        datalastev := NOW;
    END IF;

    -- setup check
    IF (clk'EVENT) AND (clk = '1') THEN
        ASSERT (NOW = 0 ns) OR ((NOW - datalastev) >= data_setup)
            REPORT "Setup error" SEVERITY warning;
        clklastrise := NOW;
    END IF;
    END PROCESS;

-- input delay processing
clrdelay  <= clr  AFTER f_delay(clr,clr_i01,clr_i10);
loaddelay <= load AFTER f_delay(load,load_i01,load_i10);
entdelay  <= ent  AFTER f_delay(ent,ent_i01,ent_i10);
enpdelay  <= enp  AFTER f_delay(enp,enp_i01,enp_i10);
clkdelay  <= clk  AFTER f_delay(clk,clk_i01,clk_i10);
adelay    <= a    AFTER f_delay(a,a_i01,a_i10);
bdelay    <= b    AFTER f_delay(b,b_i01,b_i10);
cdelay    <= c    AFTER f_delay(c,c_i01,c_i10);
ddelay    <= d    AFTER f_delay(d,d_i01,d_i10);

-- counter operation
PROCESS (clkdelay)
    VARIABLE istate : integer := 0;
    VARIABLE bstate : t_logarray (1 TO 4);
    VARIABLE unknown : boolean := true;
    BEGIN
    -- check for clear
    IF (clrdelay = '0') THEN
```

4.3. COUNTERS

```
                istate := 0;
                unknown := true;

            -- check for unknown control
            ELSIF (clrdelay = 'X') THEN unknown := true;
            ELSIF (clkdelay = '1') AND (loaddelay = 'X') THEN unknown := true;
            ELSIF (clkdelay = '1') AND ((entdelay = 'X') OR
                (enpdelay = 'X')) THEN unknown := true;

            -- check for clocked load
            ELSIF (clkdelay = '1') AND (loaddelay = '0') THEN
                bstate(1) := ddelay;
                bstate(2) := cdelay;
                bstate(3) := bdelay;
                bstate(4) := adelay;
                f_logictoint(bstate,unknown,istate);
                IF NOT unknown THEN istate := istate MOD 10; END IF;

            -- check for clocked count
            ELSIF (clkdelay = '1') AND (entdelay = '1') AND
                (enpdelay = '1') AND (istate >= 0) THEN
                IF NOT unknown THEN istate := (istate + 1) MOD 10; END IF;
            END IF;

            -- unknown state
            IF unknown THEN
                qa <= f_ttl('X') AFTER f_delay(f_ttl('X'), qa_o01, qa_o10);
                qb <= f_ttl('X') AFTER f_delay(f_ttl('X'), qb_o01, qb_o10);
                qc <= f_ttl('X') AFTER f_delay(f_ttl('X'), qc_o01, qc_o10);
                qd <= f_ttl('X') AFTER f_delay(f_ttl('X'), qd_o01, qd_o10);
                rco <= f_ttl('X') AFTER f_delay(f_ttl('X'), rco_o01, rco_o10);
            -- assign outputs
            ELSE
                f_inttologic(istate,bstate,ttl);
                qa <= bstate(4) AFTER f_delay(bstate(1), qa_o01, qa_o10);
                qb <= bstate(3) AFTER f_delay(bstate(2), qb_o01, qb_o10);
                qc <= bstate(2) AFTER f_delay(bstate(3), qc_o01, qc_o10);
                qd <= bstate(1) AFTER f_delay(bstate(4), qd_o01, qd_o10);
                IF istate = 9 THEN
                    rco <= f_ttl('1')AFTER f_delay(f_ttl('1'),rco_o01,rco_o10);
                ELSE
                    rco <= f_ttl('0')AFTER f_delay(f_ttl('0'),rco_o01,rco_o10);
                END IF;
            END IF;
        END PROCESS;
END full;
```

The ports for this model are

- **clr** - clear input
- **load** - load control
- **ent** - count enable

- **enp** - count enable
- **clk** - clock input
- **a** - parallel data input
- **b** - parallel data input
- **c** - parallel data input
- **d** - parallel data input
- **qa** - parallel data output
- **qb** - parallel data output
- **qc** - parallel data output
- **qd** - parallel data output
- **rco** - ripple carry output

Generic parameters to this model are summarized here

- **clr_i01, clr_i10** - low to high and high to low **clr** input port delays
- **load_i01, load_i10** - low to high and high to low **load** input port delays
- **ent_i01, ent_i10** - low to high and high to low **ent** input port delays
- **enp_i01, enp_i10** - low to high and high to low **enp** input port delays
- **clk_i01, clk_i10** - low to high and high to low **clk** input port delays
- **a_i01, a_i10** - low to high and high to low **a** input port delays
- **b_i01, b_i10** - low to high and high to low **b** input port delays
- **c_i01, c_i10** - low to high and high to low **c** input port delays
- **d_i01, d_i10** - low to high and high to low **d** input port delays
- Minimum **clr** pulse width **clr_min**
- Minimum **clk** pulse width **clk_min**
- Data data setup **data_setup** and hold **data_hold** minimums
- **rco_o01, rco_o10** - low to high and high to low **rco** output port delays
- **qa_o01, qa_o10** - low to high and high to low **qa** output port delays
- **qb_o01, qb_o10** - low to high and high to low **qb** output port delays
- **qc_o01, qc_o10** - low to high and high to low **qc** output port delays
- **qd_o01, qd_o10** - low to high and high to low **qd** output port delays

This model features four error checks

- **clk** frequency/spike detection

4.3. COUNTERS

- **clr** spike detection
- Serial data setup time checking
- Serial data hold time checking

as shown in the error checking section of the architecture. The spike error checks each utilize a separate process statement. An additional single process performs the hold and setup checks.

The spike detection processes use a local variable to save the previous event time for the signal being checked. This is somewhat more efficient than using the delayed attribute since it does not require the simulator to create an additional signal.

The setup/hold checking process uses two local variables to save the time of the last data and clock event respectively. This process assumes that the data must be settled before the latch control becomes active high. Adjustment to the following expression is required in order to change the latching control characteristics.

$$(clk'EVENT)AND(clk = '1')$$

Events from the data and control inputs **load, ent, enp, a, b, c** and **d** are treated as a single data event. A more sophisticated version of this error check (with additional generic parameters) could be crafted which tests for setup/hold errors on any of the data inputs separately through the use of multiple local variables, one for each data input. Most data books do not specify this level of detail in timing and therefore this approach has not been taken.

The input delay processing section of the model incorporates the appropriate delay for each of the input ports. An internal signal for each input port is declared. The main device function process utilizes the delayed signals rather than the primary inputs.

The counter operation is handled by a single process which is sensitive to the input delayed signals **clkdelay**. The following conditions are checked in order whenever the asynchronous control or address inputs have an event

- Synchronous clear
- Unknown control or clock
- Clocked parallel load
- Clocked count

Once the state of the device is determined (and stored in the local array variable **bstate**, the output signal assignment statements for **qa, qb, qc** and **qd** are executed. The carry out assignment to **rco** is also executed. Output state changes are executed with appropriate delay introduced through the use of the f_delay function as discussed in section 7.6. This function utilizes the output rise and fall delays and based on the new state value chooses the proper delay.

4.3.4 Synchronous 4 Bit Binary Counter

This device is a synchronous presettable binary counter with 4 bits. Synchronous operation is provided by having all outputs clocked (rising edge) simultaneously. The outputs may each be preset to any number between 0 and 15. Presetting is synchronous, setting a low at the input disables the counter and causes the outputs to match the data inputs after the next clock pulse regardless of the values of the enable inputs. The clear function for this device is asynchronous and a low at the **clear** input sets all outputs low regardless of the values of the **clk**, **load** or **enable** inputs. Both count enable inputs **enp** and **ent** must be low to count. Input **ent** enables the carry output. The ripple carry output **rco** will produce a low pulse while the count is zero (all inputs low) counting down or maximum (10) counting up. Changes at control inputs **enp**, **ent** and **load** that will modify the operating mode have no effect on the contents of the counter until clocking occurs. The function of the counter (whether enabled, disabled, loading, or counting) will be determined solely by the conditions meeting the stable setup and hold times. carry pulse can be used to enable successive cascaded stages. Transitions at **enp** and **ent** are allowed regardless of the value of the clock input. This device corresponds to the Texas Instruments 163 series TTL parts [TI86]. Figure 4.14 shows the symbolic representation of this device. The following shows the full model for a counter with the characteristics

- 4 bit
- Binary count
- Synchronous/clocked clear
- Synchronous/clocked parallel load
- Synchronous/clocked count
- Dual count enable control
- Parallel output
- Carry output

4.3. COUNTERS

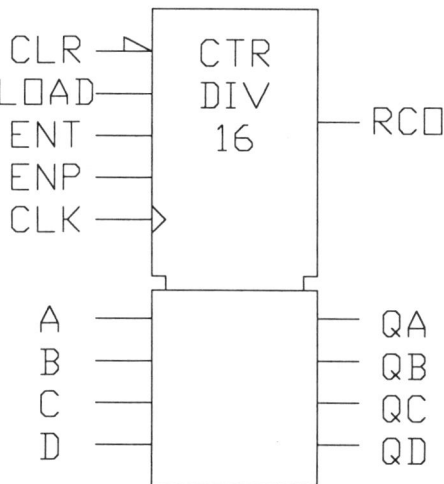

Figure 4.14: Logic Diagram 4 Bit Sync Binary Counter

Synchronous 4 Bit Binary Counter

```
USE std.std_logic.ALL;
USE std.std_ttl.ALL;
ENTITY syn4binsync IS
    GENERIC (clr_i01,clr_i10,
             load_i01,load_i10,
             ent_i01,ent_i10,
             enp_i01,enp_i10,
             clk_i01,clk_i10,
             a_i01,a_i10,
             b_i01,b_i10,
             c_i01,c_i10,
             d_i01,d_i10,
             clk_min, clr_min, data_hold, data_setup,
             rco_o01,rco_o10,
             qa_o01,qa_o10,
             qb_o01,qb_o10,
             qc_o01,qc_o10,
             qd_o01,qd_o10 : TIME := 2 ns);
    PORT    (clr,load,ent,enp,clk,a,b,c,d: IN t_wlogic;
             rco, qa,qb,qc,qd: OUT t_wlogic);
END syn4binsync;

ARCHITECTURE full OF syn4binsync IS
    SIGNAL clrdelay,loaddelay,entdelay,enpdelay,clkdelay,
           adelay,bdelay,cdelay,ddelay : t_wlogic;
BEGIN
    -- check clk frequency/spike detection
```

```
PROCESS (clk)
   VARIABLE clklastev : TIME := 0 ns;
   BEGIN
   ASSERT (NOW = 0 ns) OR ((NOW - clklastev) >= clk_min)
       REPORT "Spike detected on clock" SEVERITY warning;
   clklastev := NOW;
   END PROCESS;

-- spike detection
PROCESS (clr)
   VARIABLE clrlastev : TIME := 0 ns;
   BEGIN
   ASSERT (NOW = 0 ns) OR ((NOW - clrlastev) >= clr_min)
       REPORT "Spike detected on clr" SEVERITY warning;
   clrlastev := NOW;
   END PROCESS;

-- check for setup/hold violations
PROCESS (clk,load,ent,enp,a,b,c,d)
   VARIABLE datalastev : TIME := 0 ns;
   VARIABLE clklastrise : TIME := 0 ns;
   BEGIN
   -- hold check
   IF (load'EVENT) OR (ent'EVENT) OR (enp'EVENT) OR
      (a'EVENT) OR (b'EVENT) OR (c'EVENT) OR (d'EVENT) THEN
       ASSERT (NOW = 0 ns) OR ((NOW - clklastrise) >= data_hold)
           REPORT "Hold error" SEVERITY warning;
       datalastev := NOW;
   END IF;

   -- setup check
   IF (clk'EVENT) AND (clk = '1') THEN
       ASSERT (NOW = 0 ns) OR ((NOW - datalastev) >= data_setup)
           REPORT "Setup error" SEVERITY warning;
       clklastrise := NOW;
   END IF;
   END PROCESS;

-- input delay processing
clrdelay <= clr AFTER f_delay(clr,clr_i01,clr_i10);
loaddelay <= load AFTER f_delay(load,load_i01,load_i10);
entdelay <= ent AFTER f_delay(ent,ent_i01,ent_i10);
enpdelay <= enp AFTER f_delay(enp,enp_i01,enp_i10);
clkdelay <= clk AFTER f_delay(clk,clk_i01,clk_i10);
adelay <= a AFTER f_delay(a,a_i01,a_i10);
bdelay <= b AFTER f_delay(b,b_i01,b_i10);
cdelay <= c AFTER f_delay(c,c_i01,c_i10);
ddelay <= d AFTER f_delay(d,d_i01,d_i10);

-- counter operation
PROCESS (clkdelay)
    VARIABLE istate : integer := 0;
```

4.3. COUNTERS

```
            VARIABLE bstate : t_logarray (1 TO 4);
            VARIABLE unknown : boolean := true;
            BEGIN
            -- check for clear
            IF (clrdelay = '0') THEN
                istate := 0;
                unknown := false;

            -- check for unknown control
            ELSIF (clrdelay = 'X') THEN unknown := true;
            ELSIF (clkdelay = '1') AND (loaddelay = 'X') THEN unknown := true;
            ELSIF (clkdelay = '1') AND ((entdelay = 'X') OR
                (enpdelay = 'X')) THEN unknown := true;

            -- check for clocked load
            ELSIF (clkdelay = '1') AND (loaddelay = '0') THEN
                bstate(1) := ddelay;
                bstate(2) := cdelay;
                bstate(3) := bdelay;
                bstate(4) := adelay;
                f_logictoint(bstate,unknown,istate);

            -- check for clocked count
            ELSIF (clkdelay = '1') AND (entdelay = '1') AND (enpdelay = '1')
                AND (istate >= 0) THEN
                    IF NOT unknown THEN istate := (istate + 1) MOD 16; END IF;
            END IF;

            -- unknown state
            IF unknown THEN
                qa <= f_ttl('X') AFTER f_delay(f_ttl('X'), qa_o01, qa_o10);
                qb <= f_ttl('X') AFTER f_delay(f_ttl('X'), qb_o01, qb_o10);
                qc <= f_ttl('X') AFTER f_delay(f_ttl('X'), qc_o01, qc_o10);
                qd <= f_ttl('X') AFTER f_delay(f_ttl('X'), qd_o01, qd_o10);
                rco <= f_ttl('X') AFTER f_delay(f_ttl('X'), rco_o01, rco_o10);
            -- assign outputs
            ELSE
                f_inttologic(istate,bstate,ttl);
                qa <= bstate(4) AFTER f_delay(bstate(1), qa_o01, qa_o10);
                qb <= bstate(3) AFTER f_delay(bstate(2), qb_o01, qb_o10);
                qc <= bstate(2) AFTER f_delay(bstate(3), qc_o01, qc_o10);
                qd <= bstate(1) AFTER f_delay(bstate(4), qd_o01, qd_o10);
                IF istate = 15 THEN
                    rco <= f_ttl('1')AFTER f_delay(f_ttl('1'),rco_o01,rco_o10);
                ELSE
                    rco <= f_ttl('0')AFTER f_delay(f_ttl('0'),rco_o01,rco_o10);
                END IF;
            END IF;
            END PROCESS;
END full;
```

The ports for this model are

- **clr** - clear input
- **load** - load control
- **ent** - count enable
- **enp** - count enable
- **clk** - clock input
- **a** - parallel data input
- **b** - parallel data input
- **c** - parallel data input
- **d** - parallel data input
- **qa** - parallel data output
- **qb** - parallel data output
- **qc** - parallel data output
- **qd** - parallel data output
- **rco** - ripple carry output

Generic parameters to this model are summarized here

- **clr_i01, clr_i10** - low to high and high to low **clr** input port delays
- **load_i01, load_i10** - low to high and high to low **load** input port delays
- **ent_i01, ent_i10** - low to high and high to low **ent** input port delays
- **enp_i01, enp_i10** - low to high and high to low **enp** input port delays
- **clk_i01, clk_i10** - low to high and high to low **clk** input port delays
- **a_i01, a_i10** - low to high and high to low **a** input port delays
- **b_i01, b_i10** - low to high and high to low **b** input port delays
- **c_i01, c_i10** - low to high and high to low **c** input port delays
- **d_i01, d_i10** - low to high and high to low **d** input port delays
- Minimum **clk** pulse width **clk_min**
- Minimum **clr** pulse width **clr_min**
- Data data setup **data_setup** and hold **data_hold** minimums
- **rco_o01, rco_o10** - low to high and high to low **rco** output port delays
- **qa_o01, qa_o10** - low to high and high to low **qa** output port delays
- **qb_o01, qb_o10** - low to high and high to low **qb** output port delays

4.3. COUNTERS

- **qc_o01, qc_o10** - low to high and high to low **qc** output port delays
- **qd_o01, qd_o10** - low to high and high to low **qd** output port delays

This model features four error checks

- **clk** frequency/spike detection
- **clr** spike detection
- Serial data setup time checking
- Serial data hold time checking

as shown in the error checking section of the architecture. The spike error checks each utilize a separate process statement. An additional single process performs the hold and setup checks.

The spike detection processes use a local variable to save the previous event time for the signal being checked. This is somewhat more efficient than using the delayed attribute since it does not require the simulator to create an additional signal.

The setup/hold checking process uses two local variables to save the time of the last data and clock event respectively. This process assumes that the data must be settled before the latch control becomes active high. Adjustment to the following expression is required in order to change the latching control characteristics.

$$(clk'EVENT)AND(clk ='1')$$

Events from the data and control inputs **load, ent, enp, a, b, c** and **d** are treated as a single data event. A more sophisticated version of this error check (with additional generic parameters) could be crafted which tests for setup/hold errors on any of the data inputs separately through the use of multiple local variables, one for each data input. Most data books do not specify this level of detail in timing and therefore this approach has not been taken.

The input delay processing section of the model incorporates the appropriate delay for each of the input ports. An internal signal for each input port is declared. The main device function process utilizes the delayed signals rather than the primary inputs.

The counter operation is handled by a single process which is sensitive to the input delayed signals **clkdelay**. The following conditions are checked in order whenever the asynchronous control or address inputs have an event

- Synchronous clear
- Unknown control or clock
- Clocked parallel load
- Clocked count

220   CHAPTER 4. SEQUENTIAL DEVICES

Once the state of the device is determined (and stored in the local array variable **bstate**, the output signal assignment statements for **qa**, **qb**, **qc** and **qd** are executed. The carry out assignment to **rco** is also executed. Output state changes are executed with appropriate delay introduced through the use of the **f_delay** function as discussed in section 7.6. This function utilizes the output rise and fall delays and based on the new state value chooses the proper delay.

4.3.5 Synchronous Up/Down 4-Bit Decade Counter

This device is a synchronous presettable counter with 4 bits. Synchronous operation is provided by simultaneously clocking (rising edge) all outputs. The outputs may each be preset and is performed synchronously; a low at the load input disables the counter and causes outputs to match data inputs after the next clock pulse. Both count enable inputs **enp** and **ent** must be low to count. The direction of the count is determined by the value of the **ud** input. When **ud** is high, the counter counts up; when low, it counts down. Input **ent** enables the carry output. The ripple carry output **rco** will produce a low pulse while the count is zero (all inputs low) counting down or maximum (10) counting up. Changes in **enp** and **ent** are allowed regardless of the value of the clock input. Changes in **enp**, **ent**, **load** and **ud** that will modify the operating mode have no effect on the contents of the counter until clocking occurs. The function of the counter (whether enabled, disabled, loading, or counting) will be determined solely by the conditions meeting the stable setup and hold times. This device corresponds to the Texas Instruments 168 TTL series parts [TI86]. Figure 4.15 shows the symbolic representation of this device. The following shows the full model for a counter with the characteristics

- 4 bit
- Decade count
- Up/down operation
- Dual count enable control
- Synchronous/clocked parallel load
- Synchronous/clocked count
- Parallel output
- Carry output

4.3. COUNTERS

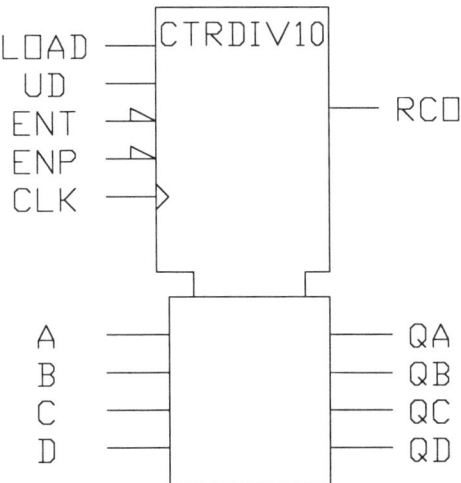

Figure 4.15: Logic Diagram Bit Up/Down Sync Decade Counter

Synchronous Up/Down 4-Bit Decade Counter

```
USE std.std_logic.ALL;
USE std.std_ttl.ALL;
ENTITY syn4decadeud IS
    GENERIC (ud_i01,ud_i10,
             load_i01,load_i10,
             ent_i01,ent_i10,
             enp_i01,enp_i10,
             clk_i01,clk_i10,
             a_i01,a_i10,
             b_i01,b_i10,
             c_i01,c_i10,
             d_i01,d_i10,
             clk_min, data_hold, data_setup,
             rco_o01,rco_o10,
             qa_o01,qa_o10,
             qb_o01,qb_o10,
             qc_o01,qc_o10,
             qd_o01,qd_o10 : TIME := 2 ns);
    PORT    (ud,load,ent,enp,clk,a,b,c,d: IN t_wlogic;
             rco, qa,qb,qc,qd: OUT t_wlogic);
END syn4decadeud;

ARCHITECTURE full OF syn4decadeud IS
    SIGNAL uddelay,loaddelay,entdelay,enpdelay,clkdelay,
           adelay,bdelay,cdelay,ddelay : t_wlogic;
BEGIN
```

```
-- check clk frequency/spike detection
PROCESS (clk)
   VARIABLE clklastev : TIME := 0 ns;
   BEGIN
   ASSERT (NOW = 0 ns) OR ((NOW - clklastev) >= clk_min)
      REPORT "Spike detected on clock" SEVERITY warning;
   clklastev := NOW;
   END PROCESS;

-- check for setup/hold violations
PROCESS (clk,load,ent,enp,a,b,c,d,ud)
   VARIABLE datalastev : TIME := 0 ns;
   VARIABLE clklastrise : TIME := 0 ns;
   BEGIN
   -- hold check
   IF (load'EVENT) OR (ent'EVENT) OR (enp'EVENT) OR (ud'EVENT) OR
      (a'EVENT) OR (b'EVENT) OR (c'EVENT) OR (d'EVENT) THEN
      ASSERT (NOW = 0 ns) OR ((NOW - clklastrise) >= data_hold)
         REPORT "Hold error" SEVERITY warning;
      datalastev := NOW;
   END IF;

   -- setup check
   IF (clk'EVENT) AND (clk = '1') THEN
      ASSERT (NOW = 0 ns) OR ((NOW - datalastev) >= data_setup)
         REPORT "Setup error" SEVERITY warning;
      clklastrise := NOW;
   END IF;
   END PROCESS;

-- input delay processing
uddelay <= ud AFTER f_delay(ud,ud_i01,ud_i10);
loaddelay <= load AFTER f_delay(load,load_i01,load_i10);
entdelay <= ent AFTER f_delay(ent,ent_i01,ent_i10);
enpdelay <= enp AFTER f_delay(enp,enp_i01,enp_i10);
clkdelay <= clk AFTER f_delay(clk,clk_i01,clk_i10);
adelay <= a AFTER f_delay(a,a_i01,a_i10);
bdelay <= b AFTER f_delay(b,b_i01,b_i10);
cdelay <= c AFTER f_delay(c,c_i01,c_i10);
ddelay <= d AFTER f_delay(d,d_i01,d_i10);

-- counter operation
PROCESS (clkdelay)
   VARIABLE istate : integer := 0;
   VARIABLE bstate : t_logarray (1 TO 4);
   VARIABLE unknown : boolean := true;
   BEGIN
   -- check for unknown control
   IF (clkdelay'EVENT) AND (clkdelay = '1') AND
      (loaddelay = 'X') THEN unknown :=,true;
   ELSIF (clkdelay'EVENT) AND (clkdelay = '1') AND
         ((entdelay = 'X') OR (enpdelay = 'X')) THEN unknown := true;
```

4.3. COUNTERS

```
        -- check for unknown clock
        ELSIF (clkdelay'EVENT) AND (clkdelay = 'X') THEN unknown := true;

        -- check for clocked load
        ELSIF (clkdelay'EVENT) AND (clkdelay='1') AND (loaddelay='0')THEN
            bstate(1) := ddelay;
            bstate(2) := cdelay;
            bstate(3) := bdelay;
            bstate(4) := adelay;
            f_logictoint(bstate,unknown,istate);
            IF NOT unknown THEN istate := istate MOD 10; END IF;

        -- check for clocked count
        ELSIF (clkdelay'EVENT) AND (clkdelay = '1') AND (entdelay ='1') AND
            (enpdelay = '1') AND (istate >= 0) THEN
            -- watch out for unknown u/d control
            IF (uddelay = 'X') THEN unknown := true;

            -- count up
            ELSIF (uddelay = '1') THEN istate := (istate + 1) MOD 10;

            -- count down
            ELSE
                istate := istate - 1;
                IF istate < 0 THEN istate := 9; END IF;
            END IF;
        END IF;

        -- unknown state
        IF unknown THEN
            qa <= f_ttl('X') AFTER f_delay(f_ttl('X'), qa_o01, qa_o10);
            qb <= f_ttl('X') AFTER f_delay(f_ttl('X'), qb_o01, qb_o10);
            qc <= f_ttl('X') AFTER f_delay(f_ttl('X'), qc_o01, qc_o10);
            qd <= f_ttl('X') AFTER f_delay(f_ttl('X'), qd_o01, qd_o10);
            rco <= f_ttl('X') AFTER f_delay(f_ttl('X'), rco_o01, rco_o10);
        -- assign outputs
        ELSE
            f_inttologic(istate,bstate,ttl);
            qa <= bstate(4) AFTER f_delay(bstate(1), qa_o01, qa_o10);
            qb <= bstate(3) AFTER f_delay(bstate(2), qb_o01, qb_o10);
            qc <= bstate(2) AFTER f_delay(bstate(3), qc_o01, qc_o10);
            qd <= bstate(1) AFTER f_delay(bstate(4), qd_o01, qd_o10);
            IF istate = 9 THEN
                rco <= f_ttl('1')AFTER f_delay(f_ttl('1'),rco_o01,rco_o10);
            ELSE
                rco <= f_ttl('0')AFTER f_delay(f_ttl('0'),rco_o01,rco_o10);
            END IF;
        END IF;
    END PROCESS;
END full;
```

224   CHAPTER 4. SEQUENTIAL DEVICES

The ports for this model are
- **ud** - up/down control
- **load** - load control
- **ent** - count enable
- **enp** - count enable
- **clk** - clock input
- **a** - parallel data input
- **b** - parallel data input
- **c** - parallel data input
- **d** - parallel data input
- **qa** - parallel data output
- **qb** - parallel data output
- **qc** - parallel data output
- **qd** - parallel data output
- **rco** - ripple carry output

Generic parameters to this model are summarized here
- ud_i01, ud_i10 - low to high and high to low **ud** input port delays
- load_i01, load_i10 - low to high and high to low **load** input port delays
- ent_i01, ent_i10 - low to high and high to low **ent** input port delays
- enp_i01, enp_i10 - low to high and high to low **enp** input port delays
- clk_i01, clk_i10 - low to high and high to low **clk** input port delays
- a_i01, a_i10 - low to high and high to low **a** input port delays
- b_i01, b_i10 - low to high and high to low **b** input port delays
- c_i01, c_i10 - low to high and high to low **c** input port delays
- d_i01, d_i10 - low to high and high to low **d** input port delays
- Minimum **clk** pulse width **clk_min**
- Data data setup **data_setup** and hold **data_hold** minimums
- rco_o01, rco_o10 - low to high and high to low **rco** output port delays
- qa_o01, qa_o10 - low to high and high to low **qa** output port delays
- qb_o01, qb_o10 - low to high and high to low **qb** output port delays
- qc_o01, qc_o10 - low to high and high to low **qc** output port delays

4.3. COUNTERS

- **qd_o01, qd_o10** - low to high and high to low **qd** output port delays

This model features three error checks

- **clk** frequency/spike detection
- Serial data setup time checking
- Serial data hold time checking

as shown in the error checking section of the architecture. The spike error check utilizes a separate process statement. An additional single process performs the hold and setup checks.

The spike detection process uses a local variable to save the previous event time for the signal being checked. This is somewhat more efficient than using the delayed attribute since it does not require the simulator to create an additional signal.

The setup/hold checking process uses two local variables to save the time of the last data and clock event respectively. This process assumes that the data must be settled before the latch control becomes active high. Adjustment to the following expression is required in order to change the latching control characteristics.

$$(clk'EVENT) AND (clk =' 1')$$

Events from the data and control inputs **load**, **ent**, **enp**, **a**, **b**, **c** and **d** are treated as a single data event. A more sophisticated version of this error check (with additional generic parameters) could be crafted which tests for setup/hold errors on any of the data inputs separately through the use of multiple local variables, one for each data input. Most data books do not specify this level of detail in timing and therefore this approach has not been taken.

The input delay processing section of the model incorporates the appropriate delay for each of the input ports. An internal signal for each input port is declared. The main device function process utilizes the delayed signals rather than the primary inputs.

The counter operation is handled by a single process which is sensitive to the input delayed signals **clkdelay**. The following conditions are checked in order whenever the asynchronous control or address inputs have an event

- Unknown control
- Unknown clock
- Clocked parallel load
- Clocked count

Once the state of the device is determined (and stored in the local variable **istate**, this value is converted to a logic array **bstate** using the **f_inttologic** function (see section 7.7 for detailed description of this function). The logic array value **bstate** is then used in the output signal assignment statements for **qa, qb, qc** and **qd** are executed. The carry out assignment to **rco** is also executed. Output state changes are executed with appropriate delay introduced through the use of the **f_delay** function as discussed in section 7.6. This function utilizes the output rise and fall delays and based on the new state value chooses the proper delay.

Chapter 5
Memory Devices

This chapter discusses memories, programmable logic arrays, and related devices. These types of devices have the characteristic that they require considerably more internal state to represent their behavior than other types of device models.

In order to ease the modelling tasks for these types of models, a set of utility functions are utilized throughout this chapter as described here. The following shows the definitions for **memory** which is used to define a memory array, **productline** and **inputline** which are both used to represent PAL and PLD programmed arrays.

Memory modelling types

```
-- memory array type definition
TYPE memory IS ARRAY(INTEGER RANGE <>) OF integer;

-- pal array type definitions
TYPE inputline IS ARRAY(1 TO 32) OF boolean;
TYPE productline IS ARRAY(INTEGER RANGE <>) OF inputline;

-- alternative:
--   TYPE productline IS ARRAY(INTEGER RANGE <>,INTEGER RANGE <>) OF boolean;
```

The definition shown for **inputline** uses a fixed size, although VHDL will support the alternate more general approach shown in comments above.

5.1 Memory Initialization

Initialization of memory models is best handled through the use of data files. Since more than one instance of the same memory model can occur in a design, and since each instance can have different memory values, the use of a separate data file for each memory device during circuit power-up is used. The text I/O facilities of VHDL are effective for this purpose. The following

Initializing ROM Memory

```
PROCEDURE f_rominit(rom : INOUT memory; romdef : IN TEXT) IS
    VARIABLE l : line;              -- current input line from file
    VARIABLE j : integer;           -- rom index
    BEGIN
        FOR j IN rom'RANGE LOOP     -- initialize the rom
            READLINE(romdef,l);
            READ(l,rom(j));
        END LOOP;
    END f_rominit;
```

shows the definition of the **f_rominit** procedure which initializes a ROM model memory to its power-up state. This procedure reads from an open text file which contains 1 line per word, 1 integer value per line, representing the word value in memory. For example:

ROM Memory Data File

0
1
2
3
4
5
6
7

shows the memory data file for an 8 word by 3 bit memory. The variables for a ROM model are shown here

Memory variable declarations

```
VARIABLE rom : memory(1 TO 256);
FILE romdef : TEXT IS IN "tbp24s10.def";   -- rom def file
VARIABLE startup : boolean := true;        -- rom need initializing?
```

The **rom** variable contains the memory state and constraints the array type **memory** with the appropriate number of words. The **romdef** variable is declared as a text file with its associated data file name. The use of a string generic parameter allows different instances of this model to have different data files associated with them. Finally, the **startup** variable is used as follows

5.1. MEMORY INITIALIZATION

Memory modelling types

```
IF startup THEN                 -- initialize rom, once only
   f_rominit(rom,romdef);
   startup := false;
END IF;
```

to insure that the **f_rominit** procedure is called only once during simulator initialization.

The following shows the definition of the **f_palinit** which is similar to the **f_rominit** procedure but customized for initializing PAL and PLD devices.

Initializing PAL/PLD Memory

```
PROCEDURE f_palinit(pal : INOUT productline;
                   paldef : IN TEXT) IS
    VARIABLE l : line;           -- current input line from file
    VARIABLE j,k : integer;      -- pal indexes
    VARIABLE c : character;      -- current input character
    BEGIN
        -- initialize the pal
        FOR j IN pal'RANGE LOOP
            -- get next product line
            READLINE(paldef,l);

            -- loop through input values
            FOR k IN pal(j)'RANGE LOOP
                READ(l,c);
                IF c = '1' THEN pal(j)(k) := true;
                ELSE pal(j)(k) := false;
                END IF;
            END LOOP;
        END LOOP;
    END f_palinit;
```

The data file format for this procedure is

- Assumes n (words) by m (bits per word) logic array
- n lines in file, 1 per word
- m characters per line, 1 per bit
- '1' represents wired array connection
- '0' represents open array connection

as shown in the following example

PLA Data File Example

```
10000000000000000000000000000000      enable o2
00100000000000000000000000000000      o2 <= i1
00000000000000000000000000000010      grounded
00000000000000000000000000000010      grounded
00000000000000000000000000000010      grounded
00000000000000000000000000000010      grounded
00000000000000000000000000000010      grounded
00000000000000000000000000000010      grounded
```

which shows part of a PLA data file. The word size contains 32 bits, often referred to as the PLA input lines. In this case there are 64 lines in the file, representing the product lines for the PLA. Comments are allowed beyond the 32 characters used for array entries as shown. The variables for a PLA model are shown here

PLA/PLD variable declarations

```
VARIABLE pal : productline(1 TO 64);
-- Alternative: VARIABLE pal : productline(1 TO 64,1 TO 32);
FILE paldef : TEXT IS IN "pal1618.def";    -- pal def file
VARIABLE startup : boolean := true;        -- pal need initializing?
```

The **pal** variable contains the memory state and constraints the array type **productline** with the appropriate number of words. The alternative shown can be used with the more general two dimensional unconstrained array definition discussed above. The **paldef** variable is declared as a text file with its associated data file name. The use of a string generic parameter allows different instances of this model to have different data files associated with them. Finally, the **startup** variable is used as follows

Memory modelling types

```
-- initialize pal, once only
IF startup THEN
    f_palinit(pal,paldef);
    startup := false;
END IF;
```

to insure that the **f_palinit** procedure is called only once during simulator initialization.

5.2 Read Only Memories

The following models describe various types of ROM devices. These models follow a standard format and all are presented with full timing functionality. It is possible to reduce the simulation and model complexity in stages as follows

- Remove input delay processing
    - Remove all input timing generic parameters
    - Remove all internal delay signals
    - Remove the input delay signal assignment statements
    - Rename all references in the main behavioral process from the delayed port names to primary input port names
- Remove error checks: delete the error checking processes which occur prior to the main behavioral process
- Use simpler output delay calculations: adjust the final output assignment statements

By methodically adjusting these major components of the models it is possible to build a wide range of models which fit into the model accuracy continuum discussed in section 2.5.

The following adaptations are possible starting from these base models:

- Change size of memory by changing the ROM memory variable declaration. Adjustment of the address width will be required as well, along with word width adjustments. The initialization code does not need to change for word widths which range between 1 and 32 bits.
- Enable/disable options can be adjusted by altering the control logic of the main memory operation process. Two options are shown in the following examples, one is an asynchronous memory, the other is synchronous.

5.2.1 1024 bit (256 by 4) ROM

This programmable read-only memory provides 1024 bits of memory organized as 256 words of 4-bits each. High values at the **g1** and **g2** inputs enables all outputs. A low value at either enable input causes all outputs to be in the tristate off condition. Figure 5.1 shows the symbolic representation of this device. This device corresponds to the Texas Instruments TBP24S10 series TTL parts [TI86]. The following shows the full model for a read only memory device, with characteristics

- Total 1024 bits of memory

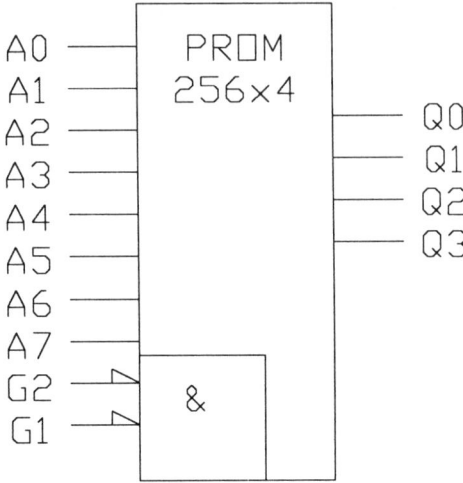

Figure 5.1: Logic Diagram 1024 Bit ROM

- 256 words of 4 bits each
- Asynchronous address
- Two asynchronous enable
- Tristate outputs

1024 bit (256 by 4) ROM

```
USE std.std_logic.ALL;
USE std.std_ttl.ALL;
USE std.textio.ALL;
USE work.palpack.ALL;
ENTITY tbp24s10 IS
    GENERIC (a0_i01,a0_i10,
             a1_i01,a1_i10,
             a2_i01,a2_i10,
             a3_i01,a3_i10,
             a4_i01,a4_i10,
             a5_i01,a5_i10,
             a6_i01,a6_i10,
             a7_i01,a7_i10,
             g1_i01,g1_i10,
             g2_i01,g2_i10,
             q0_o01,q0_o10,q0_oz0,q0_oz1,q0_o0z,q0_o1z,
             q1_o01,q1_o10,q1_oz0,q1_oz1,q1_o0z,q1_o1z,
             q2_o01,q2_o10,q2_oz0,q2_oz1,q2_o0z,q2_o1z,
             q3_o01,q3_o10,q3_oz0,q3_oz1,q3_o0z,q3_o1z : TIME := 2 ns);
```

5.2. READ ONLY MEMORIES

```
    PORT    (a0,a1,a2,a3,a4,a5,a6,a7 : IN t_wlogic;
             g2,g1 : IN t_wlogic;
             q0,q1,q2,q3 : INOUT t_wlogic);
END tbp24s10;

ARCHITECTURE full OF tbp24s10 IS
    -- input delay signals
    SIGNAL a0delay,a1delay,a2delay,a3delay,a4delay,
           a5delay,a6delay,a7delay,g1delay,g2delay : t_wlogic;
BEGIN
    -- input delay processing
    a0delay <= a0 AFTER f_delay(a0,a0_i01,a0_i10);
    a1delay <= a1 AFTER f_delay(a1,a1_i01,a1_i10);
    a2delay <= a2 AFTER f_delay(a2,a2_i01,a2_i10);
    a3delay <= a3 AFTER f_delay(a3,a3_i01,a3_i10);
    a4delay <= a4 AFTER f_delay(a4,a4_i01,a4_i10);
    a5delay <= a5 AFTER f_delay(a5,a5_i01,a5_i10);
    a6delay <= a6 AFTER f_delay(a6,a6_i01,a6_i10);
    a7delay <= a7 AFTER f_delay(a7,a7_i01,a7_i10);
    g1delay <= g1 AFTER f_delay(g1,g1_i01,g1_i10);
    g2delay <= g2 AFTER f_delay(g2,g2_i01,g2_i10);

    -- memory operation
    PROCESS (a0delay,a1delay,a2delay,a3delay,
             a4delay,a5delay,a6delay,a7delay,
             g1delay,g2delay)
        VARIABLE rom : memory(1 TO 256);
        FILE romdef : TEXT IS IN "tbp24s10.def";   -- rom def file
        VARIABLE startup : boolean := true;        -- rom need initializing?
        VARIABLE baddress : t_logarray(1 TO 8);
        VARIABLE iaddress : integer;
        VARIABLE unknown : boolean;
        VARIABLE word : t_logarray(1 TO 4);
    BEGIN
    -- initialize rom, once only
    IF startup THEN
        f_rominit(rom,romdef);
        startup := false;
    END IF;

    -- device enabled?
    IF (g1delay = '1') OR (g2delay = '1') THEN
        -- calculate address
        baddress := (a7delay,a6delay,a5delay,a4delay,
                     a3delay,a2delay,a1delay,a0delay);
        f_logictoint(baddress,unknown,iaddress);

        -- unknown address?
        IF unknown THEN
            -- unknown output
            q0 <= FX AFTER
```

```
              f_zdelay(q0,FX,q0_o01,q0_o10,q0_oz0,q0_oz1,q0_o0z,q0_o1z);
            q1 <= FX AFTER
              f_zdelay(q1,FX,q1_o01,q1_o10,q1_oz0,q1_oz1,q1_o0z,q1_o1z);
            q2 <= FX AFTER
              f_zdelay(q2,FX,q2_o01,q2_o10,q2_oz0,q2_oz1,q2_o0z,q2_o1z);
            q3 <= FX AFTER
              f_zdelay(q3,FX,q3_o01,q3_o10,q3_oz0,q3_oz1,q3_o0z,q3_o1z);
        ELSE
            -- pickup the memory value
            f_inttologic(rom(iaddress),word,ttl);
            q0 <= word(4) AFTER f_zdelay(
              q0,word(4),q0_o01,q0_o10,q0_oz0,q0_oz1,q0_o0z,q0_o1z);
            q1 <= word(3) AFTER f_zdelay(
              q1,word(3),q1_o01,q1_o10,q1_oz0,q1_oz1,q1_o0z,q1_o1z);
            q2 <= word(2) AFTER f_zdelay(
              q2,word(2),q2_o01,q2_o10,q2_oz0,q2_oz1,q2_o0z,q2_o1z);
            q3 <= word(1) AFTER f_zdelay(
              q3,word(1),q3_o01,q3_o10,q3_oz0,q3_oz1,q3_o0z,q3_o1z);
        END IF;

    -- device disabled
    ELSE
        q0 <= ZX AFTER f_zdelay(
          q0,ZX,q0_o01,q0_o10,q0_oz0,q0_oz1,q0_o0z,q0_o1z);
        q1 <= ZX AFTER f_zdelay(
          q1,ZX,q1_o01,q1_o10,q1_oz0,q1_oz1,q1_o0z,q1_o1z);
        q2 <= ZX AFTER f_zdelay(
          q2,ZX,q2_o01,q2_o10,q2_oz0,q2_oz1,q2_o0z,q2_o1z);
        q3 <= ZX AFTER f_zdelay(
          q3,ZX,q3_o01,q3_o10,q3_oz0,q3_oz1,q3_o0z,q3_o1z);
    END IF;
    END PROCESS;
END full;

CONFIGURATION parts OF tbp24s10 IS
FOR full
END FOR;
END parts;
```

The ports for this model are

- a0, a1, a2, a3, a4, a5, a6, a7 - address inputs
- g1, g2 - memory enable, both must be high for memory to operate, otherwise tristate outputs are disabled
- q0, q1, q2, q3 - tristate outputs from memory

Generic parameters to this model are summarized here

- a0_i01, a0_i10 - low to high and high to low a0 input port delays
- a1_i01, a1_i10 - low to high and high to low a1 input port delays
- a2_i01, a2_i10 - low to high and high to low a2 input port delays

5.2. READ ONLY MEMORIES

- a3_i01, a3_i10 - low to high and high to low **a3** input port delays
- a4_i01, a4_i10 - low to high and high to low **a4** input port delays
- a5_i01, a5_i10 - low to high and high to low **a5** input port delays
- a6_i01, a6_i10 - low to high and high to low **a6** input port delays
- a7_i01, a7_i10 - low to high and high to low **a7** input port delays
- g1_i01, g1_i10 - low to high and high to low **g1** input port delays
- g2_i01, g2_i10 - low to high and high to low **g2** input port delays
- q0_o01, q0_o10, q0_oz0, q0_oz1, q0_o0z, q0_o1z - low to high, high to low, off to low, off to high, low to off, high to off **q0** output port delays
- q1_o01, q1_o10, q1_oz0, q1_oz1, q1_o0z, q1_o1z - low to high, high to low, off to low, off to high, low to off, high to off **q1** output port delays
- q2_o01, q2_o10, q2_oz0, q2_oz1, q2_o0z, q2_o1z - low to high, high to low, off to low, off to high, low to off, high to off **q2** output port delays
- q3_o01, q3_o10, q3_oz0, q3_oz1, q3_o0z, q3_o1z - low to high, high to low, off to low, off to high, low to off, high to off **q3** output port delays

This model has no error checks since it operates entirely asynchronously.

The input delay processing section of the model incorporates the appropriate delay for each of the input ports. An internal signal for each input port is declared. The main device function process utilizes the delayed signals rather than the primary inputs.

The memory operation is handled by a single process which is sensitive to the input delayed signals **a0delay, a1delay, a2delay, a3delay, a4delay, a5delay, a6delay, a7delay, g1delay** and **g2delay**. The **startup** variable is used to insure that the ROM initialization occurs only once by calling **f_rominit** (as discussed in section 5.1). The two enable inputs **g1delay** and **g2delay** are checked whenever an asynchronous input changes, and if both high, the following actions are taken:

- Calculate address as an integer value using **f_logictoint** (this function is described in more detail in section 7.7).
- If any of the address inputs are unknown, generate unknowns for all outputs.
- Otherwise use the integer address to lookup memory value. Convert this memory value to individual bits using the **f_inttologic** procedure (as described in section 7.7).

If the device is not enabled, then high-impedance outputs are generated. All of the above state changes are executed with appropriate delay introduced through the use of the **f_delay** function as discussed in section 7.6. This function utilizes the output rise and fall delays and based on the new state value chooses the proper delay.

5.2.2 16,384 bit (4096 by 4) register PROM

This programmable read-only memory provides 16,384 bits of memory organized as 4096 words of 4-bits each. The output register receives data from the PROMP array on the rising edge of **rclk**. A high value at the **g** input enables all outputs. A low value causes all outputs to be in the tristate off condition. Figure 5.2 shows the symbolic representation of this device.

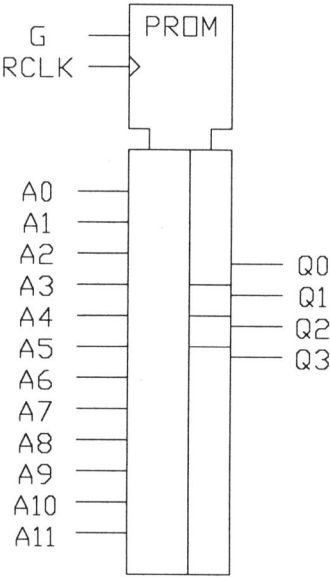

Figure 5.2: Logic Diagram 16384 Bit PROM

This device corresponds to the Texas Instruments TBP34R162 series TTL parts [TI86]. The following shows the full model for a read only memory device, with characteristics

- Total 16,384 bits of memory
- 4096 words of 4 bits each
- Asynchronous address input

5.2. READ ONLY MEMORIES

- One asynchronous enable
- Tristate outputs
- Clocked output

16,384 bit (4096 by 4) register PROM

```
USE std.std_logic.ALL;
USE std.std_ttl.ALL;
USE std.textio.ALL;
USE work.palpack.ALL;
ENTITY tbp34r162 IS
    GENERIC (a0_i01,a0_i10,
             a1_i01,a1_i10,
             a2_i01,a2_i10,
             a3_i01,a3_i10,
             a4_i01,a4_i10,
             a5_i01,a5_i10,
             a6_i01,a6_i10,
             a7_i01,a7_i10,
             a8_i01,a8_i10,
             a9_i01,a9_i10,
             a10_i01,a10_i10,
             a11_i01,a11_i10,
             g_i01,g_i10,
             rclk_i01,rclk_i10,
             rclk_min,
             addr_setup,addr_hold,
             q0_o01,q0_o10,q0_oz0,q0_oz1,q0_o0z,q0_o1z,
             q1_o01,q1_o10,q1_oz0,q1_oz1,q1_o0z,q1_o1z,
             q2_o01,q2_o10,q2_oz0,q2_oz1,q2_o0z,q2_o1z,
             q3_o01,q3_o10,q3_oz0,q3_oz1,q3_o0z,q3_o1z : TIME := 2 ns);

    PORT    (a0,a1,a2,a3,a4,a5,a6,a7,a8,a9,a10,a11 : IN t_wlogic;
             g,rclk : IN t_wlogic;
             q0,q1,q2,q3 : INOUT t_wlogic);
END tbp34r162;

ARCHITECTURE full OF tbp34r162 IS
    -- input delay signals
    SIGNAL a0delay,a1delay,a2delay,a3delay,
           a4delay,a5delay,a6delay,a7delay,
           a8delay,a9delay,a10delay,a11delay,
           gdelay,rclkdelay : t_wlogic;
BEGIN
    -- check rclk frequency/spike detection
    PROCESS (rclk)
        VARIABLE rclklastev : TIME := 0 ns;
        BEGIN
        ASSERT (NOW = 0 ns) OR ((NOW - rclklastev) >= rclk_min)
            REPORT "Spike detected on clock" SEVERITY warning;
```

```
            rclklastev := NOW;
        END PROCESS;

    -- check for setup/hold violations
    PROCESS (rclk,a0,a1,a2,a3,a4,a5,a6,a7,a8,a9,a10,a11)
        VARIABLE addrlastev : TIME := 0 ns;
        VARIABLE rclklastev : TIME := 0 ns;
        BEGIN
        -- Hold check
        IF (a0'EVENT)OR(a1'EVENT)OR(a2'EVENT)OR(a3'EVENT)OR
           (a4'EVENT)OR(a5'EVENT)OR(a6'EVENT)OR(a7'EVENT)OR
           (a8'EVENT)OR(a9'EVENT)OR(a10'EVENT)OR(a11'EVENT) THEN
            ASSERT (NOW = 0 ns) OR ((NOW - rclklastev) >= addr_hold)
                REPORT "Hold error" SEVERITY warning;
            addrlastev := NOW;
        END IF;

        -- Setup check
        IF (rclk'EVENT) AND (rclk = '1') THEN
            ASSERT (NOW = 0 ns) OR ((NOW - addrlastev) >= addr_setup)
                REPORT "Setup error" SEVERITY warning;
            rclklastev := NOW;
        END IF;
        END PROCESS;

    -- input delay processing
    a0delay <= a0 AFTER f_delay(a0,a0_i01,a0_i10);
    a1delay <= a1 AFTER f_delay(a1,a1_i01,a1_i10);
    a2delay <= a2 AFTER f_delay(a2,a2_i01,a2_i10);
    a3delay <= a3 AFTER f_delay(a3,a3_i01,a3_i10);
    a4delay <= a4 AFTER f_delay(a4,a4_i01,a4_i10);
    a5delay <= a5 AFTER f_delay(a5,a5_i01,a5_i10);
    a6delay <= a6 AFTER f_delay(a6,a6_i01,a6_i10);
    a7delay <= a7 AFTER f_delay(a7,a7_i01,a7_i10);
    a8delay <= a8 AFTER f_delay(a8,a8_i01,a8_i10);
    a9delay <= a9 AFTER f_delay(a9,a9_i01,a9_i10);
    a10delay <= a10 AFTER f_delay(a10,a10_i01,a10_i10);
    a11delay <= a11 AFTER f_delay(a11,a11_i01,a11_i10);
    gdelay <= g AFTER f_delay(g,g_i01,g_i10);
    rclkdelay <= rclk AFTER f_delay(rclk,rclk_i01,rclk_i10);

    -- memory operation
    PROCESS (g,rclk)
        VARIABLE rom : memory(1 TO 4096);
        FILE romdef : TEXT IS IN "tbp34r162.def";   -- rom def file
        VARIABLE startup : boolean := true;         -- rom need initializing?
        VARIABLE baddress : t_logarray(1 TO 12);
        VARIABLE iaddress : integer;
        VARIABLE unknown : boolean;
        VARIABLE word : t_logarray(1 TO 4);
    BEGIN
    -- initialize rom, once only
```

5.2. READ ONLY MEMORIES

```
      IF startup THEN
         f_rominit(rom,romdef);
         startup := false;
      END IF;

      -- device enabled?
      IF (gdelay = '1') THEN
         -- calculate address
         baddress := (a11delay,a10delay,a9delay,a8delay,
                      a7delay,a6delay,a5delay,a4delay,
                      a3delay,a2delay,a1delay,a0delay);
         f_logictoint(baddress,unknown,iaddress);

         -- unknown address?
         IF unknown THEN
            -- unknown output
            q0 <= FX AFTER f_zdelay(
               q0,FX,q0_o01,q0_o10,q0_oz0,q0_oz1,q0_o0z,q0_o1z);
            q1 <= FX AFTER f_zdelay(
               q1,FX,q1_o01,q1_o10,q1_oz0,q1_oz1,q1_o0z,q1_o1z);
            q2 <= FX AFTER f_zdelay(
               q2,FX,q2_o01,q2_o10,q2_oz0,q2_oz1,q2_o0z,q2_o1z);
            q3 <= FX AFTER f_zdelay(
               q3,FX,q3_o01,q3_o10,q3_oz0,q3_oz1,q3_o0z,q3_o1z);
         ELSE
            -- pickup the memory value
            f_inttologic(rom(iaddress),word,ttl);
            q0 <= word(4) AFTER f_zdelay(
               q0,word(4),q0_o01,q0_o10,q0_oz0,q0_oz1,q0_o0z,q0_o1z);
            q1 <= word(3) AFTER f_zdelay(
               q1,word(3),q1_o01,q1_o10,q1_oz0,q1_oz1,q1_o0z,q1_o1z);
            q2 <= word(2) AFTER f_zdelay(
               q2,word(2),q2_o01,q2_o10,q2_oz0,q2_oz1,q2_o0z,q2_o1z);
            q3 <= word(1) AFTER f_zdelay(
               q3,word(1),q3_o01,q3_o10,q3_oz0,q3_oz1,q3_o0z,q3_o1z);
         END IF;

      -- device disabled
      ELSE
         q0 <= ZX AFTER f_zdelay(
            q0,ZX,q0_o01,q0_o10,q0_oz0,q0_oz1,q0_o0z,q0_o1z);
         q1 <= ZX AFTER f_zdelay(
            q1,ZX,q1_o01,q1_o10,q1_oz0,q1_oz1,q1_o0z,q1_o1z);
         q2 <= ZX AFTER f_zdelay(
            q2,ZX,q2_o01,q2_o10,q2_oz0,q2_oz1,q2_o0z,q2_o1z);
         q3 <= ZX AFTER f_zdelay(
            q3,ZX,q3_o01,q3_o10,q3_oz0,q3_oz1,q3_o0z,q3_o1z);
      END IF;
   END PROCESS;
END full;

CONFIGURATION parts OF tbp34r162 IS
```

```
FOR full
END FOR;
END parts;
```

The ports for this model are

- **a0, a1, a2, a3, a4, a5, a6, a7, a8, a9, a10, a11** - address inputs
- **g** - memory enable, must be high for memory to operate, otherwise tristate outputs are disabled
- **q0, q1, q2, q3** - tristate outputs from memory
- **rclk** - input clock for updating outputs

Generic parameters to this model are summarized here

- **a0_i01, a0_i10** - low to high and high to low **a0** input port delays
- **a1_i01, a1_i10** - low to high and high to low **a1** input port delays
- **a2_i01, a2_i10** - low to high and high to low **a2** input port delays
- **a3_i01, a3_i10** - low to high and high to low **a3** input port delays
- **a4_i01, a4_i10** - low to high and high to low **a4** input port delays
- **a5_i01, a5_i10** - low to high and high to low **a5** input port delays
- **a6_i01, a6_i10** - low to high and high to low **a6** input port delays
- **a7_i01, a7_i10** - low to high and high to low **a7** input port delays
- **a8_i01, a8_i10** - low to high and high to low **a8** input port delays
- **a9_i01, a9_i10** - low to high and high to low **a9** input port delays
- **a10_i01, a10_i10** - low to high and high to low **a10** input port delays
- **a11_i01, a11_i10** - low to high and high to low **a11** input port delays
- **g_i01, g_i10** - low to high and high to low **g** input port delays
- **rclk_i01, rclk_i10** - low to high and high to low **rclk** input port delays
- Minimum **rclk** pulse width **rclk_min**
- Address setup **addr_setup** and hold **addr_hold** minimums
- **q0_o01, q0_o10, q0_oz0, q0_oz1, q0_o0z, q0_o1z** - low to high, high to low, off to low, off to high, low to off, high to off **q0** output port delays
- **q1_o01, q1_o10, q1_oz0, q1_oz1, q1_o0z, q1_o1z** - low to high, high to low, off to low, off to high, low to off, high to off **q1** output port delays

5.2. READ ONLY MEMORIES

- q2_o01, q2_o10, q2_oz0, q2_oz1, q2_o0z, q2_o1z - low to high, high to low, off to low, off to high, low to off, high to off **q2** output port delays

- q3_o01, q3_o10, q3_oz0, q3_oz1, q3_o0z, q3_o1z - low to high, high to low, off to low, off to high, low to off, high to off **q3** output port delays

This model features three error checks

- Clock frequency/spike detection
- Address setup time checking
- Address hold time checking

as shown in the error checking section of the architecture. Each of the error checks utilizes a separate process statement (the exception is the hold/setup checks which utilizes a single process).

The spike detection process uses a local variable to save the previous event time for the signal being checked. This is somewhat more efficient than using the delayed attribute since it does not require the simulator to create an additional signal.

The setup/hold checking process uses two local variables **addrlastev** and **rclklastev** to save the time of the last data and clock event respectively. This process assumes a positive edge triggered clock. Adjustment to the expression

$$(rclk'EVENT)AND(rclk ='1')$$

is required in order to change the clock characteristics.

The input delay processing section of the model incorporates the appropriate delay for each of the input ports. An internal signal for each input port is declared. The main device function process utilizes the delayed signals rather than the primary inputs.

The memory operation is handled by a single process which is sensitive to the input delayed signals **g** and **rclk**. The **startup** variable is used to insure that the ROM initialization occurs only once by calling **f_rominit** (as discussed in section 5.1). The enable input **gdelay** is checked whenever it or the clock changes, and if high, the following actions are taken:

- Calculate address as an integer value using **f_logictoint** (this function is described in more detail in section 7.7). The **baddress** temporary variable is used to pass the address bit values to this call. The **iaddress** variable returns the address integer value.

- If any of the address inputs are unknown (the **unknown** variable is returned as true), generate unknowns for all outputs.

- Otherwise use the integer address to lookup memory value. Convert this memory value to individual bits using the **f_inttologic** procedure (as described in section 7.7). The memory value is determined with the expression *rom(iaddress)* and the bit values of the word are returned in the variable **word**. The conversion is performed using the TTL technology rules.

If the device is not enabled, then high-impedance outputs are generated. All of the above state changes are executed with appropriate delay introduced through the use of the **f_delay** function as discussed in section 7.6. This function utilizes the output rise and fall delays and based on the new state value chooses the proper delay.

5.3 Random Access Memories

The following models describe various types of random access memory devices. These models follow a standard format and all are presented with full timing functionality. It is possible to reduce the simulation and model complexity in stages as follows

- Remove input delay processing
    - Remove all input timing generic parameters
    - Remove all internal delay signals
    - Remove the input delay signal assignment statements
    - Rename all references in the main behavioral process from the delayed port names to primary input port names
- Remove error checks: delete the error checking processes which occur prior to the main behavioral process
- Use simpler output delay calculations: adjust the final output assignment statements

By methodically adjusting these major components of the models it is possible to build a wide range of models which fit into the model accuracy continuum discussed in section 2.5.

The following adaptations are possible starting from these base models:

- Change size of memory by changing the RAM memory variable declaration. Adjustment of the address width will be required as well, along with word width adjustments. The initialization code does not need to change.
- Enable/disable/read/write options can be adjusted by altering the control logic of the main memory operation process.

5.3. RANDOM ACCESS MEMORIES

5.3.1 64 bit RAM

This random access memory device provides 64 bits of memory organized as 16 words of 4-bits each. During the write cycle, input data is written into the selected address when the chip-select **s** and the write-enable **rw** are low. During this mode the outputs are placed in tristate off condition (high-impedance). During the read cycle, information in memory selected by address is placed on outputs. This mode is enabled by placing the write-enable **rw** to high, and the chip-select to low. When the chip select **s** is low the outputs will be placed in the tristate off condition. Figure 5.3

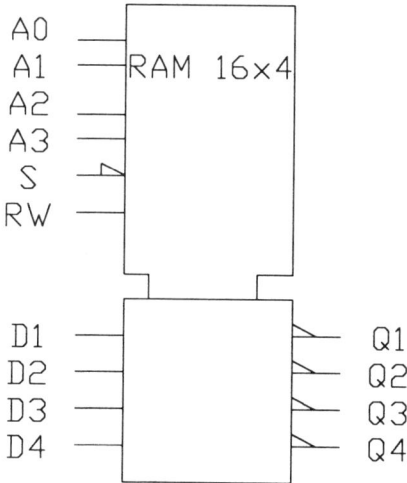

Figure 5.3: Logic Diagram 64 Bit RAM

shows the symbolic representation of this device. The following shows the function table for this device.

Function Table

function	inputs		output
	chip select	write enable	
write	L	L	Z
read	L	H	complement of data entered
inhibit	H	X	Z

The following shows the full model for a random access memory device, with characteristics

- Total 64 bits of memory
- 16 words of 4 bits each
- Asynchronous address
- Asynchronous data
- Asynchronous enable
- Asynchronous read/write control
- Tristate outputs

64 bit RAM

```
USE std.std_logic.ALL;
USE std.std_ttl.ALL;
USE std.textio.ALL;
USE work.palpack.ALL;
ENTITY ram64 IS
    GENERIC (a0_i01,a0_i10,
             a1_i01,a1_i10,
             a2_i01,a2_i10,
             a3_i01,a3_i10,
             addr_setup,addr_hold,

             s_i01,s_i10,
             s_setup,s_hold,
             rw_i01,rw_i10,
             rw_min,

             d1_i01,d1_i10,
             d2_i01,d2_i10,
             d3_i01,d3_i10,
             d4_i01,d4_i10,
             data_setup,data_hold,

             q1_o01,q1_o10,q1_oz0,q1_oz1,q1_o0z,q1_o1z,
             q2_o01,q2_o10,q2_oz0,q2_oz1,q2_o0z,q2_o1z,
             q3_o01,q3_o10,q3_oz0,q3_oz1,q3_o0z,q3_o1z,
             q4_o01,q4_o10,q4_oz0,q4_oz1,q4_o0z,q4_o1z : TIME := 2 ns);
    PORT   (a0,a1,a2,a3 : IN t_wlogic;
            s, rw : IN t_wlogic;
            d1,d2,d3,d4 : IN t_wlogic;
            q1,q2,q3,q4 : INOUT t_wlogic);
END ram64;

ARCHITECTURE full OF ram64 IS
    -- input delay signals
    SIGNAL a0delay,a1delay,a2delay,a3delay,
           sdelay,rwdelay,
           d1delay,d2delay,d3delay,d4delay : t_wlogic;
```

5.3. RANDOM ACCESS MEMORIES

```
BEGIN
   -- check rw frequency/spike detection
   PROCESS (rw)
      VARIABLE rwlastev : TIME := 0 ns;
      BEGIN
      ASSERT (NOW = 0 ns) OR ((NOW - rwlastev) >= rw_min)
         REPORT "Spike detected on write enable" SEVERITY warning;
      rwlastev := NOW;
      END PROCESS;

   -- check for setup/hold violations
   PROCESS (rw,a0,a1,a2,a3)
      VARIABLE addrlastev : TIME := 0 ns;
      VARIABLE rwlastev : TIME := 0 ns;
      BEGIN
      -- Hold check
      IF (a0'EVENT)OR(a1'EVENT)OR(a2'EVENT)OR(a3'EVENT) THEN
         ASSERT (NOW = 0 ns) OR ((NOW - rwlastev) >= addr_hold)
            REPORT "Address hold error" SEVERITY warning;
         addrlastev := NOW;
      END IF;

      -- Setup check
      IF (rw'EVENT) AND (rw = '0') THEN
         ASSERT (NOW = 0 ns) OR ((NOW - addrlastev) >= addr_setup)
            REPORT "Address setup error" SEVERITY warning;
      -- Get rw rise for hold check
      ELSIF (rw'EVENT) AND (rw = '1') THEN rwlastev := NOW;
      END IF;
      END PROCESS;

   -- check for setup/hold violations
   PROCESS (rw,d1,d2,d3,d4)
      VARIABLE datalastev : TIME := 0 ns;
      VARIABLE rwlastev : TIME := 0 ns;
      BEGIN
      -- Hold check
      IF (d1'EVENT)OR(d2'EVENT)OR(d3'EVENT)OR(d4'EVENT) THEN
         ASSERT (NOW = 0 ns) OR ((NOW - rwlastev) >= data_hold)
            REPORT "Data hold error" SEVERITY warning;
         datalastev := NOW;
      END IF;

      -- Setup check
      IF (rw'EVENT) AND (rw = '1') THEN
         ASSERT (NOW = 0 ns) OR ((NOW - datalastev) >= data_setup)
            REPORT "Data setup error" SEVERITY warning;
         rwlastev := NOW;
      END IF;
      END PROCESS;

   -- check for setup/hold violations
```

```
PROCESS (rw,s)
   VARIABLE slastev : TIME := 0 ns;
   VARIABLE rwlastev : TIME := 0 ns;
   BEGIN
   -- Hold check
   IF (s'EVENT) THEN
      ASSERT (NOW = 0 ns) OR ((NOW - rwlastev) >= s_hold)
         REPORT "Enable hold error" SEVERITY warning;
      slastev := NOW;
   END IF;

   -- Setup check
   IF (rw'EVENT) AND (rw = '1') THEN
      ASSERT (NOW = 0 ns) OR ((NOW - slastev) >= s_setup)
         REPORT "Enable setup error" SEVERITY warning;
      rwlastev := NOW;
   END IF;
   END PROCESS;

-- input delay processing
a0delay <= a0 AFTER f_delay(a0,a0_i01,a0_i10);
a1delay <= a1 AFTER f_delay(a1,a1_i01,a1_i10);
a2delay <= a2 AFTER f_delay(a2,a2_i01,a2_i10);
a3delay <= a3 AFTER f_delay(a3,a3_i01,a3_i10);
sdelay <= s AFTER f_delay(s,s_i01,s_i10);
rwdelay <= rw AFTER f_delay(rw,rw_i01,rw_i10);
d1delay <= d1 AFTER f_delay(d1,d1_i01,d1_i10);
d2delay <= d2 AFTER f_delay(d2,d2_i01,d2_i10);
d3delay <= d3 AFTER f_delay(d3,d3_i01,d3_i10);
d4delay <= d4 AFTER f_delay(d4,d4_i01,d4_i10);

-- memory operation
PROCESS (a0delay,a1delay,a2delay,a3delay,
         sdelay,rwdelay,
         d1delay,d2delay,d3delay,d4delay)
   TYPE rammemory IS ARRAY(1 TO 16) OF integer;
   VARIABLE ram : rammemory;
   VARIABLE startup : boolean := true;

   VARIABLE bwork : t_logarray(1 TO 4);
   VARIABLE iwork : integer;
   VARIABLE iaddr : integer;
   VARIABLE unknown : boolean;
BEGIN
-- initialize ram, once only
IF startup THEN
   FOR i IN ram'RANGE LOOP ram(i) := -1; END LOOP;
   startup := false;
END IF;

-- write mode?
IF (sdelay = '0') AND (rwdelay = '0') THEN
```

5.3. RANDOM ACCESS MEMORIES

```
            q1 <= ZX AFTER f_zdelay(
                q1,ZX,q1_o01,q1_o10,q1_oz0,q1_oz1,q1_o0z,q1_o1z);
            q2 <= ZX AFTER f_zdelay(
                q2,ZX,q2_o01,q2_o10,q2_oz0,q2_oz1,q2_o0z,q2_o1z);
            q3 <= ZX AFTER f_zdelay(
                q3,ZX,q3_o01,q3_o10,q3_oz0,q3_oz1,q3_o0z,q3_o1z);
            q4 <= ZX AFTER f_zdelay(
                q4,ZX,q4_o01,q4_o10,q4_oz0,q4_oz1,q4_o0z,q4_o1z);

            -- pickup address
            bwork := (a3delay,a2delay,a1delay,a0delay);
            f_logictoint(bwork,unknown,iaddr);

            -- unknown address?
            IF unknown THEN
                -- set all of memory to unknowns
                FOR i IN ram'RANGE LOOP ram(i) := -1; END LOOP;
            ELSE
                -- pickup data
                bwork := (d4delay,d3delay,d2delay,d1delay);
                f_logictoint(bwork,unknown,iwork);

                -- set memory location to new value
                ram(iaddr) := iwork;
            END IF;

    -- read mode?
    ELSIF (sdelay = '0') AND (rwdelay = '1') THEN
            -- pickup address
            bwork := (a3delay,a2delay,a1delay,a0delay);
            f_logictoint(bwork,unknown,iaddr);

            -- unknown address?
            IF unknown THEN
                -- return all unknowns
                q1 <= FX AFTER f_zdelay(
                    q1,FX,q1_o01,q1_o10,q1_oz0,q1_oz1,q1_o0z,q1_o1z);
                q2 <= FX AFTER f_zdelay(
                    q2,FX,q2_o01,q2_o10,q2_oz0,q2_oz1,q2_o0z,q2_o1z);
                q3 <= FX AFTER f_zdelay(
                    q3,FX,q3_o01,q3_o10,q3_oz0,q3_oz1,q3_o0z,q3_o1z);
                q4 <= FX AFTER f_zdelay(
                    q4,FX,q4_o01,q4_o10,q4_oz0,q4_oz1,q4_o0z,q4_o1z);
            ELSE
                -- unknown memory value?
                IF ram(iaddr) < 0 THEN
                    q1 <= FX AFTER f_zdelay(
                        q1,FX,q1_o01,q1_o10,q1_oz0,q1_oz1,q1_o0z,q1_o1z);
                    q2 <= FX AFTER f_zdelay(
                        q2,FX,q2_o01,q2_o10,q2_oz0,q2_oz1,q2_o0z,q2_o1z);
                    q3 <= FX AFTER f_zdelay(
                        q3,FX,q3_o01,q3_o10,q3_oz0,q3_oz1,q3_o0z,q3_o1z);
```

```
              q4 <= FX AFTER f_zdelay(
                  q4,FX,q4_o01,q4_o10,q4_oz0,q4_oz1,q4_o0z,q4_o1z);
            ELSE
              -- pickup memory value
              f_inttologic(ram(iaddr),bwork,ttl);
              q1 <= NOT bwork(4) AFTER f_zdelay(q1,NOT bwork(4),
                  q1_o01,q1_o10,q1_oz0,q1_oz1,q1_o0z,q1_o1z);
              q2 <= NOT bwork(3) AFTER f_zdelay(q1,NOT bwork(3),
                  q1_o01,q1_o10,q1_oz0,q1_oz1,q1_o0z,q1_o1z);
              q3 <= NOT bwork(2) AFTER f_zdelay(q2,NOT bwork(2),
                  q2_o01,q2_o10,q2_oz0,q2_oz1,q2_o0z,q2_o1z);
              q4 <= NOT bwork(1) AFTER f_zdelay(q3,NOT bwork(1),
                  q3_o01,q3_o10,q3_oz0,q3_oz1,q3_o0z,q3_o1z);
            END IF;
          END IF;

      -- inhibit mode
      ELSE
          q1 <= ZX AFTER f_zdelay(
              q1,ZX,q1_o01,q1_o10,q1_oz0,q1_oz1,q1_o0z,q1_o1z);
          q2 <= ZX AFTER f_zdelay(
              q2,ZX,q2_o01,q2_o10,q2_oz0,q2_oz1,q2_o0z,q2_o1z);
          q3 <= ZX AFTER f_zdelay(
              q3,ZX,q3_o01,q3_o10,q3_oz0,q3_oz1,q3_o0z,q3_o1z);
          q4 <= ZX AFTER f_zdelay(
              q4,ZX,q4_o01,q4_o10,q4_oz0,q4_oz1,q4_o0z,q4_o1z);
      END IF;
    END PROCESS;
END full;

CONFIGURATION parts OF ram64 IS
  FOR full
  END FOR;
END parts;
```

The ports for this model are

- a0, a1, a2, a3 - address inputs
- s - memory enable
- rw - read/write mode control
- d1, d2, d3, d4 - data inputs
- q1, q2, q3, q4 - tristate outputs

Generic parameters to this model are summarized here

- a1_i01, a1_i10 - low to high and high to low **a1** input port delays
- a2_i01, a2_i10 - low to high and high to low **a2** input port delays
- a3_i01, a3_i10 - low to high and high to low **a3** input port delays
- a4_i01, a4_i10 - low to high and high to low **a4** input port delays

5.3. RANDOM ACCESS MEMORIES

- Address setup **addr_setup** and hold **addr_hold** minimums
- **s_i01, s_i10** - low to high and high to low **s** input port delays
- Enable setup **s_setup** and hold **s_hold** minimums
- **rw_i01, rw_i10** - low to high and high to low **rw** input port delays
- Mode change setup **rw_setup** and hold **rw_hold** minimums
- Minimum **rw** pulse width **rw_min**
- **d1_i01, d1_i10** - low to high and high to low **d1** input port delays
- **d2_i01, d2_i10** - low to high and high to low **d2** input port delays
- **d3_i01, d3_i10** - low to high and high to low **d3** input port delays
- **d4_i01, d4_i10** - low to high and high to low **d4** input port delays
- Data change setup **data_setup** and hold **data_hold** minimums
- **q1_o01, q1_o10, q1_oz0, q1_oz1, q1_o0z, q1_o1z** - low to high, high to low, off to low, off to high, low to off, high to off **q1** output port delays
- **q2_o01, q2_o10, q2_oz0, q2_oz1, q2_o0z, q2_o1z** - low to high, high to low, off to low, off to high, low to off, high to off **q2** output port delays
- **q3_o01, q3_o10, q3_oz0, q3_oz1, q3_o0z, q3_o1z** - low to high, high to low, off to low, off to high, low to off, high to off **q3** output port delays
- **q4_o01, q4_o10, q4_oz0, q4_oz1, q4_o0z, q4_o1z** - low to high, high to low, off to low, off to high, low to off, high to off **q4** output port delays

This model features seven error checks

- Read/write mode control frequency/spike detection
- Address setup time checking
- Address hold time checking
- Mode setup time checking
- Mode hold time checking
- Enable setup time checking
- Enable hold time checking

as shown in the error checking section of the architecture. Each of the error checks utilizes a separate process statement (the exception is the hold/setup checks which utilize a single process).

Each of the spike detection processes uses a local variable to save the previous event time for the signal being checked. This is somewhat more efficient than using the delayed attribute since it does not require the simulator to create an additional signal.

The setup/hold checking process uses two local variables to save the time of the last data and clock event respectively. For address checking, the mode event triggers a clock event, and a change in the address lines triggers a data event. The setup check occurs in relation to the rising edge of the **rw** mode signal, while the hold check occurs in relation to the falling edge of the **rw** signal. The other two setup/hold processes are triggered by a rising edge on the mode **rw** signal.

The input delay processing section of the model incorporates the appropriate delay for each of the input ports. An internal signal for each input port is declared. The main device function process utilizes the delayed signals rather than the primary inputs.

The memory operation is handled by a single process which is sensitive to the input delayed signals **a1delay**, **a2delay**, **a3delay**, **a4delay**, **sdelay**, **rwdelay d1delay**, **d2delay**, **d3delay** and **d4delay**. The **startup** variable is used to insure that the RAM initialization occurs only once. Negative values are stored in the memory to represent uninitialized or unknown data. The enable **sdelay** or mode **rwdelay** are checked whenever an asynchronous input changes and the following actions are taken:

- If write mode and device is enabled, set all outputs to tristate off (high-impedance). The input address is converted to an integer representation using the **f_logictoint** procedure; a temporary variable **bwork** is used to pass the address value as bits, and the **iaddr** variable returns the integer value; if an unknown value is detected in the address the **unknown** variable will be true. For an unknown address, all of RAM is set to unknown, otherwise the specific RAM address is set to the value of the incoming data. Similar to the address calculations, the **f_logictoint** procedure is used to convert the bit representation **bwork** of the data into an integer value **iwork**.

- If read mode and device is enabled, the address is converted to an integer value **iaddr** by passing the bit values through **bwork** into the **f_logictoint** procedure. If the address has unknowns (see **unknown**) then the outputs are set to unknown, otherwise the value associated with the memory address are generated using the **f_inttologic** which returns the bit values in **bwork** using the ttl technology rules (in this case).

- If the device is disabled, the outputs are set to tristate-off (high impedance).

In all cases, the outputs are complemented. All of the above state changes are executed with appropriate delay introduced through the use of the f_delay function as discussed in section 7.6. This function utilizes the output rise and fall delays and based on the new state value chooses the proper delay.

5.4 PALs, PLDs

The following models describe various types of programmable array devices. These models follow a standard format and all are presented with full timing functionality. It is possible to reduce the simulation and model complexity in stages as follows

- Remove input delay processing

    – Remove all input timing generic parameters

    – Remove all internal delay signals

    – Remove the input delay signal assignment statements

    – Rename all references in the main behavioral process from the delayed port names to primary input port names

- Remove error checks: delete the error checking processes which occur prior to the main behavioral process

- Use simpler output delay calculations: adjust the final output assignment statements

By methodically adjusting these major components of the models it is possible to build a wide range of models which fit into the model accuracy continuum discussed in section 2.5.

The following adaptations are possible starting from these base models:

- Adjust the dimensions of the array, both row and column by changing the **pal** memory variable declaration. The examples shown in this section assume a 32 input columns, and a 64 product rows, but these can be adjusted easily for other devices.

- The number of inputs/outputs can be adjusted, along with the input types. The following examples show direct inputs, feedback from outputs back into the inputs (input/output), and the use clocked flip-flops to store outputs.

5.4.1 Calculating Products

The following shows the definition of the **f_product** function which takes as an input, an array of bit values **inputs**, and the programmed logic array input line map **imap**.

Product calculation

```
FUNCTION f_product(inputs : IN t_logarray;
                   imap : IN inputline) RETURN t_logic IS
    VARIABLE work : t_logic;
    VARIABLE found : boolean;
BEGIN
    -- loop through each input value
    work := f_ttl('1');
    found := false;
    FOR i IN imap'RANGE LOOP
        -- active node?
        IF imap(i) THEN
            IF inputs(i) = '0' THEN work := f_ttl('0');
            ELSIF (inputs(i) = 'X')AND(work /= '0') THEN work:=f_ttl('X');
            END IF;
            found := true;
        END IF;
    END LOOP;

    -- if active input line, return resolved value
    IF found THEN RETURN work;

    -- return unknown if no active input lines
    ELSE RETURN f_ttl('X');
    END IF;
END f_product;
```

From this the anded value of the connected signals is returned. In essence this function returns a single value associated with a product line in the PAL. This function is utilized by the following procedures which consolidate these values to form a single output value for the PAL. If all array connections are open, then the function returns unknown. This may be inappropriate with certain devices in which case this can be adjusted.

The following shows the definition of the **f_lookup7** procedure which calculates the value of an output signal based on 7 product line inputs, and also returns the enable value for the tristate output based on an eighth product line input.

5.4. PALS, PLDS

PAL Output Calculation with Enable

```
PROCEDURE f_lookup7(inputs : IN t_logarray;
                   pal : IN productline;
                   entrystart : IN integer;
                   SIGNAL o : INOUT t_wlogic;
                   o_o01,o_o10,o_oz0,o_oz1,o_o0z,o_o1z : IN time) IS
    VARIABLE ovalue : t_logic;
BEGIN
    -- check the enable input
    IF f_product(inputs,pal(entrystart)) = '1' THEN
        ovalue := NOT (f_product(inputs,pal(entrystart+1)) OR
                       f_product(inputs,pal(entrystart+2)) OR
                       f_product(inputs,pal(entrystart+3)) OR
                       f_product(inputs,pal(entrystart+4)) OR
                       f_product(inputs,pal(entrystart+5)) OR
                       f_product(inputs,pal(entrystart+6)) OR
                       f_product(inputs,pal(entrystart+7))  );
        o <= ovalue AFTER f_zdelay(
            o,ovalue,o_o01,o_o10,o_oz0,o_oz1,o_o0z,o_o1z);

    -- output disabled
    ELSE o <= ZX AFTER f_zdelay(o,ZX,o_o01,o_o10,o_oz0,o_oz1,o_o0z,o_o1z);
    END IF;
END f_lookup7;
```

The input values are passed in **inputs**, the array programming is given in **pal**, and the **entrystart** gives the index for the product lines to use as follows:

- Entry *entrystart* enables product line
- Entries *entrystart* + 1 through *entrystart* + 7 ored together to determine output value

The signal o is the output signal to which the calculated value is assigned with the appropriate delay as determined by the o_o01, o_o10, o_oz0, o_oz1, o_o0z and o_o1z parameters using the **f_zdelay** function (as described in section 7.6. The following shows how the **f_lookup7** procedure is used.

Use of Enable Product Lookup

```
-- setup state array
state := (i2delay, NOT i2delay, i1delay, NOT i1delay,
          i3delay, NOT i3delay, io6,     NOT io6,
          i4delay, NOT i4delay, io5,     NOT io5,
          i5delay, NOT i5delay, io4,     NOT io4,
          i6delay, NOT i6delay, io3,     NOT io3,
          i7delay, NOT i7delay, io2,     NOT io2,
```

```
           i8delay,  NOT i8delay,  io1,       NOT io1,
           i9delay,  NOT i9delay,  i10delay,  NOT i10delay);

-- update working states and outputs
f_lookup7(state,pal,1,o2,o2_o01,o2_o10,o2_oz0,o2_oz1,o2_o0z,o2_o1z);
```

A logic array **state** is used to collect all of the input lines for the PAL. This value is passed to the procedure along with the pal programming **pal** and the entry point (1 in this case). The calculated value is assigned to the output **o2** according to the given delays.

The following shows the definition of the **f_lookup8** procedure which calculates the value of an output signal based on 8 product line inputs.

PAL Output Calculation

```
PROCEDURE f_lookup8(inputs : IN t_logarray;
                    pal : IN productline;
                    entrystart : IN integer;
                    laststate : INOUT t_logic;
                    en : IN t_wlogic;
                    SIGNAL o : OUT t_wlogic;
                    o_o01,o_o10,o_oz0,o_oz1,o_o0z,o_o1z : IN time) IS
    VARIABLE ovalue : t_logic;
BEGIN
    -- calculate new state
    ovalue := NOT (
                    f_product(inputs,pal(entrystart+0)) OR
                    f_product(inputs,pal(entrystart+1)) OR
                    f_product(inputs,pal(entrystart+2)) OR
                    f_product(inputs,pal(entrystart+3)) OR
                    f_product(inputs,pal(entrystart+4)) OR
                    f_product(inputs,pal(entrystart+5)) OR
                    f_product(inputs,pal(entrystart+6)) OR
                    f_product(inputs,pal(entrystart+7))      );

    -- output new state if enabled
    IF en = '0' THEN o <= ovalue AFTER
        f_zdelay(laststate,ovalue,o_o01,o_o10,o_oz0,o_oz1,o_o0z,o_o1z);

    -- output disabled
    ELSE o <= ZX AFTER
        f_zdelay(laststate,ZX,o_o01,o_o10,o_oz0,o_oz1,o_o0z,o_o1z);
    END IF;
    laststate := ovalue;
END f_lookup8;
```

The input values are passed in **inputs**, the array programming is given in **pal**, and the **entrystart** gives the index for the product lines to use as follows:

5.4. PALS, PLDS

- Entries *entrystart* through *entrystart* + 7 ored together to determine output value

The current value of the output **laststate** is used to determine the new output for the **o** signal. If the enable **en** is high then the output is set to high-impedance, otherwise the calculated output is assigned. Delay is determined by the **o_o01, o_o10, o_oz0, o_oz1, o_o0z** and **o_o1z** parameters using the **f_zdelay** function (as described in section 7.6. The following shows how the **f_lookup8** procedure is used.

Use of Product Lookup

```
-- setup state array
state := (i2delay, NOT i2delay, i1delay,  NOT i1delay,
         i3delay, NOT i3delay, io6,       NOT io6,
         i4delay, NOT i4delay, io5,       NOT io5,
         i5delay, NOT i5delay, io4,       NOT io4,
         i6delay, NOT i6delay, io3,       NOT io3,
         i7delay, NOT i7delay, io2,       NOT io2,
         i8delay, NOT i8delay, io1,       NOT io1,
         i9delay, NOT i9delay, i10delay,  NOT i10delay);
-- update working states and outputs
IF (clkdelay'EVENT) AND (clkdelay = '1') THEN
   f_lookup8(state,pal,9,d1,oedelay,q1,
            q1_o01,q1_o10,q1_oz0,q1_oz1,q1_o0z,q1_o1z);
```

A logic array **state** is used to collect all of the input lines for the PAL. This value is passed to the procedure along with the pal programming **pal** and the entry point (9 in this case). The calculated value is assigned to the output **q1** according to the given delays. The old value of the output is given by **d1** along with the enable **oedelay**.

5.4.2 10 input, 2 output, 6 I/O PAL

The following shows the full model for a PAL with characteristics

- 10 input lines
- 8 output lines - 2 output only, 6 bidirectional
- Asynchronous operation
- Tristate outputs

and as shown in figure 5.4.

10 input, 2 output, 6 I/O PAL

```
USE std.std_logic.ALL;
USE std.std_ttl.ALL;
USE std.textio.ALL;
USE work.palpack.ALL;
ENTITY pal1618 IS
   GENERIC (i1_io1,i1_i10,
            i2_io1,i2_i10,
            i3_io1,i3_i10,
            i4_io1,i4_i10,
            i5_io1,i5_i10,
            i6_io1,i6_i10,
            i7_io1,i7_i10,
            i8_io1,i8_i10,
            i9_io1,i9_i10,
            i10_io1,i10_i10,
            o1_oo1,o1_o10,o1_oz0,o1_oz1,o1_o0z,o1_o1z,
            o2_oo1,o2_o10,o2_oz0,o2_oz1,o2_o0z,o2_o1z,
            io1_oo1,io1_o10,io1_oz0,io1_oz1,io1_o0z,io1_o1z,
            io2_oo1,io2_o10,io2_oz0,io2_oz1,io2_o0z,io2_o1z,
            io3_oo1,io3_o10,io3_oz0,io3_oz1,io3_o0z,io3_o1z,
            io4_oo1,io4_o10,io4_oz0,io4_oz1,io4_o0z,io4_o1z,
            io5_oo1,io5_o10,io5_oz0,io5_oz1,io5_o0z,io5_o1z,
            io6_oo1,io6_o10,io6_oz0,io6_oz1,io6_o0z,io6_o1z:TIME:=2 ns);
   PORT   (i1,i2,i3,i4,i5,i6,i7,i8,i9,i10 : IN t_wlogic;
           o1,o2 : INOUT t_wlogic;
           io1,io2,io3,io4,io5,io6 : INOUT t_wlogic);
END pal1618;

ARCHITECTURE full OF pal1618 IS
   -- input delay signals
   SIGNAL i1delay,i2delay,i3delay,i4delay,i5delay,
          i6delay,i7delay,i8delay,i9delay,i10delay : t_wlogic;
BEGIN
   -- input delay processing
   i1delay  <= i1  AFTER f_delay(i1,i1_io1,i1_i10);
   i2delay  <= i2  AFTER f_delay(i2,i2_io1,i2_i10);
   i3delay  <= i3  AFTER f_delay(i3,i3_io1,i3_i10);
   i4delay  <= i4  AFTER f_delay(i4,i4_io1,i4_i10);
   i5delay  <= i5  AFTER f_delay(i5,i5_io1,i5_i10);
   i6delay  <= i6  AFTER f_delay(i6,i6_io1,i6_i10);
   i7delay  <= i7  AFTER f_delay(i7,i7_io1,i7_i10);
   i8delay  <= i8  AFTER f_delay(i8,i8_io1,i8_i10);
   i9delay  <= i9  AFTER f_delay(i9,i9_io1,i9_i10);
   i10delay <= i10 AFTER f_delay(i10,i10_io1,i10_i10);

   -- PAL operation
   PROCESS (i1delay,i2delay,i3delay,i4delay,i5delay,
            i6delay,i7delay,i8delay,i9delay,i10delay,
            io1,io2,io3,io4,io5,io6)
      VARIABLE pal : productline(1 TO 64);
```

5.4. PALS, PLDS

```
            -- should be: VARIABLE pal : productline(1 TO 64,1 TO 32);
            FILE paldef : TEXT IS IN "pal1618.def";     -- pal def file
            VARIABLE startup : boolean := true;         -- need initializing?
            VARIABLE state : t_logarray (1 TO 32);      -- pal i/o state
        BEGIN
        -- initialize pal, once only
        IF startup THEN
            f_palinit(pal,paldef);
            startup := false;
        END IF;

        -- setup state array
        state := (i2delay, NOT i2delay, i1delay,  NOT i1delay,
                  i3delay, NOT i3delay, io6,      NOT io6,
                  i4delay, NOT i4delay, io5,      NOT io5,
                  i5delay, NOT i5delay, io4,      NOT io4,
                  i6delay, NOT i6delay, io3,      NOT io3,
                  i7delay, NOT i7delay, io2,      NOT io2,
                  i8delay, NOT i8delay, io1,      NOT io1,
                  i9delay, NOT i9delay, i10delay, NOT i10delay);

        -- update working states and outputs
        f_lookup7(state,pal,1,
            o2,o2_o01,o2_o10,o2_oz0,o2_oz1,o2_o0z,o2_o1z);
        f_lookup7(state,pal,9,
            io6,io6_o01,io6_o10,io6_oz0,io6_oz1,io6_o0z,io6_o1z);
        f_lookup7(state,pal,17,
            io5,io5_o01,io5_o10,io5_oz0,io5_oz1,io5_o0z,io5_o1z);
        f_lookup7(state,pal,25,
            io4,io4_o01,io4_o10,io4_oz0,io4_oz1,io4_o0z,io4_o1z);
        f_lookup7(state,pal,33,
            io3,io3_o01,io3_o10,io3_oz0,io3_oz1,io3_o0z,io3_o1z);
        f_lookup7(state,pal,41,
            io2,io2_o01,io2_o10,io2_oz0,io2_oz1,io2_o0z,io2_o1z);
        f_lookup7(state,pal,49,
            io1,io1_o01,io1_o10,io1_oz0,io1_oz1,io1_o0z,io1_o1z);
        f_lookup7(state,pal,57,
            o1,o1_o01,o1_o10,o1_oz0,o1_oz1,o1_o0z,o1_o1z);
        END PROCESS;
END full;


CONFIGURATION parts OF pal1618 IS
FOR full
END FOR;
END parts;
```

The ports for this model are

- i1, i2, i3, i4, i5, i6, i7, i8, i9, i10 - input lines
- o1, o2 - output lines

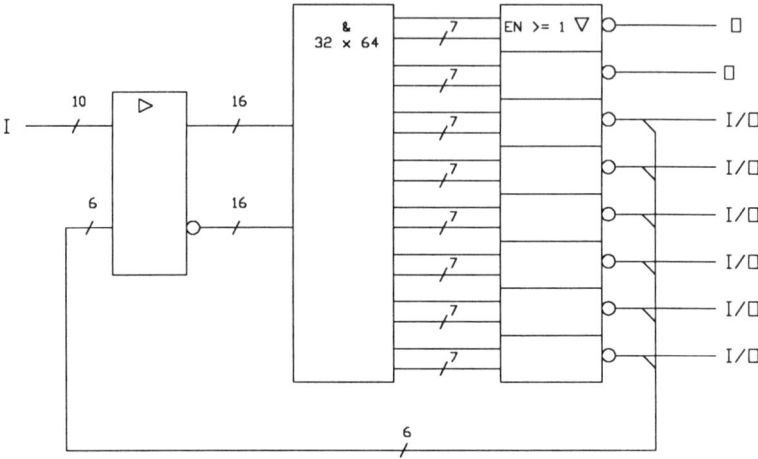

Figure 5.4: PAL16L8 Block Diagram

- io1, io2, io3, io4, io5, io6 - input and output lines

Generic parameters to this model are summarized here

- i1_i01, i1_i10 - low to high and high to low **i1** input port delays
- i2_i01, i2_i10 - low to high and high to low **i2** input port delays
- i3_i01, i3_i10 - low to high and high to low **i3** input port delays
- i4_i01, i4_i10 - low to high and high to low **i4** input port delays
- i5_i01, i5_i10 - low to high and high to low **i5** input port delays
- i6_i01, i6_i10 - low to high and high to low **i6** input port delays
- i7_i01, i7_i10 - low to high and high to low **i7** input port delays
- i8_i01, i8_i10 - low to high and high to low **i8** input port delays
- i9_i01, i9_i10 - low to high and high to low **i9** input port delays
- i10_i01, i10_i10 - low to high and high to low **i10** input port delays
- o1_o01, o1_o10, o1_oz0, o1_oz1, o1_o0z, o1_o1z - low to high, high to low, off to low, off to high, low to off, high to off **o1** output port delays
- o2_o01, o2_o10, o2_oz0, o2_oz1, o2_o0z, o2_o1z - low to high, high to low, off to low, off to high, low to off, high to off **o2** output port delays
- io1_o01, io1_o10, io1_oz0, io1_oz1, io1_o0z, io1_o1z - low to high, high to low, off to low, off to high, low to off, high to off **io1** output port delays

5.4. PALS, PLDS

- io2_o01, io2_o10, io2_oz0, io2_oz1, io2_o0z, io2_o1z - low to high, high to low, off to low, off to high, low to off, high to off **io2** output port delays
- io3_o01, io3_o10, io3_oz0, io3_oz1, io3_o0z, io3_o1z - low to high, high to low, off to low, off to high, low to off, high to off **io3** output port delays
- io4_o01, io4_o10, io4_oz0, io4_oz1, io4_o0z, io4_o1z - low to high, high to low, off to low, off to high, low to off, high to off **io4** output port delays
- io5_o01, io5_o10, io5_oz0, io5_oz1, io5_o0z, io5_o1z - low to high, high to low, off to low, off to high, low to off, high to off **io5** output port delays
- io6_o01, io6_o10, io6_oz0, io6_oz1, io6_o0z, io6_o1z - low to high, high to low, off to low, off to high, low to off, high to off **io6** output port delays

This model has no error checks.

The input delay processing section of the model incorporates the appropriate delay for each of the input ports. An internal signal for each input port is declared. The main device function process utilizes the delayed signals rather than the primary inputs.

The PAL operation is handled by a single process which is sensitive to the input delayed signals **i1delay, i2delay, i3delay, i4delay, i5delay, i6delay, i7delay, i8delay, i9delay, i10delay**. In addition, feedback in the PAL requires the process to be sensitive to the outputs **io1, io2, io3, io4, io5** and **io6**. The **startup** variable is used to insure that the PLA initialization occurs only once through the use of the **f_palinit** procedure as described in section 5.1. For each change in the inputs the following actions are taken:

- All input lines of the array are saved in the **state** variable.
- For each output, the **f_lookup7** procedure is used to calculate the ored results of the 7 input products, and to determine the enable/disable status using the eighth input product. This procedure performs the assignment directly to the output signal.

5.4.3 8 input, 2 I/O, 6 clocked output PAL

The following shows the full model for a PAL with characteristics

- 8 input lines
- Asynchronous enable for all outputs

- 6 clocked outputs
- 2 bidirectional outputs
- Tristate outputs

and as shown in figure 5.5.

8 input, 2 I/O, 6 clocked output PAL

```
USE std.std_logic.ALL;
USE std.std_ttl.ALL;
USE std.textio.ALL;
USE work.palpack.ALL;
ENTITY pal16r6a IS
    GENERIC (i1_i01,i1_i10,
             i2_i01,i2_i10,
             i3_i01,i3_i10,
             i4_i01,i4_i10,
             i5_i01,i5_i10,
             i6_i01,i6_i10,
             i7_i01,i7_i10,
             i8_i01,i8_i10,

             clk_i01,clk_i10,
             oe_i01,oe_i10,

             q1_o01,q1_o10,q1_oz0,q1_oz1,q1_o0z,q1_o1z,
             q2_o01,q2_o10,q2_oz0,q2_oz1,q2_o0z,q2_o1z,
             q3_o01,q3_o10,q3_oz0,q3_oz1,q3_o0z,q3_o1z,
             q4_o01,q4_o10,q4_oz0,q4_oz1,q4_o0z,q4_o1z,
             q5_o01,q5_o10,q5_oz0,q5_oz1,q5_o0z,q5_o1z,
             q6_o01,q6_o10,q6_oz0,q6_oz1,q6_o0z,q6_o1z,

             io1_o01,io1_o10,io1_oz0,io1_oz1,io1_o0z,io1_o1z,
             io2_o01,io2_o10,io2_oz0,io2_oz1,io2_o0z,io2_o1z :TIME:=2 ns);
    PORT    (i1,i2,i3,i4,i5,i6,i7,i8 : IN t_wlogic;
             clk,oe : IN t_wlogic;
             q1,q2,q3,q4,q5,q6 : OUT t_wlogic;
             io1,io2 : INOUT t_wlogic);
END pal16r6a;

ARCHITECTURE full OF pal16r6a IS
    -- input delay signals
    SIGNAL i1delay,i2delay,i3delay,i4delay,i5delay,i6delay,i7delay,i8delay,
           clkdelay,oedelay : t_wlogic;
BEGIN
    -- input delay processing
    i1delay <= i1 AFTER f_delay(i1,i1_i01,i1_i10);
    i2delay <= i2 AFTER f_delay(i2,i2_i01,i2_i10);
    i3delay <= i3 AFTER f_delay(i3,i3_i01,i3_i10);
    i4delay <= i4 AFTER f_delay(i4,i4_i01,i4_i10);
```

5.4. PALS, PLDS

```
        i5delay <= i5 AFTER f_delay(i5,i5_io1,i5_i10);
        i6delay <= i6 AFTER f_delay(i6,i6_io1,i6_i10);
        i7delay <= i7 AFTER f_delay(i7,i7_io1,i7_i10);
        i8delay <= i8 AFTER f_delay(i8,i8_io1,i8_i10);
        clkdelay <= clk AFTER f_delay(clk,clk_io1,clk_i10);
        oedelay <= oe AFTER f_delay(oe,oe_io1,oe_i10);

        -- register operation
        PROCESS (i1delay,i2delay,i3delay,i4delay,
                 i5delay,i6delay,i7delay,i8delay,
                 clkdelay,oedelay)
            VARIABLE d1,d2,d3,d4,d5,d6 : t_logic := U;   -- latch states
            VARIABLE pal : productline(1 TO 64);
            -- should be: VARIABLE pal : productline(1 TO 64,1 TO 32);
            VARIABLE startup : boolean := true;          -- need initializing?
            FILE paldef : TEXT IS IN "pal16r6a.def";     -- pal def file
            VARIABLE state : t_logarray (1 TO 32);       -- pal i/o state
        BEGIN
        -- initialize pal, once only
        IF startup THEN
            f_palinit(pal,paldef);
            startup := false;
        END IF;

        -- setup state array
        state := (i1delay, NOT i1delay, io1, NOT io1,
                  i2delay, NOT i2delay, d1,  NOT d1,
                  i3delay, NOT i3delay, d2,  NOT d2,
                  i4delay, NOT i4delay, d3,  NOT d3,
                  i5delay, NOT i5delay, d4,  NOT d4,
                  i6delay, NOT i6delay, d5,  NOT d5,
                  i7delay, NOT i7delay, d6,  NOT d6,
                  i8delay, NOT i8delay, io2, NOT io2);

        -- update working states and outputs
        f_lookup7(state,pal,1,
           io1,io1_oO1,io1_o10,io1_ozO,io1_oz1,io1_oOz,io1_o1z);

        -- rising edge on clock?
        IF (clkdelay'EVENT) AND (clkdelay = '1') THEN
            f_lookup8(state,pal,9,
                 d1,oedelay,q1,q1_oO1,q1_o10,q1_ozO,q1_oz1,q1_oOz,q1_o1z);
            f_lookup8(state,pal,17,
                 d2,oedelay,q2,q2_oO1,q2_o10,q2_ozO,q2_oz1,q2_oOz,q2_o1z);
            f_lookup8(state,pal,25,
                 d3,oedelay,q3,q3_oO1,q3_o10,q3_ozO,q3_oz1,q3_oOz,q3_o1z);
            f_lookup8(state,pal,33,
                 d4,oedelay,q4,q4_oO1,q4_o10,q4_ozO,q4_oz1,q4_oOz,q4_o1z);
            f_lookup8(state,pal,41,
                 d5,oedelay,q5,q5_oO1,q5_o10,q5_ozO,q5_oz1,q5_oOz,q5_o1z);
            f_lookup8(state,pal,49,
                 d6,oedelay,q6,q6_oO1,q6_o10,q6_ozO,q6_oz1,q6_oOz,q6_o1z);
```

```
        END IF;

    f_lookup7(state,pal,57,
        io2,io2_o01,io2_o10,io2_oz0,io2_oz1,io2_o0z,io2_o1z);
    END PROCESS;
END full;


CONFIGURATION parts OF pal16r6a IS
FOR full
END FOR;
END parts;
```

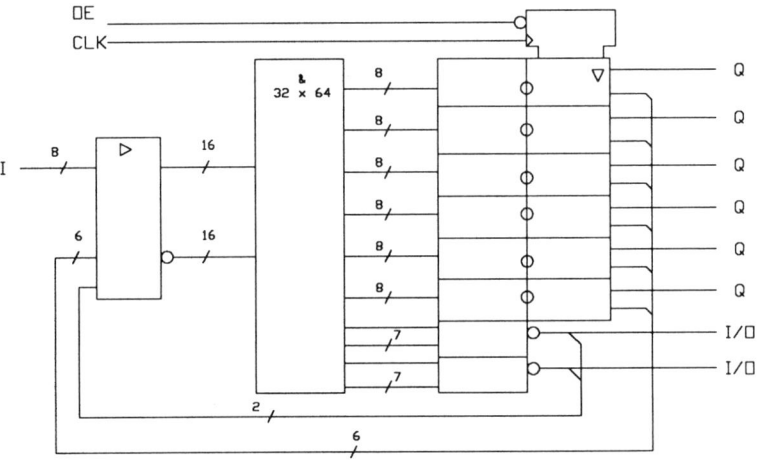

Figure 5.5: PAL16R6 Block Diagram

The ports for this model are

- i1, i2, i3, i4, i5, i6, i7, i8 - input lines
- clk - output clock
- oe - output enable
- q1, q2, q3, q4, q5, q6 - clocked outputs
- io1, io2 - input and output lines

Generic parameters to this model are summarized here

- i1_i01, i1_i10 - low to high and high to low i1 input port delays
- i2_i01, i2_i10 - low to high and high to low i2 input port delays
- i3_i01, i3_i10 - low to high and high to low i3 input port delays

5.4. PALS, PLDS

- i4_i01, i4_i10 - low to high and high to low **i4** input port delays
- i5_i01, i5_i10 - low to high and high to low **i5** input port delays
- i6_i01, i6_i10 - low to high and high to low **i6** input port delays
- i7_i01, i7_i10 - low to high and high to low **i7** input port delays
- i8_i01, i8_i10 - low to high and high to low **i8** input port delays
- clk_i01, clk_i10 - low to high and high to low **clk** input port delays
- oe_i01, oe_i10 - low to high and high to low **oe** input port delays
- q1_o01, q1_q10, q1_oz0, q1_oz1, q1_o0z, q1_q1z - low to high, high to low, off to low, off to high, low to off, high to off **q1** output port delays
- q2_o01, q2_q20, q2_oz0, q2_oz1, q2_o0z, q2_q2z - low to high, high to low, off to low, off to high, low to off, high to off **q2** output port delays
- q3_o01, q3_q30, q3_oz0, q3_oz1, q3_o0z, q3_q3z - low to high, high to low, off to low, off to high, low to off, high to off **q3** output port delays
- q4_o01, q4_q40, q4_oz0, q4_oz1, q4_o0z, q4_q4z - low to high, high to low, off to low, off to high, low to off, high to off **q4** output port delays
- q5_o01, q5_q50, q5_oz0, q5_oz1, q5_o0z, q5_q5z - low to high, high to low, off to low, off to high, low to off, high to off **q5** output port delays
- q6_o01, q6_q60, q6_oz0, q6_oz1, q6_o0z, q6_q6z - low to high, high to low, off to low, off to high, low to off, high to off **q6** output port delays
- io1_o01, io1_o10, io1_oz0, io1_oz1, io1_o0z, io1_o1z - low to high, high to low, off to low, off to high, low to off, high to off **io1** output port delays
- io2_o01, io2_o10, io2_oz0, io2_oz1, io2_o0z, io2_o1z - low to high, high to low, off to low, off to high, low to off, high to off **io2** output port delays

This model has no error checks.

The input delay processing section of the model incorporates the appropriate delay for each of the input ports. An internal signal for each input port is declared. The main device function process utilizes the delayed signals rather than the primary inputs.

The PAL operation is handled by a single process which is sensitive to the input delayed signals **i1delay, i2delay, i3delay, i4delay, i5delay,** i6delay, i7delay, i8delay, clkdelay and oedelay. The **startup** variable is used to insure that the PLA initialization occurs only once through the use of the **f_palinit** procedure as described in section 5.1. For each change in the inputs the following actions are taken:

- All input lines of the array are saved in the **state** variable.
- For the two bidirectional outputs, the **f_lookup7** procedure is used to calculate the ored results of the 7 input products, and to determine the enable/disable status using the eighth input product. This procedure performs the assignment directly to the output signal.
- For the clocked outputs, the **f_lookup8** procedure is used to calculate the ored results of the 8 input products taking the **oedelay** enable into account for tristate output.

5.4.4 8 input, 8 clocked output PAL

The following shows the full model for a PAL with characteristics

- 8 input lines
- Asynchronous enable for all outputs
- 8 clocked outputs
- Tristate outputs

and as shown in figure 5.6.

8 input, 8 clocked output PAL

```
USE std.std_logic.ALL;
USE std.std_ttl.ALL;
USE std.textio.ALL;
USE work.palpack.ALL;
ENTITY pal16r8a IS
    GENERIC (i1_i01,i1_i10,
             i2_i01,i2_i10,
             i3_i01,i3_i10,
             i4_i01,i4_i10,
             i5_i01,i5_i10,
             i6_i01,i6_i10,
             i7_i01,i7_i10,
             i8_i01,i8_i10,

             clk_i01,clk_i10,
             oe_i01,oe_i10,

             q1_o01,q1_o10,q1_oz0,q1_oz1,q1_o0z,q1_o1z,
             q2_o01,q2_o10,q2_oz0,q2_oz1,q2_o0z,q2_o1z,
             q3_o01,q3_o10,q3_oz0,q3_oz1,q3_o0z,q3_o1z,
```

5.4. PALS, PLDS

```
                q4_oO1,q4_o10,q4_oz0,q4_oz1,q4_o0z,q4_o1z,
                q5_oO1,q5_o10,q5_oz0,q5_oz1,q5_o0z,q5_o1z,
                q6_oO1,q6_o10,q6_oz0,q6_oz1,q6_o0z,q6_o1z,
                q7_oO1,q7_o10,q7_oz0,q7_oz1,q7_o0z,q7_o1z,
                q8_oO1,q8_o10,q8_oz0,q8_oz1,q8_o0z,q8_o1z : TIME := 2 ns);

     PORT   (i1,i2,i3,i4,i5,i6,i7,i8 : IN t_wlogic;
            clk,oe : IN t_wlogic;
            q1,q2,q3,q4,q5,q6,q7,q8 : OUT t_wlogic);
END pal16r8a;

ARCHITECTURE full OF pal16r8a IS
    -- input delay signals
    SIGNAL i1delay,i2delay,i3delay,i4delay,i5delay,i6delay,i7delay,i8delay,
           clkdelay,oedelay : t_wlogic;
BEGIN
    -- input delay processing
    i1delay <= i1 AFTER f_delay(i1,i1_iO1,i1_i10);
    i2delay <= i2 AFTER f_delay(i2,i2_iO1,i2_i10);
    i3delay <= i3 AFTER f_delay(i3,i3_iO1,i3_i10);
    i4delay <= i4 AFTER f_delay(i4,i4_iO1,i4_i10);
    i5delay <= i5 AFTER f_delay(i5,i5_iO1,i5_i10);
    i6delay <= i6 AFTER f_delay(i6,i6_iO1,i6_i10);
    i7delay <= i7 AFTER f_delay(i7,i7_iO1,i7_i10);
    i8delay <= i8 AFTER f_delay(i8,i8_iO1,i8_i10);
    clkdelay <= clk AFTER f_delay(clk,clk_iO1,clk_i10);
    oedelay <= oe AFTER f_delay(oe,oe_iO1,oe_i10);

    -- register operation
    PROCESS (i1delay,i2delay,i3delay,i4delay,
             i5delay,i6delay,i7delay,i8delay,
             clkdelay,oedelay)
        VARIABLE d1,d2,d3,d4,d5,d6,d7,d8 : t_logic := U;  -- latch states
        VARIABLE pal : productline(1 TO 64);
        -- should be: VARIABLE pal : productline(1 TO 64,1 TO 32);
        VARIABLE startup : boolean := true;       -- need initializing?
        FILE paldef : TEXT IS IN "pal16r8a.def";  -- pal def file
        VARIABLE state : t_logarray (1 TO 32);    -- pal i/o state
    BEGIN
    -- initialize pal, once only
    IF startup THEN
        f_palinit(pal,paldef);
        startup := false;
    END IF;

    -- rising edge on clock?
    IF (clkdelay'EVENT) AND (clkdelay = '1') THEN
        -- setup state array
        state := (i1delay, NOT i1delay, d1,  NOT d1,
                  i2delay, NOT i2delay, d2,  NOT d2,
                  i3delay, NOT i3delay, d3,  NOT d3,
                  i4delay, NOT i4delay, d4,  NOT d4,
```

```
                    i5delay, NOT i5delay, d5, NOT d5,
                    i6delay, NOT i6delay, d6, NOT d6,
                    i7delay, NOT i7delay, d7, NOT d7,
                    i8delay, NOT i8delay, d8, NOT d8);

        -- calculate new output states
        f_lookup8(state,pal,1,
            d1,oedelay,q1,q1_o01,q1_o10,q1_oz0,q1_oz1,q1_o0z,q1_o1z);
        f_lookup8(state,pal,9,
            d2,oedelay,q2,q2_o01,q2_o10,q2_oz0,q2_oz1,q2_o0z,q2_o1z);
        f_lookup8(state,pal,17,
            d3,oedelay,q3,q3_o01,q3_o10,q3_oz0,q3_oz1,q3_o0z,q3_o1z);
        f_lookup8(state,pal,25,
            d4,oedelay,q4,q4_o01,q4_o10,q4_oz0,q4_oz1,q4_o0z,q4_o1z);
        f_lookup8(state,pal,33,
            d5,oedelay,q5,q5_o01,q5_o10,q5_oz0,q5_oz1,q5_o0z,q5_o1z);
        f_lookup8(state,pal,41,
            d6,oedelay,q6,q6_o01,q6_o10,q6_oz0,q6_oz1,q6_o0z,q6_o1z);
        f_lookup8(state,pal,49,
            d7,oedelay,q7,q7_o01,q7_o10,q7_oz0,q7_oz1,q7_o0z,q7_o1z);
        f_lookup8(state,pal,57,
            d8,oedelay,q8,q8_o01,q8_o10,q8_oz0,q8_oz1,q8_o0z,q8_o1z);
    END IF;
    END PROCESS;
END full;

CONFIGURATION parts OF pal16r8a IS
FOR full
END FOR;
END parts;
```

The ports for this model are

- i1, i2, i3, i4, i5, i6, i7, i8 - input lines
- clk - output clock
- oe - output enable
- q1, q2, q3, q4, q5, q6, q7, q8 - clocked outputs

Generic parameters to this model are summarized here

- i1_i01, i1_i10 - low to high and high to low i1 input port delays
- i2_i01, i2_i10 - low to high and high to low i2 input port delays
- i3_i01, i3_i10 - low to high and high to low i3 input port delays
- i4_i01, i4_i10 - low to high and high to low i4 input port delays
- i5_i01, i5_i10 - low to high and high to low i5 input port delays
- i6_i01, i6_i10 - low to high and high to low i6 input port delays
- i7_i01, i7_i10 - low to high and high to low i7 input port delays

5.4. PALS, PLDS

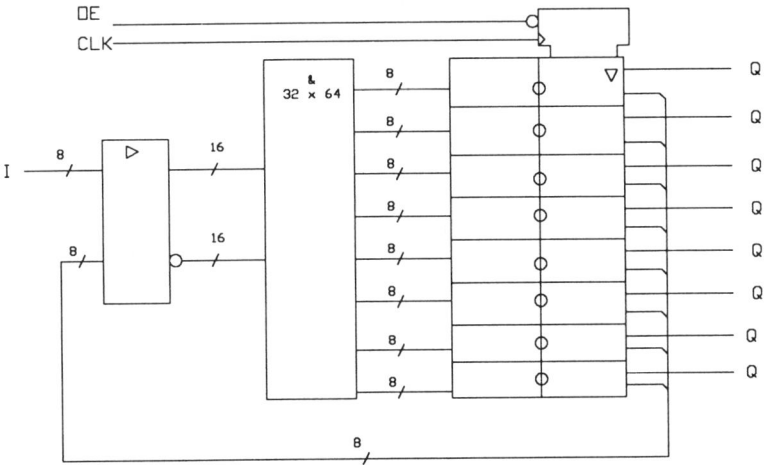

Figure 5.6: PAL16R8 Block Diagram

- i8_i01, i8_i10 - low to high and high to low i8 input port delays
- clk_i01, clk_i10 - low to high and high to low clk input port delays
- oe_i01, oe_i10 - low to high and high to low oe input port delays
- q1_o01, q1_q10, q1_oz0, q1_oz1, q1_o0z, q1_q1z - low to high, high to low, off to low, off to high, low to off, high to off q1 output port delays
- q2_o01, q2_q20, q2_oz0, q2_oz1, q2_o0z, q2_q2z - low to high, high to low, off to low, off to high, low to off, high to off q2 output port delays
- q3_o01, q3_q30, q3_oz0, q3_oz1, q3_o0z, q3_q3z - low to high, high to low, off to low, off to high, low to off, high to off q3 output port delays
- q4_o01, q4_q40, q4_oz0, q4_oz1, q4_o0z, q4_q4z - low to high, high to low, off to low, off to high, low to off, high to off q4 output port delays
- q5_o01, q5_q50, q5_oz0, q5_oz1, q5_o0z, q5_q5z - low to high, high to low, off to low, off to high, low to off, high to off q5 output port delays
- q6_o01, q6_q60, q6_oz0, q6_oz1, q6_o0z, q6_q6z - low to high, high to low, off to low, off to high, low to off, high to off q6 output port delays

- q7_o01, q7_q70, q7_oz0, q7_oz1, q7_o0z, q7_q7z - low to high, high to low, off to low, off to high, low to off, high to off **q7** output port delays

- q8_o01, q8_o10, q8_oz0, q8_oz1, q8_o0z, q8_o1z - low to high, high to low, off to low, off to high, low to off, high to off **q8** output port delays

This model has no error checks.

The input delay processing section of the model incorporates the appropriate delay for each of the input ports. An internal signal for each input port is declared. The main device function process utilizes the delayed signals rather than the primary inputs.

The PAL operation is handled by a single process which is sensitive to the input delayed signals **i1delay, i2delay, i3delay, i4delay, i5delay, i6delay, i7delay, i8delay, clkdelay** and **oedelay**. The **startup** variable is used to insure that the PLA initialization occurs only once through the use of the **f_palinit** procedure as described in section 5.1. For each change in the inputs the following actions are taken:

- All input lines of the array are saved in the **state** variable.

- For the clocked outputs, the **f_lookup8** procedure is used to calculate the ored results of the 8 input products taking the **oedelay** enable into account for tristate output.

Chapter 6

Complex Devices

Modeling off-the-shelf complex VLSI digital devices, such as microprocessors, memory controllers, and floating point units is a challenging experience. This chapter directly addresses how to model such devices through the use of illustrated techniques which help to break down a complex modeling problem into manageable pieces. Advanced modeling techniques are used to describe the behavior of a complex device in VHDL.

A model of the SPARC 32-bit RISC microprocessor (Fairchild MB86901) will be used throughout this chapter to illustrate the concepts. A full understanding will be aided by review of the databook for this device.

6.1 Getting Started

Datasheets for the more recent 32-bit microprocessors contain 100-300 pages of information in the form of text and graphs which describe the behavior of the device. Some databooks are split into a hardware reference manual (for the hardware designer) and a programmers guide (primarily for the software developer, although the hardware designer needs to be aware of a great many of the software issues as well). Such complexity is far beyond the 10-20 page datasheets of simple 8-bit microprocessors in the past.

Knowing where to begin modeling such a device is a major step toward an accurate model. But before we begin to decompose the modeling problem, the goals of the effort must be well-defined. The following section outlines the key choices which must be made.

6.1.1 Partial versus Full Functional Models

In the course of simulating a design, occasions may arise when a model of a complex device is not available in time for the project schedule. The designer has two choices:

- Leave the complex device out of the circuit and feed the pins of the empty IC socket with values that have been generated by hand or from the actual device using the simulator stimulus language
- Build a complete model which performs the above operation driven from a data file

Models of this type are called *bus-functional* models. Bus functional models are partially complete models of microcontrollers, and microprocessor-like devices which have an active bus interface protocol. A model of such a device would then consist of the state machine which defines the bus protocol along with a user interface which links the state-machine to the user input facility, usually a file as shown in figure 6.1. Building a user interface for

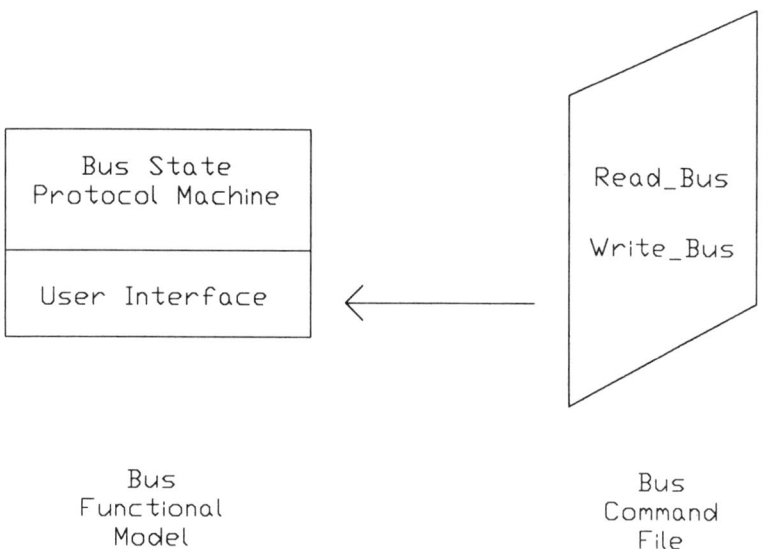

Figure 6.1: Bus Functional Model

bus functional models is not trivial. It generally consists of a compiler or interpreter which reads in commands to the bus functional model (BFM) in a language called the Bus Command Language (BCL). Commands such as when to read or write the bus, and on what address and with which

6.1. GETTING STARTED

data are read by the BFM front end command processor. The BFM synchronizes the BCL commands to the simulator event queue and causes the appropriate bus interactions to take place.

Bus functional models do not relieve the designer from the role of playing "human microprocessor" since the designer needs to understand the actions of the missing microprocessor model and feed those responses to the bus functional model. He does however avoid dealing with the detailed timing interaction of the bus in order to accomplish this task.

In doing so, the functionality of the circuit bus design can be validated and the timing of the bus can be analyzed even though the full-functional model is not available. BFM's are often best used as an interim approach until the full-functional model becomes available.

It is possible to build a model of a complex device which simply exercises the bus on which the device is connected (bus-functional model) or to build a model which only executes that portion of the device which you are particularly concerned with in your design (partial-functional model). Only one philosophy of modeling creates a model which performs all of the operations of the actual device and in time synchronized order with the actual device. This type of modeling is called *full-functional modeling* and is the type of modeling discussed in the remainder of this chapter.

Full-functional models:

1. perform all of the operations of the actual device, and

2. perform those operations with the same timing relationships as does the actual device.

We will see later in this chapter that full-functionality can be verified using a VLSI tester or hardware modeller as a quality control check. With the the goal of accuracy in mind, we proceed to the architecture of the model.

6.1.2 Architecture

The architecture of a device can sometimes be an aid to the modeling process in those cases where the breakdown of functionality is well known and easily defined.

There are two approaches to writing a model. One is to decompose the model into *autonomous* behavioral blocks, develop each block and then wire them together with signals. The other method is to write the model as having only one block and use non-signal, data objects (variables) to tie the model together.

Hybrid Structural/Behavioral Model Decomposition

Sometimes the architecture of the device is so straightforward that the description of the model can be easily broken down according to the device block diagram. But in most cases the interrelationships between the blocks of logic shown on a block diagram illustration are far more complex than the simplified block diagram.

An example of a device whose architecture lends itself to easy decomposition is the TRW TDC1028 Digital Filter/Correlator chip [TRW84]. A series of tapped delay lines and multiply-accumulators form the structure of the device as shown in figure 6.2. From this architecture, it is easy to see

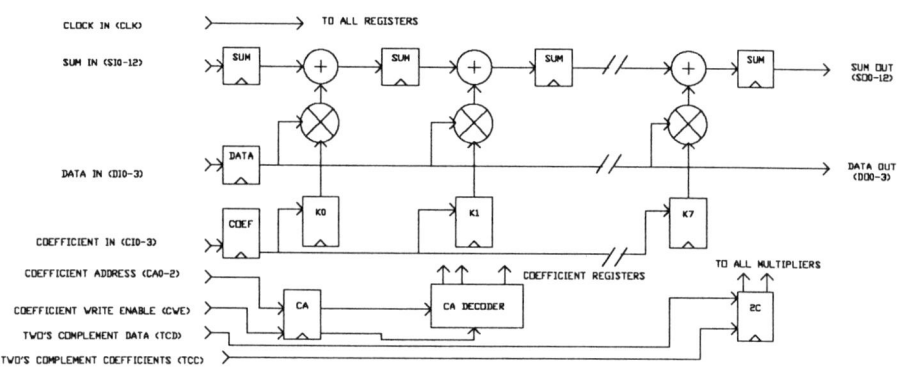

Figure 6.2: TDC1028 Block Diagram

that the model could be described as a series of multiply-accumulate-delay taps which connect to one another serially to form the device. Each tap can be represented by a behavioral model in rather simplistic form. The multiplications and additions are done over a twelve bit field. Therefore, instead of performing multiplications on a bit by bit basis, you could first convert the data streams into integers and then simply add and multiply them using the VHDL arithmetic operators. This would speed up the execution of the tap considerably.

But first, lets evaluate the impact of this structural decomposition on the number of signals required to interconnect each of the elements. Each tap has two data inputs, a **sum_in** line, a **data_in** line, and a coefficient line for a total of 20 signals. Similarly, twenty signals are passed onto the next

6.1. GETTING STARTED

stage. There are eight stages. Hence, the hybrid structural-behavioral form of this model requires 320 signals through which communication occurs. So many signals in a model can create a runtime burden on the simulator, with the general rule that the fewer signals the better.

A much better approach here is to do the following:

1. Define one process block for the entire device.
2. Convert all in-coming data into integers within the process.
3. Write a procedure which performs the operation of a tap.
4. Place this procedure in a loop and exercise the tap over the eight data/coefficient pairs.
5. Convert the integer data items back to signals.

Doing so will yield a model which runs faster, is easier to debug, and uses less memory.

Unless you can decompose the model into *autonomous* building blocks which do not unnecessarily add too many signals to the model, and operate completely separate from the rest of the model, it is best to model the device as a single architecture.

In the general case, for highly complex VLSI models whose internal architecture is not known to the modeller (which is usually the case unless the modeller is a member of the chip design team), the choice is to use one architectural unit for the entire device and utilize non-signal data objects for intra-process communication.

Entity Definition

We begin the definition of a model by specifying its interface to the outside world through a set of ports and generic parameters. Since most devices are available from their manufacturer in a variety of speeds, now is the time to take into consideration the model timing interface.

An entity declaration of the Fujitsu SPARC processor is shown below.

MB86901 Interface

```
--  ---------------------------------------------------------------
--     File name  :  MB86901.entity.vhdl
--     Title      :  SPARC MB86901 (S-25) 32-bit RISC Processor
--     Purpose    :  VHDL Behavioral Model of Fujitsu chip
--     Author(s)  :  W. Billowitch
--     Notes      :  880925BBGG databook
--  ---------------------------------------------------------------
--     Modification History :
--  ---------------------------------------------------------------
```

```
--   Version No:| Author:| Mod. Date:| Changes Made:
--      v1.000  |  wdb   |  12/31/88  | Original coding
-- ------------------------------------------------------------
USE std.std_logic.ALL;
USE std.std_cmos.ALL ;
USE std.timpak.ALL   ;  -- procedures defined in this chapter

ENTITY mb86901_generic IS
    GENERIC (
        t_1:time;  -- sys clock cycle time
        t_2:time;  -- sys clock rise time
        t_3:time;  -- sys clock fall time
        t_4:time;  -- sys clock skew time
        t_5:time;  -- sys clock (clk1) hi time
        t_6:time;  -- sys clock (clk1) lo time
        t_7:time;  -- sys clock (clk2) hi time
        t_8:time;  -- sys clock (clk2) lo time
        t_9:time;  -- ADR valid delay, from clk1 rising
        t_10:time; -- ADR hold, from clk1 rising
        t_11:time; -- ASI valid delay, from clk1 rising
        t_12:time; -- ASI hold, from clk1 rising
        t_13:time; -- Read Data Setup, before clk1 rising
        t_14:time; -- Read Data Hold, after clk1 rising
        t_15:time; -- write data valid delay from clk1 rising
        t_16:time; -- write data hold, after clk1 rising
        t_17:time; -- write data turn off from clk1
        t_17a:time; -- write data turn on  from clk1
        t_18:time; -- /AOE turn off time
        t_19:time; -- /AOE turn on  time
        t_20:time; -- /ADROE turn off time
        t_21:time; -- /ADROE turn on  time
        t_22:time; -- /DOE turn off time
        t_23:time; -- /DOE turn on  time
        t_24:time; -- SIZE, from clk1 rising
        t_25:time; -- RD, from clk1 rising
        t_26:time; -- /WE, from clk1 rising
        t_27:time; -- LDST, from clk1 rising
        t_28:time; -- NULL_CYC, from clk1 rising
        t_29:time; -- NULL_CYC, from /MHOLD falling
        t_30:time; -- /HAL, from clk1 rising
        t_31:time; -- /HAL, from /MHOLD falling
        t_32:time; -- LOCK, from clk1 rising
        t_33:time; -- DFETCH, from clk1 rising
        t_34:time; -- /ERROR, from clk1 rising
        t_35:time; -- /MDS, before CLK1 falling
        t_36:time; -- /MDS, before CLK1 rising
        t_37:time; -- /MEXC, before CLK1 falling
        t_38:time; -- /MXEC, before CLK1 rising
        t_39:time; -- /MHOLDA, before CLK1 falling
        t_40:time; -- /MHOLDB, before CLK1 falling
        t_41:time; -- /MHOLDC, before CLK1 falling
        t_42:time; -- /SHOLD, before CLK1 falling
```

6.1. GETTING STARTED

```
        t_43:time;        -- /BHOLD, before CLK1 falling
        t_44:time;        -- IRL, before CLK1 rising
        t_45:time;        -- /RESET, before CLK1 falling
        t_46_active:time; -- F<31:0>, from CLK1 rising
        t_46_hold:time;   -- F<31:0>, from CLK1 rising
        t_47_active:time; -- FINS, from CLK1 rising
        t_47_hold:time;   -- FINS, from CLK1 rising
        t_48_active:time; -- FADR, from CLK1 rising
        t_48_hold:time;   -- FADR, from CLK1 rising
        t_49_active:time; -- FEND, from CLK1 rising
        t_49_hold:time;   -- FEND, from CLK1 rising
        t_50_active:time; -- FLUSH, from CLK1 rising
        t_50_hold:time;   -- FLUSH, from CLK1 rising
        t_51_active:time; -- FXACK, from CLK1 rising
        t_51_hold:time;   -- FXACK, from CLK1 rising
        t_52_setup:time;  -- /FHOLD, before CLK1 rising
        t_52_hold:time;   -- /FHOLD, before CLK1 rising
        t_53_setup:time;  -- /FEXC, before CLK1 rising
        t_53_hold:time;   -- /FEXC, before CLK1 rising
        t_54_setup:time;  -- /FCC, before CLK1 rising
        t_54_hold:time;   -- /FCC, before CLK1 rising
        t_55_setup:time;  -- FCCV, before CLK1 rising
        t_55_hold:time;   -- FCCV, before CLK1 rising
        x_force_val:t_state -- value forced when X found.
        );
PORT    (
        -- address bus
        adr:OUT t_wlogic_vector(31 DOWNTO 0);
        asi:OUT t_wlogic_vector( 7 DOWNTO 0);

        -- data bus
        d:INOUT t_wlogic_vector(31 DOWNTO 0);

        -- bus cycle definition
        lock:OUT t_wlogic;   -- bus lock
        rd:OUT t_wlogic;     -- read
        dfetch:OUT t_wlogic; -- data fetch
        size:OUT t_wlogic_vector(1 DOWNTO 0); -- data size
        ldst:OUT t_wlogic;   -- atomic load/store

        -- bus control
        hal:OUT t_wlogic;    -- hold address latch
        we:OUT t_wlogic;     -- write enable
        mds:IN t_wlogic;     -- memory data strobe
        aoe:IN t_wlogic;     -- alternate output enable
        adroe:IN t_wlogic;   -- address output enable
        doe:IN t_wlogic;     -- data output enable
        mholda:IN t_wlogic;  -- memory hold
        mholdb:IN t_wlogic;  -- memory hold
        mholdc:IN t_wlogic;  -- memory hold
        shold:IN t_wlogic;   --
        bhold:IN t_wlogic;   -- bus hold
```

```
        nullcyc:OUT t_wlogic; -- nullify current cycle
        in_null:IN t_wlogic; -- inhibit null cycle enable
        tc:IN t_wlogic; --
        mexc:IN t_wlogic; -- memory exception

        -- floating point coprocessor interface
        f:OUT t_wlogic_vector(31 DOWNTO 0); -- FP instr/addr bus
        fexc:IN t_wlogic; -- fl. pt. exception
        fhold:IN t_wlogic; -- fl. pt. hold
        fcc:IN t_wlogic_vector(1 DOWNTO 0); -- FP cond code
        fccv:IN t_wlogic; -- fl. pt. condition code valid
        fp:IN t_wlogic; -- fl. pt. unit present
        fxack:OUT t_wlogic; -- fl. pt. exception acknowledge
        flush:OUT t_wlogic; -- fl. pt. instruction flush
        fend:OUT t_wlogic; -- fl. pt. end
        fadr:OUT t_wlogic; -- fl. pt. address
        fins:OUT t_wlogic; -- fl. pt. instruction
        -- interrupt level
        irl:IN t_wlogic_vector(3 DOWNTO 0); -- intrpt req level

        -- special
        reset:IN t_wlogic; -- processor reset
        error:OUT t_wlogic; -- processor error

        -- clocks
        clk1:IN t_wlogic; -- 75 % / 25 % duty cycle
        clk2:IN t_wlogic; -- 25 % / 75 % duty cycle

        -- unused pins
        vdd:IN t_wlogic_vector(15 DOWNTO 0); -- unused
        gnd:IN t_wlogic_vector(19 DOWNTO 0); -- unused

        -- unused test interface pins
        xsm:IN t_wlogic; -- scan clocks enable
        xtest:IN t_wlogic; -- scan clocks enable
        xack:IN t_wlogic; -- scan clocks
        bck:IN t_wlogic; -- scan clocks
        sdi:IN t_wlogic; -- scan data input
        sdo:OUT t_wlogic -- scan data output
    );
BEGIN
    -------------------------------------------------------------------------
    -- ENTITY STATEMENT PART: This part of the model essentially defines
    --     an interface shell for the model and contains statements which
    --     are common to all architectures of this model. Passive (no
    --     signals are updated) assertion statements, procedure calls, or
    --     passive process statements can go here.
    -------------------------------------------------------------------------


    ---------------------------
    -- CLOCK WAVEFORM CHECKS --
    ---------------------------
```

6.1. GETTING STARTED

```
    clk1_chk : PROCESS (clk1)
        VARIABLE timekeeper : waveform_record;
    BEGIN
    assert_waveform_check (
                            pin         => clk1,       -- check clk1 waveform
                            period_min  => t_1,
                            period_max  => t_5 + t_6,
                            pw_hi_min   => t_5,
                            pw_hi_max   => t_5 + t_2,
                            pw_lo_min   => t_6,
                            pw_lo_max   => t_6 + t_3,
                            tr_min      => 0 ns,
                            tr_max      => t_2,
                            tf_min      => 0 ns,
                            tf_max      => t_3,
                            keeper      => timekeeper
                          );
    END PROCESS;

    clk2_chk : PROCESS (clk2)
        VARIABLE timekeeper : waveform_record;
    BEGIN
    assert_waveform_check (
                            pin         => clk2,       -- check clk2 waveform
                            period_min  => t_1,
                            period_max  => t_7 + t_8,
                            pw_hi_min   => t_7,
                            pw_hi_max   => t_7 + t_2,
                            pw_lo_min   => t_8,
                            pw_lo_max   => t_8 + t_3,
                            tr_min      => 0 ns,
                            tr_max      => t_2,
                            tf_min      => 0 ns,
                            tf_max      => t_3,
                            keeper      => timekeeper
                          );
    END PROCESS;
END mb86901_generic;
```

Notice that the entity name has a **_generic** suffix attached to it. This will be used later to bind a specific timing instance of the model to this generic form of the device. Each entry in the interface list is commented with its functionality in the actual device. The names of the generic items were obtained from the datasheet by adding **t_** to each of the timing variables in the AC timing information section of the datasheet. Throughout the model, the identifiers used should be descriptive of their function as this will aid in maintaining the model at a later time. Notice in the entity statement part of the model, there are two calls to a library routine named **assert_waveform_check**. Those two concurrent passive procedures monitor ports **clk1** and **clk2** respectively. And in the event that either of the clocks violates their specification for

- period
- pulse_width hi
- pulse_width lo
- rise time
- fall time

an assertion is created informing the user that the clock signal has failed to meet the spec which could result in the device malfunctioning in an actual circuit.

An entity is an ideal place to execute concurrent passive procedure calls such as these which do not depend on the state of the model during execution. Similarly, concurrent assertion statements which reveal user errors can also be placed in this part of the model.

6.1.3 Behavior

Behavioral modeling involves a translation process from the language and drawings of the databook to the hardware description language, in this case VHDL. To begin, we establish a framework for model development.

Establishing a foundation for the model

In VHDL, the process statement is the main construct used to describe devices behaviorally. In a process, the associated variables retain their state from one activation of the model to another. Therefore, registers, flags, data variables, bus cycle counters and the like should be defined and kept in a process.

Process variables are not visible outside of the process. Therefore you cannot write one process for the bus handler and another for the microsequencer and have the two communicate through variables. The only way for the two processes to communicate is through signals. The use of internal signals should be minimized, although there are times in which internal signals are of great help in clarifying a model as we will see in section 6.1.3.

Using Processes

A process can be thought of as a parallel procedure whose internal state is independently managed and retained within the scope of the process. There are a number of guidelines to consider when dealing with processes:

6.1. GETTING STARTED

1. A process will execute once during simulator initialization. Keep this in mind because if there is any code which you do not want to have executed, now is the time to create branches around that portion of the code. Initial value expressions can be effectively used to set up state values within the process.

2. The variables defined within the declarative region of a process retain their state from one simulation cycle to another. This allows you to manage data, registers, flags, etc without having to declare those items as signals.

3. If multiple processes exist in the same model, no assumptions can be made about execution order. If the model depends upon the specific order of process execution, then the model is in error and will have unpredictable results.

4. Processes can be made to be sensitive to specific signals. This is useful in a variety of situations including signal synchronization.

5. In behavioral modeling, except in sophisticated situations requiring bus resolution functions internal to the model, no two processes should assign values to the same signal.

In determining the number of processes to use consider the following rule: *Use a separate process only when the actions of that process are autonomous to the actions of any other process.* An example of a model of the 54/74F138 1-to-8 decoder illustrates this example as shown below

54/74F138 1-to-8 Decoder

```
USE std.std_logic.ALL;
ENTITY sn74138 IS
    GENERIC (
        tplh_a_to_o:time:= 3500 ps; -- tplh:An -> On
        tphl_a_to_o:time:= 4000 ps; -- tphl:An -> On
        tplh_e12_to_o:time:= 3500 ps; -- tplh:E1 or E2 -> On
        tphl_e12_to_o:time:= 3000 ps; -- tphl:E1 or E2 -> On
        tplh_e3_to_o:time:= 4000 ps; -- tplh:E3 -> On
        tphl_e3_to_o:time:= 3500 ps  -- tphl:E3 -> On
        );
    PORT (
        a0:IN t_wlogic;
        a1:IN t_wlogic;
        a2:IN t_wlogic;
        e1:IN t_wlogic;
        e2:IN t_wlogic;
        e3:IN t_wlogic;
        o0:OUT t_wlogic;
        o1:OUT t_wlogic;
```

```vhdl
            o2:OUT t_wlogic;
            o3:OUT t_wlogic;
            o4:OUT t_wlogic;
            o5:OUT t_wlogic;
            o6:OUT t_wlogic;
            o7:OUT t_wlogic
            );
END sn74138;

USE std.std_logic.ALL;
ARCHITECTURE behavioral OF sn74138 IS
    PROCEDURE decode3_to_8 (i2,i1,i0 : IN t_wlogic;
        out7,out6,out5,out4,
        out3,out2,out1,out0 : OUT t_wlogic) IS
    BEGIN
        out7 := F1; out6 := F1; out5 := F1; out4 := F1;
        out3 := F1; out2 := F1; out1 := F1; out0 := F1;
        IF    ((i2 = F1)AND(i1 = F1)AND(i0 = F1)) THEN out7:=F0;
        ELSIF ((i2 = F1)AND(i1 = F1)AND(i0 = F0)) THEN out6:=F0;
        ELSIF ((i2 = F1)AND(i1 = F0)AND(i0 = F1)) THEN out5:=F0;
        ELSIF ((i2 = F1)AND(i1 = F0)AND(i0 = F0)) THEN out4:=F0;
        ELSIF ((i2 = F0)AND(i1 = F1)AND(i0 = F1)) THEN out3:=F0;
        ELSIF ((i2 = F0)AND(i1 = F1)AND(i0 = F0)) THEN out2:=F0;
        ELSIF ((i2 = F0)AND(i1 = F0)AND(i0 = F1)) THEN out1:=F0;
        ELSIF ((i2 = F0)AND(i1 = F0)AND(i0 = F0)) THEN out0:=F0;
        END IF;
    END decode3_to_8;
    SIGNAL  e : integer;
BEGIN
    PROCESS (a0,a1,a2,e1,e2,e3)
        VARIABLE q7,q6,q5,q4,q3,q2,q1,q0 : t_wlogic;
        VARIABLE enable : boolean;
        VARIABLE tplh : time := 1 ns;
        VARIABLE tphl : time := 1 ns;
    BEGIN
        enable := (e1 = F0) AND (e2 = F0) AND (e3 = F1);
        IF (enable) THEN
            e <= 1;
            IF (e1'STABLE)     AND -- Only the addr changed
               (e2'STABLE)     AND
               (e3'STABLE)     THEN
                tplh := tplh_a_to_o;
                tphl := tphl_a_to_o;
                e <= 2;
            ELSIF (a0'STABLE)  AND -- Only the e3 changed
                  (a1'STABLE)  AND
                  (a2'STABLE)  AND
                  (e1'STABLE)  AND
                  (e2'STABLE)  AND
                  (e3'EVENT )  THEN
                tplh := tplh_e3_to_o;
                tphl := tphl_e3_to_o;
```

6.1. GETTING STARTED

```
            e <= 3;
      ELSIF (a0'STABLE)       AND -- Only the e1/e2 changed
            (a1'STABLE)       AND
            (a2'STABLE)       AND
            (e3'STABLE)       AND
            ((e2'EVENT )      OR
             (e1'EVENT ))     THEN
         tplh := tplh_e12_to_o;
         tphl := tphl_e12_to_o;
         e <= 4;
      ELSE e <= 5;
      END IF;
      decode3_to_8 ( a2,a1,a0,q7,q6,q5,q4,q3,q2,q1,q0 );
      IF (q7 = F1) THEN o7 <= F1 AFTER tplh;
      ELSE o7 <= F0 AFTER tphl; END IF;
      IF (q6 = F1) THEN o6 <= F1 AFTER tplh;
      ELSE o6 <= F0 AFTER tphl; END IF;
      IF (q5 = F1) THEN o5 <= F1 AFTER tplh;
      ELSE o5 <= F0 AFTER tphl; END IF;
      IF (q4 = F1) THEN o4 <= F1 AFTER tplh;
      ELSE o4 <= F0 AFTER tphl; END IF;
      IF (q3 = F1) THEN o3 <= F1 AFTER tplh;
      ELSE o3 <= F0 AFTER tphl; END IF;
      IF (q2 = F1) THEN o2 <= F1 AFTER tplh;
      ELSE o2 <= F0 AFTER tphl; END IF;
      IF (q1 = F1) THEN o1 <= F1 AFTER tplh;
      ELSE o1 <= F0 AFTER tphl; END IF;
      IF (q0 = F1) THEN o0 <= F1 AFTER tplh;
      ELSE o0 <= F0 AFTER tphl; END IF;
ELSE   -- enable = F0 ...
      e <= 6;
      IF (e3'EVENT) AND ((e1'EVENT) OR (e2'EVENT)) THEN
            -- e1, e2, and e3 all changed to shut it down...
            -- pick the fastest shut down time...
            IF (tplh_e3_to_o < tplh_e12_to_o) THEN tplh:=tplh_e3_to_o;
            ELSE tplh  := tplh_e12_to_o;
            END IF;
            e <= 7;
      ELSIF (e3'EVENT)                THEN
            tplh := tplh_e3_to_o;
            -- e3 disabled the decoder...
            e <= 8;
      ELSIF (e1'EVENT) OR (e2'EVENT) THEN
            tplh := tplh_e12_to_o;
            -- e1 or e2 disabled the decoder...
            e <= 9;
      ELSE e <= 10;
      END IF;
      o7 <= F1 AFTER tplh; o3 <= F1 AFTER tplh;
      o6 <= F1 AFTER tplh; o2 <= F1 AFTER tplh;
      o5 <= F1 AFTER tplh; o1 <= F1 AFTER tplh;
      o4 <= F1 AFTER tplh; o0 <= F1 AFTER tplh;
```

```
    END IF;
  END PROCESS;
END BEHAVIORAL;

CONFIGURATION sn74F138 OF sn74138 IS
    FOR behavioral
    END FOR;
END sn74F138;
```

This example is a full production quality model of the device and takes into consideration that the propagation delays are path dependent and level dependent. In other words, you cannot simply state that, for example, output pin o5 has a delay of 15 ns under all conditions. Output pin o5's delay can be one of three different delays depending on whether the address lines caused the output to change or whether the change was caused by the e1, e2 or e3 enable lines and whether the output pin changes to '0' or '1'. Notice that all of the delays are generically mapped and provided default expressions.

In choosing an approach to modeling, an initial thought was to break the model into three processes. One for the address lines, one for the e1 and e2 lines and a third for the e3 line. But then each process would have to assign values to each of the outputs violating one of our suggested rules above. In addition, it would require the breakdown of delays into sub-delay elements. This becomes a difficult and error prone task for a model of this complexity.

In this model, there is only one process which is sensitive to all of the input pins. To determine which pins changed, you can use the s'EVENT attribute. In this model, the functionality is trivial. The majority of the code deals with which delay to use.

Another example of the MB86901 SPARC processor illustrates the use of multiple processes as shown below

Multiple Processes in MB86901 Model

```
ARCHITECTURE behavioral OF mb86901_generic IS
    SIGNAL mds_sync:t_state;
    SIGNAL mexc_sync:t_state;
    SIGNAL mholda_sync:t_state;
    SIGNAL mholdb_sync:t_state;
    SIGNAL mholdc_sync:t_state;
    SIGNAL shold_sync:t_state;
    SIGNAL bhold_sync:t_state;
    SIGNAL IRL_sync:t_state_vector(3 DOWNTO 0);
    SIGNAL reset_sync:t_state;
    SIGNAL fexc_sync:t_state;
    SIGNAL fhold_sync:t_state;
```

6.1. GETTING STARTED

```
    SIGNAL fcc_sync:t_state_vector(1 DOWNTO 0);
    SIGNAL fccv_sync:t_state;
    SIGNAL aoe_sync:t_state;
    SIGNAL adroe_sync:t_state;
    SIGNAL doe_sync:t_state;
    SIGNAL in_null_sync:t_state;
    SIGNAL tc_sync:t_state;
    SIGNAL fp_sync:t_state;
    SIGNAL xsm_sync:t_state;
    SIGNAL xtest_sync:t_state;
    SIGNAL xack_sync:t_state;
    SIGNAL bck_sync:t_state;
    SIGNAL sdi_sync:t_state;
    SIGNAL d_sync:t_state;
    CONSTANT x_force_val:t_state:= '0'; -- force all X's to '0'
BEGIN
    -- synchronize asynchronous lines to the clk1 rising/falling edges
    sync_clk1 : PROCESS (clk1)
    BEGIN
        IF f_falling_edge (clk1) THEN
            mds_sync_falling <= mds;     -- record the values now
            mexc_sync_falling <= mexc;   -- record the values now
            mholda_sync <= netval (mholda, '0'); -- sync to clk1 falling
            mholdb_sync <= netval (mholdb, '0'); -- sync to clk1 falling
            mholdc_sync <= netval (mholdc, '0'); -- sync to clk1 falling
            shold_sync <= netval (shold, '0');   -- sync to clk1 falling
            bhold_sync <= netval (bhold, '0');   -- sync to clk1 falling
        ELSIF f_rising_edge (clk1) THEN
            -- mds and mexc need to be asserted during both clk1 edges
            IF mds_sync_falling = mds THEN
                mds_sync <= netval (mds, '0');  END IF;
            IF mexc_sync_falling = mexc THEN
                mexc_sync <= netval (mexc, '0'); END IF;

            irl_sync <= netval (irl, '0');     -- sync to clk1 rising
            reset_sync <= netval (reset, '0'); -- sync to clk1 rising
            fexc_sync <= netval (fexc, '0');   -- sync to clk1 rising
            fhold_sync <= netval (fhold, '0'); -- sync to clk1 rising
            fcc_sync <= netval (fcc, '0');     -- sync to clk1 rising
            fccv_sync <= netval (fccv, '0');   -- sync to clk1 rising
        END IF;
    END PROCESS sync_clk1;

    -- Unknown handling
    aoe_sync       <= netval(aoe,     x_force_value); -- fast asynchronous
    doe_sync       <= netval(doe,     x_force_value); -- fast asynchronous
    in_null_sync   <= netval(in_null, x_force_value); -- fast asynchronous
    adroe_sync     <= netval(adroe,   x_force_value);
    tc_sync        <= netval(tc,      x_force_value);
    fp_sync        <= netval(fp,      x_force_value);
    xsm_sync       <= netval(xsm,     x_force_value);
    xtest_sync     <= netval(xtest,   x_force_value);
```

```
    xack_sync    <= netval(xack,    x_force_value);
    bck_sync     <= netval(bck,     x_force_value);
    sdi_sync     <= netval(sdi,     x_force_value);
    d_sync       <= netval(d,       x_force_value);

    -- main behavioral process
    subprocess : PROCESS ( aoe_sync,adroe_sync,doe_sync,in_null_sync,
                           tc_sync,fp_sync,xsm_sync,xtest_sync,
                           xack_sync,bck_sync,sdi_sync,d_sync,
                           clk1,clk2
                         )
      ...
    BEGIN
      ...
    END PROCESS subprocess;
END behavioral;
```

Most microprocessor class devices synchronize control lines to a clock edge. Each of these lines will be recognized internal to the device, if they satisfy their setup and hold times with respect to the sampling edge (in the code segment shown, the setup/hold checks were removed for this discussion). This is one of the first areas to investigate when modeling a complex device, since synchronizing signals to the clock makes the rest of the modeling task much easier.

In the declarative part of the architecture a number of signals are defined which are used as internal signals synchronized to the clock and X-state free. Process **sync_clk1** synchronizes the control lines to the falling or rising edge of the **clk1** port. Signals **mds** and **mexc** are not recognized internal to the chip unless they remain stable through both the falling and rising edges of **clk1**.

All of the remaining signals are processed to remove unknowns which may be present. This is done through concurrent signal assignment statements.

Modeling register sets

Register sets are most conveniently modeled as arrays of the data type of the register for register widths greater than 8. For eight bit processors, integers are an effective way to model registers. The reason for changing from integer type to arrays has to do with the word width of most popular workstation platforms and their compilers. It is rather certain that almost all contemporary workstations used for simulation are 16-bit or 32-bit based machines. An 8-bit integer representation of a register works well here. Sixteen bit integers may work, but multiplying two 16 bit integers can yield a 32-bit result with perhaps an overflow. And it is that situation

6.1. GETTING STARTED

which you then comes cannot easily control from VHDL.

For most applications, it is safe to use **t_wlogic_vector** or vectors of some other type such as (0,1,X,Z) to represent the register. In the following cases, the register data types are arrays of **t_state**.

When register groups are dealt with in a common manner such as 8 data registers and 9 address registers as of the MC68000 family, and here in the MB86901's register file and pipeline, it is useful to declare those objects as arrays of arrays, rather than individual arrays. In doing so, a given register of a register file can be addressed simply by its index in the array of registers. In addition, most arithmetic and logical operators are overloaded to handle arbitrary length vectors making it is easy to pass a register vector to the function and quickly return a value. The following demonstrates these concepts.

MB86901 Register Set

```
-- register set
VARIABLE   pipeline   : ARRAY (4 DOWNTO 0) OF register_32;
ALIAS      ibuffer_1  : register32 IS pipeline (4);
ALIAS      ibuffer_2  : register32 IS pipeline (3);
ALIAS      decode_reg : register32 IS pipeline (2);
ALIAS      exec_reg   : register32 IS pipeline (1);
ALIAS      write_reg  : register32 IS Pipeline (0);

VARIABLE   psr        : register32; -- Processor State Register
ALIAS      impl       : register4  IS psr(31 DOWNTO 28);
ALIAS      ver        : register4  IS psr(27 DOWNTO 24);
ALIAS      icc        : register4  IS psr(23 DOWNTO 20);
ALIAS      res        : register6  IS psr(19 DOWNTO 14);
ALIAS      pil        : register4  IS psr(11 DOWNTO  8);
ALIAS      cwp        : register5  IS psr( 4 DOWNTO  0);
CONSTANT   n          : natural := 23; -- negative flag in PSR
CONSTANT   z          : natural := 22; -- zero     flag in PSR
CONSTANT   v          : natural := 21; -- overflow flag in PSR
CONSTANT   c          : natural := 20; -- carry    flag in PSR
CONSTANT   ec         : natural := 13;
CONSTANT   ef         : natural := 12;
CONSTANT   s          : natural := 7;
CONSTANT   ps         : natural := 6;
CONSTANT   et         : natural := 5;

VARIABLE   wim        : register32; -- Window Invalid Mask

VARIABLE   tbr        : register32; -- Trap Base Register
ALIAS      tba        : register20 IS tbr(31 DOWNTO 12);
ALIAS      tt         : register8  IS tbr(11 DOWNTO  4);
ALIAS      nul        : register4  IS tbr( 4 DOWNTO  0);

VARIABLE   y          : register32; -- Y Register
```

```
VARIABLE    pc         : register32; -- Program Counter
VARIABLE    npc        : register32; -- Next Program Counter

-- register file
VARIABLE    r          : ARRAY (119 DOWNTO 0) OF register32;
```

Instruction Sets

Instruction sets are easily modeled as enumerated types as shown below for the MB86901.

MB86901 Instructions

```
TYPE instruction_set IS (
        FABS    , ADD     , ADDCC   , ADDX     ,
        ADDXCC  , FADDD   , FADDS   , I_AND    ,
        ANDCC   , ANDN    , LDSTUB  , LDSTUBA  ,
        BICC    , FBFCC   , CALL    , FCMPD    ,
        FCMPED  , FCMPS   , FCMPES  , FDTOI    ,
        FDTOS   , FITOD   , FITOS   , FSTOD    ,
        FSTOI   , FDIVD   , FDIVS   , XNOR     ,
        XNORCC  , XOR     , XORCC   , I_OR     ,
        ORCC    , ORN     , ORNCC   , JMPL     ,
        LDD     , LDDA    , LDDF    , LDF      ,
        LDFSR   , LDSB    , LSDBA   , LDSH     ,
        LDSHA   , LDUB    , LDUBA   , LDUH     ,
        LDUHA   , LD      , LDA     , FMOVS    ,
        FMULD   , FMULS   , MULSCC  , FNEGS    ,
        RDPSR   , RDTBR   , RDWIM   , RDY      ,
        RESTORE , RETT    , SAVE    , SETHI    ,
        SLL     , SRA     , SRL     , STB      ,
        STBA    , STDFQ   , STDF    , STD      ,
        STDA    , STF     , STFSR   , STH      ,
        STHA    , ST      , STA     , SUB      ,
        SUBCC   , SUBX    , SUBXCC  , FUSBD    ,
        FSUBS   , TADDCC  , TADDCCTV, TSUBCC   ,
        TSUBCCTV, TICC    , WRPSR   , WRTBR    ,
        WRWIM   , WRY     );
```

Notice that the instruction mnemonic **I_AND** and **I_OR** replace AND and OR respectively since they violate the reserved word list of VHDL.

Chip reset

Almost every operation which occurs in a complex device begins upon the negation of chip reset. The function of chip reset is to place the device (and model) in a known start-up state. It is from that state that the

6.1. GETTING STARTED

proper execution of the model can take place. It is also one of the first subprograms that you will need to write before testing the basic integrity of the model.

Initialization

In order to assure a properly functioning model, it is essential to have the discipline to specifically initialize each variable to a known state prior to model execution. In VHDL, variables are assigned a default initial value of T'Left. However, for completeness and visual awareness, it is recommended that initial value expressions be assigned to every variable even when the value is the same as the default value.

Step by step Approach for Complex Microprocessor model

Below is a method of attack which was found to work well in the development of many complex VLSI microprocessor class models. Review the databook as you go and highlight in yellow those sections of the databook which you have modeled. In doing so, it will be apparent what remains to be completed by looking for un-highlighted areas.

1. Define the shell of the model (the entity in VHDL)
2. Identify mode independent assertion checks and place them in the entity statement part.
3. Synchronize any signals that are sampled.
4. Remove unknowns from the interface data stream.
5. Define the register set and instruction set
6. Write the bus handler routine
7. Write the reset handler routine
8. Write the sequencer routine
9. Write the NOP instruction subprogram (if the processor doesn't have a NOP instruction then fake one)
10. Kick off the processor reset and see if it runs...make repairs as needed. At this stage congratulate yourself, since you have handled one of the most difficult parts of the modeling task. That is, defining a structure for the model and getting the model to run NOPS.
11. Write a branch instruction
12. See if you can run NOP-Branch-NOP-Branch... This checks proper operation of the sequencer and pipeline.

13. Write the interrupt handler
14. Test the interrupt handler on the branch loop... This tests returns from exception as well as proper exception handling including stack operations that may have occurred
15. Write each instruction subprogram and integrate them in with the model.
16. Execute each instruction with each addressing mode.
17. Clean up any remaining loose ends.

This basic scheme has been used for many complex behavioral models and has been quite successful.

6.2 The Timing Model

There are three general timing related issues concerning models. One is that there is, in general, more than one version of a chip, each with its own timing data. Secondly, for a given version of a device, there is a manufacturing tolerance on those timing specifications which represent minimum, typical, and maximum values. Lastly, the exact timing of a device depends on its environment, namely loading, temperature, and supply voltage.

In the next three sections, we will develop this concept to a full timing implementation which takes into account speed versions and min-typ-max timing.

6.2.1 Device Speeds

In handling multiple device speeds, it is easiest to define a generic entity whose behavior reflects the device in the general case, and then to pass to this generic model specific timing data for a given device version.

STEP 1 : Define an entity of the generic model including generics as show below:

Defining Generic Model

```
ENTITY mb86901_generic IS
    GENERIC (   t_1     : time;
                ...
                x_force_val : t_state
            );
    PORT    (
                adr:OUT t_wlogic_vector(31 DOWNTO 0);
                ...
                sdo:OUT t_wlogic
```

6.2. THE TIMING MODEL

```
              );
BEGIN
 ...
END mb86901_generic;
--------------------------------------------------
ARCHITECTURE behavioral OF mb86901_generic IS
 ...
BEGIN
 ...
END behavioral;
--------------------------------------------------
CONFIGURATION mb86901_config OF mb86901_generic IS
    FOR behavioral
    END FOR;
END mb86901_config;
```

STEP 2 : Create one entity for each of the specific timing versions that exist as below:

Defining Specific Timing Versions

```
-- 20 Mhz Version
--------------------------------------------------
ENTITY mb86901_APR_G   IS
    GENERIC (     instance_id : integer
            );
    PORT    (
                    adr:OUT t_wlogic_vector(31 DOWNTO 0);
                         ...
                    sdo:OUT t_wlogic
            );
BEGIN
 ...
END mb86901_apr_g;
--------------------------------------------------
ARCHITECTURE structural OF mb86901_APR_G IS
    COMPONENT ic_socket GENERIC (t_1: time;
                                  ...
                                 x_force_val  : t_state
                                );
                        PORT (
                           adr:OUT t_wlogic_vector(31 DOWNTO 0);
                                ...
                           sdo:OUT t_wlogic
                             );
    END COMPONENT;
BEGIN
    -- Now call the generic part while passing in timing
    instance : ic_socket
                  GENERIC MAP ( t_1 => 50 ns,
                                    ...
```

```
                              t_55_hold => 6 ns
                           )
                 PORT MAP  ( adr,
                             ...
                            sdo
                           );
END structural;
-------------------------------------------------
CONFIGURATION mhz_20 OF mb86901_apr_g IS
   USE work.ALL;
   FOR structural
      -- Now "plug" the generic part into the socket !
      FOR instance : ic_socket
         USE ENTITY work.mb86901_generic (behavioral);
      END FOR;
   END FOR;
END mhz_20;
```

STEP 3 : Repeat step 2 for each timing version you need.

There are numerous advantages to this approach including the ability to hide timing details from the user as shown in the entity **mb86901_apr_g** when referenced from a circuit.

6.2.2 Min/Max Timing

In order to assure that a given design will work under the worst case timing conditions of a given component's manufacturing timing spread, designers frequently exercise their circuits using worst case timing data. This takes the form of minimum setup times and maximum propagation delays.

Models must have provisions for the designer to select which set of timing data to use on a given simulation run.

6.2.3 Drive/Loading Dependencies

The Fujitsu SPARC Processor chip, like many other devices, has its propagation delays dependent upon the capacitive loading of its output pins. For the MB86901, the graph is shown in figure 6.3. The graph indicates that a 60 pf load will add 4.5 ns to the propagation delay of a given output pin; almost a 10 percent slow down for a 25 Mhz system clock rate. Hence, the importance of minimal loading and timing analysis are crucial to high speed performance.

In order to determine what propagation delay to use, the loading for each and every output pin in the circuit must be evaluated. For this to occur, the loading information for all of the input pins driven by a given output pin must be determined.

6.2. THE TIMING MODEL

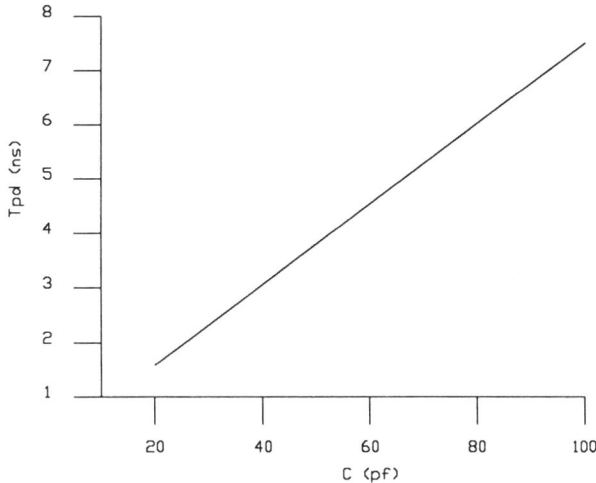

Figure 6.3: Capacitive Loading for Fujitsu MB86901

In the interest of maintaining technology independence in the model, specific timing data should not be a part of the model. Rather, specific timing information should be passed into the model through generics as shown in section 6.2.1.

6.2.4 A Uniform Approach to Device Dependent Data

In the past few years we have seen a continual increase in integration of functions within VLSI devices. Particularly in advanced microcontrollers it is not unusual to find ROM, A/D converters and EEPROM all on the same chip. An example of such a device is Motorolla's MC68HC11A4.

Other programmable devices include the more recent programmable gate-array products from companies such as Xilinx and Altera and the traditional assortment of ROMs and PROMs from a variety of manufacturers.

Each of these needs to be programmed. But in a simulation environment, there are a number of issues to be addressed:

- The program must be associated with a specific instance of the device (the same as burning a PROM and plugging it into a socket).

- During netlist expansion, the simulator assigns a unique instance identifier to the device. But what happens if you change the circuit? Does this same ic_socket receive the same instance number as it had the last time the circuit was expanded? Chances are that it doesn't. Therefore, any association scheme between a program file and a device cannot use the simulators internal instance naming scheme.

- It is undesireable to compile the program into the VHDL model as this, in general, takes a great deal of time and causes the simulator to re-expand and re-initialize the circuit. (This is true to a lesser extent with incremental-compilation type simulator products as well).
- The program needs to be easily changeable and the circuit rerun quickly. All of this points to an association scheme where the instantiated model reads a file and programs itself accordingly.
- It is desirable to have a common way of handling ROM, PROM, and other data for a given device.

One way in which to do this is to define a generic called **instance_id** in the entity which is fully bound to the generic model. Then use the **instance_id** in a procedure call to program the part as follows:

Binding VHDL Model to Data File

```
program_eprom ( "MC68HC11A4.EPROM", instance_id ); -- EPROM program
program_rom   ( "MC68HC11A4.ROM"  , instance_id ); -- ROM program
sample_a_d    ( "MC68HC11A4.ADC"  , instance_id ); -- A/D dataset
```

If there were two instances identified as 23 and 45, then in the directory of the circuit, you should create the following files:

Data File Names

```
MC68HC11A4.EPROM.23    MC68HC11A4.ROM.23    MC68HC11A4.ADC.23
MC68HC11A4.EPROM.45    MC68HC11A4.ROM.45    MC68HC11A4.ADC.45
```

These would exist along with the timing information contained in the file:

Timing File Names

```
MC68HC11A4.TIMING.23    MC68HC11A4.TIMING.45
```

This approach can be appreciated especially when the number of programmable devices in the circuit grows. The modeling is straightforward, the file names are clear as to which instance they correspond and of what class of data they contain.

6.3 Error Handling

The power of simulation becomes evident in its ability to identify designer errors in the application of a device in a circuit and to quickly inform the designer of the conditions in which the errors occurred.

Errors can occur if a setup time is violated, or if a signal is not held long enough after a clock edge, or if a clock periodicity deviates from the device specification perhaps due to a temperature rise affecting the tuning frequency of a crystal oscillator. In all of these cases, the designer should be informed of the error.

This section will discuss ways of handling each of these areas.

6.3.1 Unknowns

It is rare that the logical type Bit ('0','1') suffices for actual logic where tristate conditions and open-collector or open-emitter situations arise. In response to the need to model actual hardware, designers have developed state-strength or multi-state data types for use in describing model behavior as discussed in chapter 7.

Throughout this book, we have used **t_wlogic** as a multi-value datatype for signals. However, in actual hardware, only the real values of '0' or '1' apply. Inside the chip, there are no unknowns. Depending on the technology, an OPEN port may take on a either a '1' or a '0' value. But whatever its value, it is not unknown inside the device. When gates are concerned, it is possible to propagate an unknown through the model. The logic table for an OR gate is illustrated below

OR Gate Unknown Propagation

inputs	00	0X	X0	01	10	11	1X	X1	XX
output	0	X	X	1	1	1	1	1	X

But how does a complex microprocessor propagate unknowns? We will illustrate the concept with a few examples:

- *Memory Move Instruction*: Data is moved from one place in memory to another. Unknowns in the data present no problem.
- *Arithmetic Instruction*: If we are adding two numbers and one of them contains unknowns (from a previous load instruction), then with a bit more effort we could figure out which bits of the result were affected by the unknown and make those bits be unknown as well.
- *Instruction*: Now we have a bit of a problem, since unknowns appear on an instruction field. Perhaps the model cannot determine which of

four instructions to use since two bits are unknown. Proper handling of this situation is not clear in many cases.

- *Address*: Suppose the data containing unknowns was moved from the data registers to the address registers and then used to de-reference a data item. At what address do we then find the data?

From this discussion, it is clear that no attempt should be made to ascertain what operation to perform upon hitting an unknown in either a data stream or instruction or address. Instead, two actions should be taken:

- Map the unknown value to a '0' or '1' value in order to force a valid operation

- Issue an assertion to flag to the user that an unknown was detected and indicate what forced mapping was used

The mapping of unknown values should be handled through the use of generic parameters so that the user can adjust as appropriate for the given circuit and goals he has for his simulation run. An example of a useful function to accomplish this is given below

Unknown Handling Function

```
------------------------------------------------------------
-- Function Name : NETVAL
--
-- Parameters :
--      in   :: pin of t_wlogic which may be an X
-- Returns    : Clean binary value of 1 or 0
-- Purpose    : to clean out all X's and force the values to a given state
-- Notes      : Creates assertions to flag the existence of the X
--
-- Use        : pin_23_clean <= NETVAL (pin_23, '0');
--
------------------------------------------------------------
FUNCTION netval ( SIGNAL          pin : t_wlogic;
                  CONSTANT force_value : t_state
                ) RETURN t_state
IS
    VARIABLE state : t_state;
BEGIN
    CASE f_state (pin) IS
      WHEN '0' => state := '0';
      WHEN '1' => state := '1';
      WHEN 'X' => state := force_value; -- map to force_value
                         -- Assume ZX implies tristate value
      WHEN 'U' => state := force_value; -- map to force_value
    END CASE;
    RETURN state;
END netval;
```

6.3.2 Setup / Hold Time Techniques

Setup and hold time checks are also important to the circuit designer since problems which can affect the proper operation of the circuit are flagged. An example of a production quality setup check routine is given below

Setup/Hold Checks

```
-------------------------------------------------------------
-- Procedure Name : Assert_setup_violation
-- in            : test_port -- t_wlogic
-- in            : reference port -- t_wlogic
-- in            : condition -- boolean
-- Purpose       : To flag setup violations
-- Notes         : For use as either an passive concurrent procedure or
--                  sequential statement
--
--                  --------      ------------
--                  _____XXXXX_____    test port
--                  --------------|------
--                                |    \_____  reference port
--                  ----->|       |<---------
--
-- Use           : Assert_setup_violation ( test_port => control_line,
--                                          test_port_name => "control_line",
--                                          reference_port => clk,
--                                          ref_port_name => "clk",
--                                          condition => f_falling_edge(clk),
--                                          setup_spec => 23 ns
--                                          );
-------------------------------------------------------------
PROCEDURE assert_setup_violation
          ( SIGNAL     test_port        : IN t_wlogic;
            CONSTANT   test_port_name   : IN string;
            SIGNAL     reference_port   : IN t_wlogic;
            CONSTANT   ref_port_name    : IN string;
            CONSTANT condition          : IN boolean;
            CONSTANT setup_spec         : IN time
          )
IS
BEGIN
    -- we don't care if test_port wiggles here, only the ref_port

    IF reference_port'EVENT AND condition THEN
        ASSERT ((NOW - test_port'LAST_EVENT) > setup_spec)
            REPORT "Setup Violation on " & test_port_name &
                "w.r.t " & ref_port_name
            SEVERITY warning;
    END IF;
END assert_setup_violation;
```

This procedure is written to be used as either a concurrent passive procedure call which can be called in the entity statement part of a model or as a sequential procedure which can be called from within a process block.

In the formal parameter list of this routine, both the test and reference ports and names are passed. The routine watches for changes on the reference port. If a change has occurred and the condition satisfied, then the assertion is checked. If the last time the test port changed is more than the setup spec then the setup time specification is satisfied otherwise an assertion is flagged.

A routine for hold time checking could be developed similarly.

6.3.3 Waveform Checking

Most complex devices are run from a periodic clock which needs to maintain a certain periodicity, pulse width and transition time in order for the device which it is driving to operate properly. In modeling such devices, it is important to check for violations of these conditions and report them to the circuit designer.

Philosophically, it is suggested that the model continue to run after the error message has been reported. The designer will then have to decide whether to ignore the warning or to fix the waveform to comply with the device spec.

A useful procedure to use in testing for waveform violations is given below. First a record is defined whose purpose it is to keep track of internal data and retain its state as shown below:

Waveform Record Definition

```
TYPE waveform_record IS RECORD
                tf : time ;
                tr : time ;
                t0 : time ;
                t1 : time ;
                tX : time ;
                pw_lo : time ;
                pw_hi : time ;
                x_found : boolean ;
                period : time ;
END RECORD;
```

This procedure works by recording the transition times of the waveform whenever the state of the waveform is '0', '1' or 'X'. From the diagram below, the rise and fall times are taken as the $0->X->1$ time and $1->X->0$ time respectively. In cases where the signal does not traverse

6.3. ERROR HANDLING

through the X state, the transition time is considered to be zero. Pulse width high and low times are measured and recorded as well. The total period is determined from the addition of the pulse widths. The following demonstrates how this record is utilized in the **assert_waveform_check** procedure.

Waveform Checking

```
------------------------------------------------------------
-- Procedure Name : Assert_waveform_check
--
-- Parameters :
--          in  ::
-- Purpose     : To be used ONLY within a PROCESS
--               to monitor the waveform spec of a periodic signal.
-- Notes       : This procedure will be called each time that the
--               signal changes value.
--             : RISETIME => usually, rise time is 0 under step changes
--               of 0->1 during simulation modeling. But sophisticated
--               models may represent risetime as 0->X->1. Under this
--               situation, the risetime is assumed to be twice the X->1
--               transition time.
--             : FALLTIME => usually, fall time is 0 under step changes
--               of 1->0 during simulation modeling. But sophisticated
--               models may represent falltime as 1->X->0. Under this
--               situation, the falltime is assumed to be twice the X->0
--               transition time.
--             : PERIOD   =>
--
--          |<-------------------Period------------------->|
--          |<---------- PW_high--------->|<----PW_low---->|
--          |   _____     |                |   _____
--          |  /                     \    |                |  /
-- 50%- __|_/|-------" X state " ----|\_|__  _____ ____/----------
--       /  |  |                      |  |  _____/
--       |  |  |                      |  |  |
--       |  |  |                      |  |  |
--      -->|  tr |<--                -->|  tf  |<--
------------------------------------------------------------
PROCEDURE assert_waveform_check
       ( SIGNAL    pin         : in t_wlogic; -- signal to check
         CONSTANT  period_min  : in time;
         CONSTANT  period_max  : in time;
         CONSTANT  pw_hi_min   : in time;
         CONSTANT  pw_hi_max   : in time;
         CONSTANT  pw_lo_min   : in time;
         CONSTANT  pw_lo_max   : in time;
         CONSTANT  tr_min      : in time;
         CONSTANT  tr_max      : in time;
         CONSTANT  tf_min      : in time;
         CONSTANT  tf_max      : in time;
```

```
            VARIABLE  keeper      : inout waveform_record
        )
IS
BEGIN
    IF f_state(pin) = '1' THEN
        keeper.t1 := NOW;
        IF keeper.x_found THEN  keeper.tr := 2*(keeper.t1 - keeper.tX);
        ELSE                    keeper.tr := 0 ns;
        END IF;
        keeper.pw_lo :=
            (keeper.t1-(keeper.tr/2))-(keeper.t0-(keeper.tf/2));
        keeper.period := keeper.pw_lo + keeper.pw_hi;

    ELSIF f_state(pin) = '0' THEN
        keeper.t0 := NOW;
        IF keeper.x_found THEN keeper.tf := 2*(keeper.t0 - keeper.tX);
        ELSE                   keeper.tf := 0 ns;
        END IF;
        keeper.pw_hi :=
            (keeper.t0-(keeper.tf/2))-(keeper.t1-(keeper.tr/2));
        keeper.period := keeper.pw_lo + keeper.pw_hi;
    END IF;

    IF f_state(pin) = 'X' THEN
        keeper.x_found := true;
        keeper.tX := NOW;
    ELSE
        keeper.x_found := false;
    END IF;

    IF NOW > 0 ns THEN
        ASSERT (keeper.period <= period_max)
            REPORT " period is too long" SEVERITY warning;
        ASSERT (keeper.period >= period_min)
            REPORT " period is too short" SEVERITY warning;
        ASSERT (keeper.pw_lo <= pw_lo_max)
            REPORT " pw_lo is too long" SEVERITY warning;
        ASSERT (keeper.pw_lo >= pw_lo_min)
            REPORT " pw_lo  is too short" SEVERITY warning;
        ASSERT (keeper.pw_hi <= pw_hi_max)
            REPORT " pw_hi is too long" SEVERITY warning;
        ASSERT (keeper.pw_hi >= pw_hi_min)
            REPORT " pw_hi  is too short" SEVERITY warning;
        ASSERT (keeper.tr <= tr_max)
            REPORT " tr is too long" SEVERITY warning;
        ASSERT (keeper.tr >= tr_min)
            REPORT " tr  is too short" SEVERITY warning;
        ASSERT (keeper.tf <= tf_max)
            REPORT " tf is too long" SEVERITY warning;
        ASSERT (keeper.tf >= tf_min)
            REPORT " tf  is too short" SEVERITY warning;
    END IF;
```

```
END assert_waveform_check;
```

This procedure is used inside of a process statement as shown below.

Checking a Waveform

```
wavechk : PROCESS (clk)
    VARIABLE timekeeper : waveform_record;
BEGIN
assert_waveform_check ( clk, 120 ns, 198 ns, 75 ns,
                        101 ns, 75 ns, 101 ns,
                        0 ns, 5 ns, 0 ns, 5 ns,
                        timekeeper
                      );
END PROCESS;
```

In the process, a variable **timekeeper** is declared. This is used to retain the timing information in the procedure.

6.4 Techniques for Modeling

The following sections describe techniques which are useful in modelling complex devices.

6.4.1 Bus Handlers

In developing a microprocessor class model, one of the first routines to develop is the bus handler. Bus handlers are simply a sequential machine which controls the transfer of information to and from the bus. The bus handler subprogram must be able to take care of issues such as:

- Byte, Word, Long Word and Quad Word data transfers
- Bus Error exceptions during transfer of data and termination of the bus cycle
- Assertion and Negation of the bus control lines
- Assertion checks for data line setup/hold times
- Tri-stating of the bus during inactive periods
- Synchronization with the rest of the processor

The form of the code is similar to that of the sequencer in section 6.4.3. Each state performs a given operation and the **bus_cycle** goes back to zero on the completion of the cycle.

6.4.2 Instruction Decoders

Observation of the instruction set opcodes often reveals fields which determine the opcode and other fields which determine addressing modes. Recognizing the common thread is the first step toward writing a decoder.

First, write an opcode comparator function which takes don't-cares as a legal matching pattern. For example, the following shows the opc_matches which uses a string pattern to match the opcode.

Opcode Comparator

```
-----------------------------------------------------------
-- Function Name : Opc_matches
--
-- Parameters :
--        in   :: matching pattern - string
--        in   :: opcode           - register13
-- Returns     : boolean true if there is a match; false otherwise
-- Purpose     : To decoder opcodes
-- Notes       : example of use :
--         IF    opc_matches ( "11xxxxx001001", opcode ) THEN
--               mnemonic := LDSB;
--         ELSIF opc_matches ( "11xxxxx011001", opcode ) THEN
--               mnemonic := LDSBA;
--         END IF;
-----------------------------------------------------------
FUNCTION opc_matches ( CONSTANT pattern : xstring13;
                       CONSTANT opcode  : register13 ) RETURN boolean IS
    VARIABLE match : boolean := true;
BEGIN
    FOR i IN opcode'RANGE LOOP
        NEXT WHEN pattern (i) = 'x'; -- jump to the next pattern
        IF opcode(i) = '1' AND pattern(i) /= '1' THEN RETURN false; END IF;
        IF opcode(i) = '0' AND pattern(i) /= '0' THEN RETURN false; END IF;
    END LOOP;
    RETURN true;
END;
```

Use of a string lets us simply transcribe the opcodes from the datasheet to the if-elsif block without error.

Another way to write a decoder is to write a series of case statements. For the MB86901, the outer case statement would concentrate on bits 31..30 of the instruction. The next set of inner case statements would concentrate of bits 24..19 of the instruction. By using the case statement technique, it is easy to make mistakes in decoding the instruction, since more work is required of you in interpreting the datasheet. For almost all decoders of RISC or CISC instructions, the pattern matching scheme works best.

6.4.3 Sequencers

Sequencers are sequential state machines, and like all state machines including counters, a sequencer can be modeled as a case statement. A sequencer is shown below

Sequencer

```
Procedure Simple_Sequencer is
begin
    case seq_cycle is
        when 0 =>   Fetch_opcode;
                    seq_cycle := seq_cycle + 1;

        when 1 =>   Decode_opcode;
                    seq_cycle := seq_cycle + 1;

        when 2 =>   Fetch_data;
                    seq_cycle := seq_cycle + 1;

        when 2 =>   Execute_opcode;
                    seq_cycle := seq_cycle + 1;

        when 3 =>   Store_result;
                    seq_cycle := 0;
    end case;
end sequencer;
```

6.4.4 Instruction Sets

Instruction sets can readily be modeled as one subprogram for each instruction mnemonic. The execution of any given instruction involves the execution of the corresponding subprogram.

6.5 Quality Assurance

Developing high quality models of off-the-shelf components requires a commitment of time and money for the task to be accomplished properly. As a benchmark for determining the level of effort required to develop a model without the benefit and detailed knowledge of the actual chip designer, the following rules of thumb for model development time are useful:

- SSI / MSI complexity: 2 - 15 hours
- MSI / LSI complexity: 16 - 40 hours
- LSI complexity (ie. microcontroller): 41 - 160 hours

- USART: 160 - 320 hours
- RISC microprocessor: 500 - 1500 hours
- CISC microprocessor: 1000 - 2000 hours
- DMA controller: 500 - 1500 hours
- Complex DSP chip: 1000 - 3000 hours
- Complex Co-Processor: 1000 - 3000 hours

There is a wide variation in effort for a given class of device. The level of effort correlates well with the complexity of the device. The more functions integrated into a device, the longer it will take to model.

While the level of effort stated above may seem high, the estimates derive from extensive experience in tracking model development efforts and include time for

- model development
- test generation
- hardware model development
- test validation
- documentation

The goal is to develop a model which emulates the actual device. The information available on the device is the datasheet and an occasional phone call to the semiconductor vendor's applications group.

Assuring quality in behavioral modeling must first start with a systematic approach to building quality into the model from the outset. To make sure that every paragraph in the datasheet has been modeled, a highlighting marker becomes an effective coverage analysis tool as outlined in section 6.1.3. Simply highlight each paragraph in the datasheet as it is modeled.

6.5.1 Developing a Test Plan

The test plan should include a mechanism for addressing each and every function of the device. In addition, every timing constraint needs to be checked as well. The following techniques and suggestions provide guidance in this area.

Avoiding Double-Blind Errors

It should be kept in mind that running tests against model code is not sufficient to prove that the model is fully coded or that the code works as does the chip, particularly when the tests are written by the same person who wrote the model.

Double-Blind errors can error where:

6.5. QUALITY ASSURANCE

1. the modeller misunderstood the functionality of the part, and
2. coded it according to that understanding, and then
3. tested it according to that same understanding.

The test passes, but the model is incorrect.

To avoid such problems, a few suggestions are in order:

- Have someone other than the model writer develop the tests
- Check that every setup/hold/waveform violation is tested
- Document each model extensively as you are modeling it, not after the coding is completed so that issues are fresh in mind while the coding is taking place
- Use commercially available assembly code (that is tested) to run against the model whenever possible
- Run code coverage on the model
- Run the model against the actual device

Developing a Regression Test Suite

A regression test suite is essential to maintain for each model. The tests should each be simple enough to pinpoint specific areas of functionality or timing in the model to the exclusion of others.

Develop a test plan as you proceed through the datasheet and model the part. For every paragraph, write down an applicable test procedure to assure that the paragraph is tested. This will assure that what is described in text is tested in the model after having been coded. In addition to these measures, techniques of code coverage, assertion tests and hardware modeling should be used.

The model should be run against a complete regression suite. The regression suite should contain but not be limited to tests for the following:

- All setup violations should be tested
- All hold violations should be tested
- All waveform violations should be tested
- All functional violations should be tested
- All X's should be flagged and mapped
- All Race conditions should be tested
- All functionality should be tested
- For ports with bus-resolution functions. The port should be tested with ports connected and disconnected to the bus to insure proper operation of the bus-resolution function.

Synergy between Test and Development

The most effective method of testing a model is to have one person develop a test suite while the other develops the model. Using this technique avoids the double blind problem mentioned above but also provides an independent view toward testing.

Since the schedule for a model's development includes the time to develop tests, a significant reduction in delivery date can be accomplished by splitting the work between two parties. The reduction will not be complete though, since both parties may have to communicate about the problem and in addition, two people now need to learn about the device instead of just one. All things considered, a 35 percent reduction in schedule time can be realized with a corresponding 40 percent increase in cost as shown in figure 6.4.

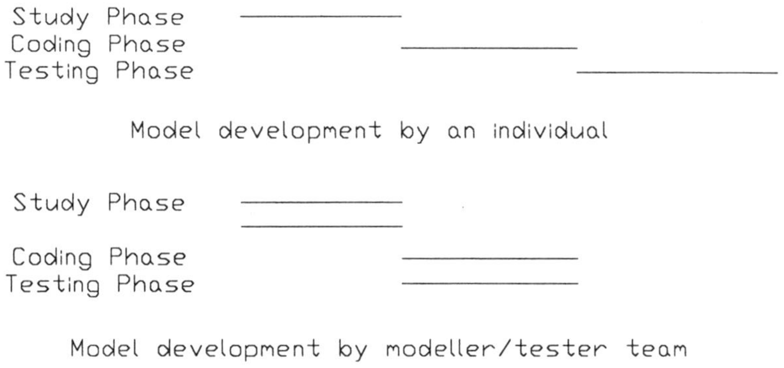

```
Study Phase      ―――――――
Coding Phase              ―――――――
Testing Phase                      ―――――――

         Model development by an individual

Study Phase      ―――――――
                 ―――――――
Coding Phase              ―――――――
Testing Phase             ―――――――

      Model development by modeller/tester team
```

Figure 6.4: Model Development Time

6.5.2 Validation of the Model

The following sections outline techniques which can be used to validate a complex behavioral model.

Code Coverage

Code coverage is an elegant way to determine how well a model is tested. At the beginning of every procedure call and branch path insert a write statement to a file which writes out a unique number or identifier for that branch if during the execution of the model, the path is traversed. The following demonstrates this approach.

6.5. QUALITY ASSURANCE

Code Coverage Diagnostics

```
PROCEDURE xyz IS
BEGIN
    ...
    WRITELINE ( code_coverage, "xyz: entry pt." );
    ...
    IF condition THEN
        ...
        WRITELINE ( code_coverage, "xyz: branch 22" );
        ...
    ELSIF condition_2 THEN
        ...
        WRITELINE ( code_coverage, "xyz: branch 23" );
        ...
    END IF;
END;
```

This writes a series of lines to a file which we'll assume, for example, is called **coverage_test_suite_1**. For each test suite run against the model, a separate coverage file can be generated.

Then when you have completed all of the tests, concatenate the code coverage files you just created together (see **catf** utility on Apollo workstations) followed by a utility to sort the file into order (see **srf**) and followed by a utility to delete duplicate lines (such as **dldupl**). To find out which branch paths were not traversed, create a file of all of the known branch path output statements (see **fpat** or **grep**). Sort this file as was done on the code coverage file. This is your reference template with which to compare the test coverage.

Use a file comparator utility (such as **cmf**) to compare the expected code coverage file against the results of all the tests. This powerful technique will quickly point out which need testing.

Assertion Tests

Figure 6.5 shows how setup and hold times are measured. Suppose a model contained a setup/hold spec for the transition of values on the DATA lines with respect to the falling edge of the **clock**. If **Ts** is the setup time specification and **Th** is the hold time specification, then as the transition of the **data** moves toward the setup time spec as shown above, an assertion will occur (if the model is performing properly).

If we move the **data** transition point 1 ns to the right every time the clock is negated, then we can walk the transition point through the setup time violation area and observe assertions being flagged on every falling

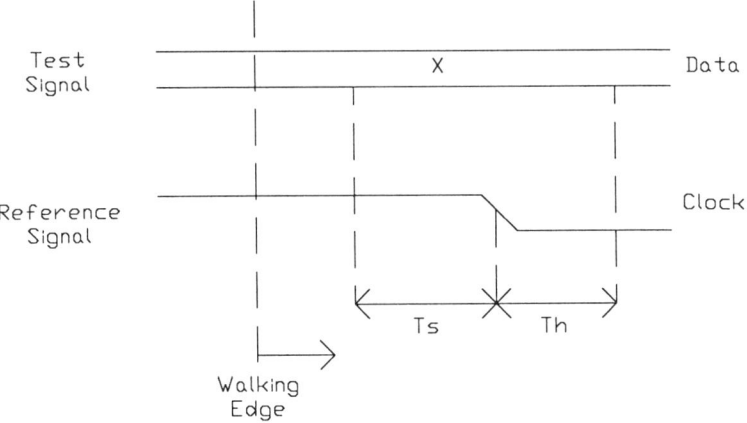

Figure 6.5: Setup/Hold Time

edge of the clock during the time in which the transition time is in the region **Ts** to **clock** falling.

Similarly, if we continue to propagate the transition 1 ns through the hold time region beyond the falling edge of the clock, we should observe hold time errors during this time until the transition is past the hold time spec.

Hardware Modeling

Hardware modellers provide the final and most conclusive tests that can be run against a behavioral model from a functionality and clock edge timing point of view. Hardware modellers should not, however, be called upon to verify precision timing since the timing information coming from the hardware modeller actually comes from the shell script written around the functional hardware model rather than the device under test itself. More advanced hardware modellers may have transient data recording capability which may provide the nanosecond precision. But keep in mind that the observed timing will be for only that chip which is plugged into the modeller and no other.

Figure 6.6 shows how a hardware modeller is used to test a VHDL model. Hardware models can tell you whether the model is doing what the actual device is doing on a functionality basis and on a clock edge to clock edge basis. It can tell you that on clock edge 13, for example, a write to address 16#FFFE occurs followed by an instruction fetch from address 16#AB9E.

6.5. QUALITY ASSURANCE

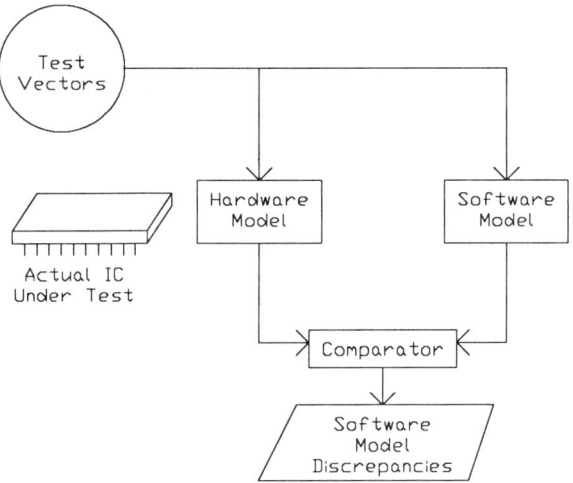

Figure 6.6: Hardware Model Testing

Chapter 7

The Standard Logic Package

The *Standard Logic Package* contains a collection of useful modelling facilities encapsulated in a VHDL package. This chapter will discuss this package in detail, highlighting the rationale behind each feature of the package and ways in which the package can be used to aid in modelling.

The following summarizes the key capabilities which this package offers to the hardware designer:

- Predefined logic value system
- Utility functions for manipulating logic values
- Logic tables
- Timing functions
- Utilize strictly VHDL IEEE standard 1076-1987 features, insuring that this package will run on a variety of VHDL environments

The value of this package (and a reflection of the elegance of the VHDL language) is that the hardware designer can utilize the power of the standard logic package without requiring detailed knowledge of its internal operation. As a result, what would otherwise be a complicated VHDL coding problem becomes simple. For example, the designer need never be concerned with writing a VHDL bus resolution function since this is provided automatically. In most cases, the designer need not be concerned with the logic value strengths and can focus on logic value states when writing models since strengths are handled automatically by overloaded operators.

This package reflects many man-years of experience with behavioral model library development. The requirements which are reflected in the

development of this package include:
- Support for switch level simulation
- Technology independence, allowing extensions for the future and providing support for all current technologies
- Provide consistent features which facilitate logic level modelling
- Hide the complexity of the package from the designer as much as possible, make models readable
- Provide for the timing accuracy continuum, giving the modeller flexibility to choose the appropriate level of model accuracy

The following sections will outline in detail the operation and use of the standard logic package.

7.1 Using the Standard Logic Package

Almost all of the models discussed in this book rely on the facilities which are defined in the standard logic package. Access to these facilities are possible through use of the VHDL use statement as shown below

Accessing Standard Logic

```
-- access standard logic facilities
USE std.std_logic.ALL;

-- determine specific technology to use
USE std.std_ttl.ALL;

-- VHDL entity declaration
--      .
--      .
--      .
```

The two statements shown should be declared immediately before defining a VHDL entity. In this way the facilities of the standard logic package are accessible from a VHDL device model. The first statement accesses all of the general facilities of the package. The second statement determines which specific technology rules will be used. The following summarizes the technology rules which are supported in the current version of the package:
- **std_cmos** - cmos technology rules
- **std_ecl** - ecl technology rules
- **std_nmos** - nmos technology rules
- **std_ttl** - ttl technology rules
- **std_ttloc** - ttl open-collector technology rules

7.2 The Logic Value System

VHDL is an unusual hardware description language with its flexibility in regard to value systems. It is possible in VHDL to define arbitrary and new value systems. In many cases this capability offers the designer flexibility which makes high level modelling of systems much more effective. For example, the designer can describe digital data in an abstract form in VHDL (by representing a bus value as an integer or a record) making it possible to ignore implementation details during simulation.

As the hardware designer gets closer to implementation of a system, this flexibility becomes less valuable and often undesirable. Furthermore, the designer likely will rely on pre-built VHDL libraries supplied by internal modelling groups or by outside vendors. Substantial experience has been gained in the use of the various value systems and in some cases in the mixing of value systems during a single simulation. Unfortunately, when value systems are mixed, significant technical challenges emerge including:

- Proper mapping from one value system to another

- Maintaining reasonable simulation efficiency

- Managing the increased complexity related to displaying simulation results to the user

- Managing the increased complexity of models which must handle mixed value systems

Mixing of logic value systems forces the use of VHDL type conversion functions which introduce complexity, inaccuracies and run-time performance penalties. In order for simulation to be effective, it is important that models from various sources utilize a common value system.

In order to avoid these conflicts and provide a uniform and accurate simulation environment for logic simulation, the standard logic package defines a single value system which anticipates the many requirements of the logic designer. The table below defines the logic values used in the Standard Logic Package.

Standard Logic System

value	F	R	W	Z	D	Z	W	R	F
U									
D					D				
Z0				Z0					
Z1						Z1			
ZDX				ZDX					
DZX						DZX			
ZX				ZX					
W0			W0						
W1							W1		
WZ0			WZ0						
WZ1							WZ1		
WDX			WDX						
DWX						DWX			
WZX			WZX						
ZWX						ZWX			
WX			WX						
R0		R0							
R1								R1	
RW0		RW0							
RW1							RW1		
RZ0		RZ0							
RZ1							RZ1		
RDX		RDX							
DRX						DRX			
RZX		RZX							
ZRX						ZRX			
RWX		RWX							
WRX						WRX			
RX		RX							
F0	F0								
F1									F1
FR0	FR0								
FR1								FR1	
FW0	FW0								
FW1							FW1		
FZ0	FZ0								
FZ1						FZ1			
FDX	FDX								
DFX						DFX			
FZX	FZX								
ZFX						ZFX			
FWX	FWX								
WFX						WFX			
FRX	FRX								
RFX		RFX							
FX	FX								

7.2. THE LOGIC VALUE SYSTEM

A total of 46 unique values are defined. Virtually every technology used today can be modelled with a reasonable level of accuracy using this value system. The states zero and one represent the logic values of false and true respectively. These when coupled with strengths result in the values shown. The strengths are summarized here:

- Forced (F).
- Strong resistive (R)
- Weak resistive (W)
- High impedance (Z)
- Disconnect (D)

Forcing strength represents a direct connection to the power supply. Strong resistive strength is associated with a resistive connection to the power supply, e.g. resistor pull-up and pull-down. Weak resistive strength represents a high resistive connection to the power supply. High impedance is the isolated node strength where the logic value is maintained by a capacitive charge (for tristate this represents the open state). The disconnect strength represents no value such as an isolated node with no capacitive charge.

The U value represents an uninitialized signal. The uninitialized value is important during simulator startup and is the initial value for all signals. By observing which signals remain uninitialized, the designer can identify failures in the circuit startup. In addition, the uninitialized value insures proper simulator startup by forcing all signals to change at least once before establishing a logic value. This is often important since the event driven nature of VHDL can suppress model execution thus leading to an invalid initial state of the simulator if the uninitialized value is not used.

Except for the special U value, all values shown in the table above are derived from the application of logic interval theory
. Using this methodology, we generate every possible range (interval) of values from the extremes of F0 and F1. By taking this approach, less pessimistic (and more accurate) simulation results are possible than the simpler usage of a single unknown value. It is possible to go one step further and build a value system which contains every possible combination of state/strength sets but in most cases this flexibility is not required and the additional overhead of manipulating the large number of values (in this case 256) is not warranted.

The values for modelling wires are based on four strengths (F, R, W, Z), two states (0, 1), a disconnect strength (D) which is neither a zero nor a one, and an uninitialized value named U that is similar to FX except that it is due to a non-computed value rather than an ambiguity predicted by a model. The interval notation allows the model to refer to sets of wire-values, provided that the values within a set are contiguous given the ordering based on strength and state: F0, R0, W0, Z0, D, Z1, W1, R1, F1. From the 9 boundary values above, we can derive 45 intervals, where 45 is obtained from $(1 + 2 + 3 + 4 + 5 + 6 + 7 + 8 + 9)$. With 45 intervals and a U value, we have a 46-value system.

The VHDL type definitions associated with the logic value system are shown below.

Value System Data Types

```
-- Base type definition for value system, 46 possible values
TYPE t_logic IS (
   U, D,
   Z0,Z1,ZDX,DZX,ZX,
   W0,W1,WZ0,WZ1,WDX,DWX,WZX,ZWX,WX,
   R0,R1,RW0,RW1,RZ0,RZ1,RDX,DRX,RZX,ZRX,RWX,WRX,RX,
   F0,F1,FR0,FR1,FW0,FW1,FZ0,FZ1,FDX,DFX,FZX,ZFX,FWX,WFX,FRX,RFX,FX
);

-- Array of the base type definition, used in bus resolution
TYPE t_logic_vector IS ARRAY(NATURAL RANGE <>) OF t_logic;

-- Bus resolution function to be used on all signals of type t_wlogic
FUNCTION f_logic_bus( s : t_logic_vector ) RETURN t_logic;

-- The logic data type to be used by modellers
SUBTYPE t_wlogic IS f_logic_bus t_logic;

-- Array of logic data type; useful for signal arrays
TYPE t_wlogic_vector IS ARRAY(NATURAL RANGE <>) OF t_wlogic;

-- The basic state values.
TYPE t_state IS ( '0', '1', 'X' );

-- The different strengths, in ascending order.
TYPE t_strength IS ( 'U', 'Z', 'W', 'R', 'F' );

-- The different technologies supported by this package.
TYPE t_technology IS ( ecl, cmos, nmos, ttl, ttloc );
```

t_logic defines the basic value system and its vector type t_logic_vector is used along with the bus resolution function f_logic_bus to define the type to be used by modellers: t_wlogic. The vector type is also available as t_wlogic_vector. When declaring variables, signals or ports in models the designer should use only the t_wlogic and t_wlogic_vector data types to insure automatic inclusion of the bus resolution function.

Finally, data types t_state, t_strength and t_technology are used by other functions in the logic package to allow easy mapping between the standard logic values and state, strength and technology.

7.3 Technology Rules

Specific technology dependent application of the various values are discussed in section 7.3.

7.3 Technology Rules

This section discusses the technology specific features of the standard logic package. Currently, the package supports the following technologies and follows the strength rules as described here.

- **CMOS** – Always generate forced/false, forced/true values on output signals.
- **TTL** – Generate forced/false, forced/true values on output signals which are not tristate. For tristate output, always generate high-impedance/false, high-impedance/true output.
- **NMOS** – Always generate resistive/false, forced/true values on output signals.
- **ECL** – Always generate forced/false, resistive/true values on output signals.
- **TTLOC** – Always generate forced/false, high-impedance/unknown values on output signals.

For technology independent applications, the modeller should use the following values:

- F0 - forced false
- F1 - forced true
- FX - forced unknown
- ZX - high impedance unknown
- U - uninitialized

By following this methodology, models which utilize the complete value system can be mixed with technology independent models. This approach also avoids the pitfalls of VHDL type conversions between two value systems and automatically invokes the standard logic package bus resolution function.

The **f_tech** lookup table as shown below is useful in converting from state values to a technology dependent value.

Technology Specific Output Value Lookup

```
-- Type definition for tables that follow
TYPE f_specific_t IS ARRAY( t_state'low TO t_state'high ) OF t_logic;

-- Type definition for table
TYPE f_tech_con IS ARRAY (t_technology'low to t_technology'high) OF t_logic;
TYPE f_tech_t IS ARRAY (t_state'low to t_state'high) OF f_tech_con;

-- Given state and technology, return appropriate output value
CONSTANT f_tech : f_tech_t := (
        ( R0,       -- '0', ecl
          F0,       -- '0', cmos
          F0,       -- '0', nmos
          F0,       -- '0', ttl
          F0 ),     -- '0', ttloc

        ( F1,       -- '1', ecl
          F1,       -- '1', cmos
          R1,       -- '1', nmos
          F1,       -- '1', ttl
          ZX ),     -- '1', ttloc

        ( RFX,      -- 'X', ecl
          FX,       -- 'X', cmos
          FRX,      -- 'X', nmos
          FX,       -- 'X', ttl
          FZX )     -- 'X', ttloc
    );
```

Generally, this function will be used for generating the output signal value for a technology dependent model.

The following

Use of Technology Lookup Table

Technology	State		
	0	1	X
ecl	R0	F1	RFX
cmos	F0	F1	FX
nmos	F0	R1	FRX
ttl	F0	F1	FX
ttloc	F0	ZX	FZX

summarizes the values generated by the **f_tech** table. The rationale behind these technology specific values are discussed in sections 7.3.1, 7.3.2, 7.3.3, 7.3.4 and 7.3.5.

The following shows an example of the use of this table.

7.3. TECHNOLOGY RULES

Use of Technology Lookup Table

```
new_state    := f_and( inp1, inp2 );
output_signal <= f_tech( new_state, ttl ) AFTER delay;
```

The technology independent AND of two signals **inp1** and **inp2** are stored in the variable **new_state** which is used along with a specific technology indicator **ttl** to generate the technology dependent value for the output signal **output_signal**.

The **f_convz** table shown below converts a logic value into its corresponding high impedance value.

Converting to High-Impedance Strength

```
-- Type definition for table
TYPE f_logic_T IS ARRAY (t_logic'low to t_logic'high) OF t_logic;

-- Given logic value, return appropriate high-impedance tristate logic value
CONSTANT f_convz : f_logic_t := (
         ZX, -- U
         D,  -- D
         Z0, -- Z0
         Z1, -- Z1
         ZX, -- ZDX
         ZX, -- DZX
         ZX, -- ZX
         Z0, -- W0
         Z1, -- W1
         Z0, -- WZ0
         Z1, -- WZ1
         ZX, -- WDX
         ZX, -- DWX
         ZX, -- WZX
         ZX, -- ZWX
         ZX, -- WX
         Z0, -- R0
         Z1, -- R1
         Z0, -- RW0
         Z1, -- RW1
         Z0, -- RZ0
         Z1, -- RZ1
         ZX, -- RDX
         ZX, -- DRX
         ZX, -- RZX
         ZX, -- ZRX
         ZX, -- RWX
         ZX, -- WRX
         ZX, -- RX
         Z0, -- F0
```

```
                Z1,  -- F1
                Z0,  -- FR0
                Z1,  -- FR1
                Z0,  -- FW0
                Z1,  -- FW1
                Z0,  -- FZ0
                Z1,  -- FZ1
                ZX,  -- FDX
                ZX,  -- DFX
                ZX,  -- FZX
                ZX,  -- ZFX
                ZX,  -- FWX
                ZX,  -- WFX
                ZX,  -- FRX
                ZX,  -- RFX
                ZX   -- FX
);
```

This table is useful in calculating the new state of a tristate node that has just had the control signal turned off with outputs as shown below

7.3. TECHNOLOGY RULES

High-Impedance Conversion

Logic Value	Z Logic Value
U	ZX
D	D
Z0	Z0
Z1	Z1
ZDX	ZDX
DZX	DZX
ZX	ZX
W0	Z0
W1	Z1
WZ0	Z0
WZ1	Z1
WDX	ZDX
DWX	DZX
WZX	ZX
ZWX	ZX
WX	ZX
R0	Z0
R1	Z1
RW0	Z0
RW1	Z1
RZ0	Z0
RZ1	Z1
RDX	ZDX
DRX	DZX
RZX	ZX
ZRX	ZX
RWX	ZX
WRX	ZX
RX	ZX
F0	Z0
F1	Z1
FR0	Z0
FR1	Z1
FW0	Z0
FW1	Z1
FZ0	Z0
FZ1	Z1
FDX	ZDX
DFX	DZX
FZX	ZX
ZFX	ZX
FWX	ZX
WFX	ZX
FRX	ZX
RFX	ZX
FX	ZX

The example shown below

Example of High-Impedance Strength Conversion

```
output_signal <= f_convz( input_signal ) AFTER delay;
```

shows how this table is utilized to generate a high-impedance output value on signal **output_signal**.

A related function **f_convu** as shown below

Calculating a Future Tristate Value

```
FUNCTION f_convu( newval  : IN t_logic;
                  lastval : IN t_logic;
                  tech    : IN t_technology ) RETURN t_logic IS
    VARIABLE oldst : t_state;
    VARIABLE newst : t_state;
BEGIN
    -- Pickup states for values
    oldst := f_state( lastval );
    newst := f_state( newval );

    CASE newst IS
        -- new value is false
        WHEN '0' =>
            CASE oldst IS
                -- from false to false
                WHEN '0' =>
                    CASE tech IS
                        WHEN ecl => RETURN RZO;
                        WHEN cmos =>RETURN FZO;
                        WHEN nmos =>RETURN FZO;
                        WHEN ttl => RETURN FZO;
                        WHEN ttloc => RETURN FZO;
                    END CASE;

                -- from true/unknown to false
                WHEN '1'|'X'       =>
                    CASE tech IS
                        WHEN ecl => RETURN RZX;
                        WHEN cmos =>RETURN FZX;
                        WHEN nmos =>RETURN FZX;
                        WHEN ttl => RETURN FZX;
                        WHEN ttloc => RETURN FZX;
                    END CASE;
            END CASE;

        -- new value is true
        WHEN '1' =>
            CASE oldst IS
```

7.3. TECHNOLOGY RULES

```
                    -- from true to true
                    WHEN '1' =>
                        CASE tech IS
                            WHEN ecl => RETURN FZ1;
                            WHEN cmos =>RETURN FZ1;
                            WHEN nmos =>RETURN RZ1;
                            WHEN ttl => RETURN FZ1;
                            WHEN ttloc => RETURN ZX;
                        END CASE;

                    -- from false/unknown to true
                    WHEN '0'|'X'    =>
                        CASE tech IS
                            WHEN ecl => RETURN ZFX;
                            WHEN cmos =>RETURN ZFX;
                            WHEN nmos =>RETURN ZRX;
                            WHEN ttl => RETURN ZFX;
                            WHEN ttloc => RETURN ZX;
                        END CASE;
                END CASE;

            -- new value is unknown
            WHEN 'X'    =>
                -- from any value to unknown
                CASE tech IS
                    WHEN ecl => RETURN RFX;
                    WHEN cmos =>RETURN FX;
                    WHEN nmos =>RETURN FRX;
                    WHEN ttl => RETURN FX;
                    WHEN ttloc => RETURN FZX;
                END CASE;
        END CASE;
END f_convu;
```

and is used to calculate the next value for a tristate output of a device. This function calculates the state that should be output when the gating signal is of an unknown value. The following

Calculating a Future Tristate Value

technology	newstate/oldstate						
	0/0	0/1	0/X	1/0	1/1	1/X	X/any
ecl	RZ0	RZX	RZX	ZFX	FZ1	ZFX	RFX
cmos	FZ0	FZX	RZX	ZFX	FZ1	ZFX	FX
nmos	FZ0	FZX	FZX	ZRX	RZ1	ZRX	FRX
ttl	FZ0	FZX	FZX	ZFX	FZ1	ZFX	FX
ttloc	FZ0	FZX	FZX	ZX	ZX	ZX	FZX

summarizes the operation of this function. The algorithm used calculates the largest interval of values which the gate could generate given the unknown tristate control input.

Finally, function **f_convx** as shown below

Pessimistic Future Tristate Value

```
FUNCTION f_convx( newval  : IN t_logic;
                  lastval : IN t_logic;
                  tech    : IN t_technology ) RETURN t_logic IS
BEGIN
           CASE tech IS
              WHEN ecl => RETURN RFX;
              WHEN cmos =>RETURN FX;
              WHEN nmos =>RETURN FRX;
              WHEN ttl => RETURN FX;
              WHEN ttloc => RETURN FZX;
           END CASE;

END f_convx;
```

can be used in place of **f_convu** but returns more pessimistic results and may be useful in certain applications. The following

Calculating a Pessimistic Future Tristate Value

technology	output
ecl	RFX
cmos	FX
nmos	FRX
ttl	FX
ttloc	FZX

summarizes the operation of this function.

7.3.1 ECL - Emitter Coupled Logic

The following

ECL Output Signal Values

State	Logic Value
0	R0
1	F1
X	RFX

7.3. TECHNOLOGY RULES

shows the output signal values which are used for ECL logic. By setting the strength of false values lower (R) than the true values (F) we are assured that false will override true. The unknown value is chosen to span R0 and F1.

The specific VHDL table declaration for the ECL technology is shown below

ECL Lookup Table

```
CONSTANT f_ecl : f_specific_t :=
    ( R0,     -- '0'
      F1,     -- '1'
      RFX     -- 'X'
    );
```

In order to ease modelling of ECL devices, the standard boolean operators are overloaded as shown below.

ECL Boolean Operators

```
-- "and" operator.
FUNCTION "and" ( l,r : t_wlogic ) RETURN t_wlogic IS
BEGIN RETURN( f_ecl( f_and( f_state( l ) ) ( f_state( r ) ) ) ); END;

-- "or" operator.
FUNCTION "or" ( l,r : t_wlogic ) RETURN t_wlogic IS
BEGIN RETURN( f_ecl( f_or( f_state( l ) ) ( f_state( r ) ) ) ); END;

-- "nand" operator.
FUNCTION "nand" ( l,r : t_wlogic ) RETURN t_wlogic IS
BEGIN RETURN( f_ecl( f_nand( f_state( l ) ) ( f_state( r ) ) ) ); END;

-- "nor" operator.
FUNCTION "nor" ( l,r : t_wlogic ) RETURN t_wlogic IS
BEGIN RETURN( f_ecl( f_nor( f_state( l ) ) ( f_state( r ) ) ) ); END;

-- "xor" operator.
FUNCTION "xor" ( l,r : t_wlogic ) RETURN t_wlogic IS
BEGIN RETURN( f_ecl( f_xor( f_state( l ) ) ( f_state( r ) ) ) ); END;

-- "not" operator.
FUNCTION "not" ( l   : t_wlogic ) RETURN t_wlogic IS
BEGIN RETURN( f_ecl( f_not( f_state( l ) ) ) ); END;
```

To access the standard logic functions for the ECL technology the statements shown below

Accessing Standard Logic with ECL Technology

```
-- access standard logic facilities
USE std.std_logic.ALL;

-- choose ECL technology
USE std.std_ecl.ALL;

-- VHDL entity declaration
--    .
--    .
--    .
```

The following gives an example of the use of the overloaded nand operator:

Overloaded NAND Operator for ECL

```
-- gate function
y <= adelay NAND bdelay
    AFTER f_delay(adelay NAND bdelay, y_o01, y_o10);
```

Several cases exist in which an unknown signal value requires special processing. These cases occur in technologies which utilize different strengths for false and true logic values on output. For example, for an ECL buffer the following rules apply to the correct modelling of this device

ECL Unknown Handling

Value Of A	Value Of B
false	resistive false (**R0**)
true	forced true (**F1**)
unknown	interval (**RFX**)

7.3.2 CMOS - Complementary MOS

The following

CMOS Output Signal Values

State	Logic Value
0	F0
1	F1
X	FX

7.3. TECHNOLOGY RULES

shows the output signal values which are used for CMOS logic. Both false and true values have equal strength. The unknown value is chosen to span F0 and F1.

The specific VHDL table declaration for the CMOS technology is shown below

CMOS Lookup Table

```
CONSTANT f_cmos : f_specific_t :=
    ( F0,     -- '0'
      F1,     -- '1'
      FX      -- 'X'
    );
```

In order to ease modelling of CMOS devices, the standard boolean operators are overloaded as shown below.

CMOS Boolean Operators

```
-- "and" operator.
FUNCTION "and" ( l,r : t_wlogic ) RETURN t_wlogic IS
BEGIN RETURN( f_cmos( f_and( f_state( l ) ) ( f_state( r ) ) ) ); END;

-- "or" operator.
FUNCTION "or" ( l,r : t_wlogic ) RETURN t_wlogic IS
BEGIN RETURN( f_cmos( f_or( f_state( l ) ) ( f_state( r ) ) ) ); END;

-- "nand" operator.
FUNCTION "nand" ( l,r : t_wlogic ) RETURN t_wlogic IS
BEGIN RETURN( f_cmos( f_nand( f_state( l ) ) ( f_state( r ) ) ) ); END;

-- "nor" operator.
FUNCTION "nor" ( l,r : t_wlogic ) RETURN t_wlogic IS
BEGIN RETURN( f_cmos( f_nor( f_state( l ) ) ( f_state( r ) ) ) ); END;

-- "xor" operator.
FUNCTION "xor" ( l,r : t_wlogic ) RETURN t_wlogic IS
BEGIN RETURN( f_cmos( f_xor( f_state( l ) ) ( f_state( r ) ) ) ); END;

-- "not" operator.
FUNCTION "not" ( l   : t_wlogic ) RETURN t_wlogic IS
BEGIN RETURN( f_cmos( f_not( f_state( l ) ) ) ); END;
```

To access the standard logic functions for the CMOS technology the statements shown below

Accessing Standard Logic with CMOS Technology

```
-- access standard logic facilities
USE std.std_logic.ALL;

-- choose CMOS technology
USE std.std_cmos.ALL;

-- VHDL entity declaration
--     .
--     .
--     .
```

The following gives an example of the use of the overloaded nand operator:

Overloaded NAND Operator for CMOS

```
-- gate function
y <= adelay NAND bdelay
    AFTER f_delay(adelay NAND bdelay, y_o01, y_o10);
```

7.3.3 NMOS - n-Channel MOS

The following

NMOS Output Signal Values

State	Logic Value
0	F0
1	R1
X	FRX

shows the output signal values which are used for NMOS logic. By setting the strength of true values lower (R) than the false values (F) we are assured that false will override true. The unknown value is chosen to span F0 and R1.

The specific VHDL table declaration for the NMOS technology is shown below.

7.3. TECHNOLOGY RULES

NMOS Lookup Table

```
CONSTANT f_nmos : f_specific_t :=
    ( F0,      -- '0'
      R1,      -- '1'
      FRX      -- 'X'
    );
```

In order to ease modelling of NMOS devices, the standard boolean operators are overloaded as shown below.

NMOS Boolean Operators

```
-- "and" operator.
FUNCTION "and" ( l,r : t_wlogic ) RETURN t_wlogic IS
BEGIN RETURN( f_nmos( f_and( f_state( l ) ) ( f_state( r ) ) ) ); END;

-- "or" operator.
FUNCTION "or" ( l,r : t_wlogic ) RETURN t_wlogic IS
BEGIN RETURN( f_nmos( f_or( f_state( l ) ) ( f_state( r ) ) ) ); END;

-- "nand" operator.
FUNCTION "nand" ( l,r : t_wlogic ) RETURN t_wlogic IS
BEGIN RETURN( f_nmos( f_nand( f_state( l ) ) ( f_state( r ) ) ) ); END;

-- "nor" operator.
FUNCTION "nor" ( l,r : t_wlogic ) RETURN t_wlogic IS
BEGIN RETURN( f_nmos( f_nor( f_state( l ) ) ( f_state( r ) ) ) ); END;

-- "xor" operator.
FUNCTION "xor" ( l,r : t_wlogic ) RETURN t_wlogic IS
BEGIN RETURN( f_nmos( f_xor( f_state( l ) ) ( f_state( r ) ) ) ); END;

-- "not" operator.
FUNCTION "not" ( l   : t_wlogic ) RETURN t_wlogic IS
BEGIN RETURN( f_nmos( f_not( f_state( l ) ) ) ); END;
```

To access the standard logic functions for the NMOS technology the statements shown below.

Accessing Standard Logic with NMOS Technology

```
-- access standard logic facilities
USE std.std_logic.ALL;

-- choose NMOS technology
USE std.std_nmos.ALL;

-- VHDL entity declaration
```

```
--      .
--      .
--      .
```

The following gives an example of the use of the overloaded nand operator:

Overloaded NAND Operator for NMOS

```
-- gate function
y <= adelay NAND bdelay
    AFTER f_delay(adelay NAND bdelay, y_o01, y_o10);
```

Several cases exist for unknown signal values which require special processing. These cases occur in technologies which utilize different strengths for false and true logic values on output. See section 7.3.1 for an example of this type of processing.

7.3.4 TTL - Transistor transistor logic

The following

TTL Output Signal Values

State	Logic Value
0	F0
1	F1
X	FX

shows the output signal values which are used for TTL logic. Both false and true values have the forced strength (F). The unknown value is chosen to span F0 to F1.

The specific VHDL table declaration for the TTL technology is shown below.

TTL Lookup Table

```
CONSTANT f_ttl : f_specific_t :=
    ( F0,      -- '0'
      F1,      -- '1'
      FX       -- 'X'
    );
```

In order to ease modelling of TTL devices, the standard boolean operators are overloaded as shown below.

7.3. TECHNOLOGY RULES

TTL Boolean Operators

```
-- "and" operator.
FUNCTION "and" ( l,r : t_wlogic ) RETURN t_wlogic IS
BEGIN RETURN( f_ttl( f_and( f_state( l ) ) ( f_state( r ) ) ) ); END;

-- "or" operator.
FUNCTION "or" ( l,r : t_wlogic ) RETURN t_wlogic IS
BEGIN RETURN( f_ttl( f_or( f_state( l ) ) ( f_state( r ) ) ) ); END;

-- "nand" operator.
FUNCTION "nand" ( l,r : t_wlogic ) RETURN t_wlogic IS
BEGIN RETURN( f_ttl( f_nand( f_state( l ) ) ( f_state( r ) ) ) ); END;

-- "nor" operator.
FUNCTION "nor" ( l,r : t_wlogic ) RETURN t_wlogic IS
BEGIN RETURN( f_ttl( f_nor( f_state( l ) ) ( f_state( r ) ) ) ); END;

-- "xor" operator.
FUNCTION "xor" ( l,r : t_wlogic ) RETURN t_wlogic IS
BEGIN RETURN( f_ttl( f_xor( f_state( l ) ) ( f_state( r ) ) ) ); END;

-- "not" operator.
FUNCTION "not" ( l    : t_wlogic ) RETURN t_wlogic IS
BEGIN RETURN( f_ttl( f_not( f_state( l ) ) ) ); END;
```

To access the standard logic functions for the TTL technology the statements shown below.

Accessing Standard Logic with TTL Technology

```
USE std.std_logic.ALL;    -- access standard logic facilities
USE std.std_ttl.ALL;      -- choose TTL technology

-- VHDL entity declaration
--     .
--     .
```

The following shows the use of the overloaded nand operator.

Overloaded NAND Operator for TTL

```
y <= adelay NAND bdelay
     AFTER f_delay(adelay NAND bdelay, y_o01, y_o10);
```

Several cases exist in which an unknown signal value requires special processing. These cases occur in technologies which utilize different strengths for false and true logic values on output. For example, for a tristate buffer the following rules apply to the correct modelling of this device

ECL Unknown Handling

Value Of B	C State	C Strength
false	C_{prev}	hi–impedance
true	state of A	technology dependent
unknown	– C_{prev}, if A = C_{prev} – unknown, if A <> C_{prev}	an interval, between hi–impedance and the technology

where C_{prev} is the previous state of C.

7.3.5 TTLOC - Open-collector TTL

The following

TTLOC Output Signal Values

State	Logic Value
0	F0
1	ZX
X	FZX

shows the output signal values which are used for TTLOC logic.

The specific VHDL table declaration for the TTLOC technology is shown below

TTLOC Lookup Table

```
CONSTANT f_ttloc : f_specific_t :=
    ( F0,     -- '0'
      ZX,     -- '1'
      FZX     -- 'X'
    );
```

In order to ease modelling of TTLOC devices, the standard boolean operators are overloaded as shown below.

7.3. TECHNOLOGY RULES

TTLOC Boolean Operators

```
-- "and" operator.
FUNCTION "and" ( l,r : t_wlogic ) RETURN t_wlogic IS
BEGIN RETURN( f_ttloc( f_and( f_state( l ) ) ( f_state( r ) ) ) ); END;

-- "or" operator.
FUNCTION "or" ( l,r : t_wlogic ) RETURN t_wlogic IS
BEGIN RETURN( f_ttloc( f_or( f_state( l ) ) ( f_state( r ) ) ) ); END;

-- "nand" operator.
FUNCTION "nand" ( l,r : t_wlogic ) RETURN t_wlogic IS
BEGIN RETURN( f_ttloc( f_nand( f_state( l ) ) ( f_state( r ) ) ) ); END;

-- "nor" operator.
FUNCTION "nor" ( l,r : t_wlogic ) RETURN t_wlogic IS
BEGIN RETURN( f_ttloc( f_nor( f_state( l ) ) ( f_state( r ) ) ) ); END;

-- "xor" operator.
FUNCTION "xor" ( l,r : t_wlogic ) RETURN t_wlogic IS
BEGIN RETURN( f_ttloc( f_xor( f_state( l ) ) ( f_state( r ) ) ) ); END;

-- "not" operator.
FUNCTION "not" ( l    : t_wlogic ) RETURN t_wlogic IS
BEGIN RETURN( f_ttloc( f_not( f_state( l ) ) ) ); END;
```

To access the standard logic functions for the TTLOC technology the statements shown below.

Accessing Standard Logic with TTLOC Technology

```
USE std.std_logic.ALL;     -- access standard logic facilities
USE std.std_ttloc.ALL;     -- choose TTLOC technology

-- VHDL entity declaration
--    .
--    .
```

The following gives an example of the use of the overloaded nand operator:

Overloaded NAND Operator for TTLOC

```
y <= adelay NAND bdelay
        AFTER f_delay(adelay NAND bdelay, y_o01, y_o10);
```

Several cases exist for unknown signal values which require special processing. These cases occur in technologies which utilize different strengths for false and true logic values on output. See section 7.3.1 for an example of this type of processing.

7.4 Bus Resolution

The standard logic package utilizes a single technology independent bus resolution function. See section 7.2 for a discussion of how the bus resolution function is associated with the type definition for **t_wlogic**.

The following shows the body of the **f_logic_bus** bus resolution function.

Bus Resolution Function

```
FUNCTION f_logic_bus( s : t_logic_vector ) RETURN t_logic IS
    VARIABLE result : t_logic;       -- result so far
BEGIN
    -- If no inputs then default is D
    IF (s'LENGTH = 0) THEN RETURN D;

    -- For one input return value directly
    ELSIF (s'LENGTH = 1) THEN RETURN s(s'LOW);

    -- Calculate value based on inputs
    ELSE
        result := D;

        -- Iterate through all inputs
        FOR i IN s'LOW TO s'HIGH LOOP
            if ( s(i) = U ) then
                return U;
            end if;
            if ( s(i) = D ) then
                NEXT;
            end if;
            result := f_busres(result)(s(i));
        END LOOP;

        -- Return the resultant value
        RETURN result;
    END IF;

END f_logic_bus;
```

The function defaults to a value of D. Each input signal **s(i)** is polled in sequence. A working result **result** is maintained. The current signal value and the working result are merged using the **f_busres** table lookup which returns the least pessimistic value interval which represents both values.

The following code shows the definition of **f_busres** lookup table.

7.4. BUS RESOLUTION

Bus Value Lookup

```
-- Type definitions for table
TYPE t_bus IS ARRAY(t_logic'LOW TO t_logic'HIGH) OF t_logic;
TYPE t_bus_tab IS ARRAY(t_logic'LOW TO t_logic'HIGH) OF t_bus;

CONSTANT f_busres : t_bus_tab := (
  ( U, U,
    U, U, U, U, U,
    U, U, U, U, U, U, U, U, U,
    U, U, U, U, U, U, U, U, U, U, U, U, U,
    U, U, U, U, U, U, U, U, U, U, U, U, U, U, U, U, U ),-- U
  ( U,D,
    ZO,Z1,ZDX,DZX,ZX,
    WO,W1,WZO,WZ1,WDX,DWX,WZX,ZWX,WX,
    RO,R1,RWO,RW1,RZO,RZ1,RDX,DRX,RZX,ZRX,RWX,WRX,RX,     -- D
    FO,F1,FRO,FR1,FWO,FW1,FZO,FZ1,FDX,DFX,FZX,ZFX,FWX,WFX,FRX,RFX,FX),
  ( U,ZO,
    ZO,ZX,ZO,ZX,ZX,
    WO,W1,WZO,ZWX,WZO,ZWX,WZX,ZWX,WX,
    RO,R1,RWO,RW1,RZO,ZRX,RZO,ZRX,RZX,ZRX,RWX,WRX,RX,     -- ZO
    FO,F1,FRO,FR1,FWO,FW1,FZO,ZFX,FZO,ZFX,FZX,ZFX,FWX,WFX,FRX,RFX,FX),
  ( U,Z1,
    ZX,Z1,ZX,Z1,ZX,
    WO,W1,WZX,WZ1,WZX,WZ1,WZX,ZWX,WX,
    RO,R1,RWO,RW1,RZX,RZ1,RZX,RZ1,RZX,ZRX,RWX,WRX,RX,     -- Z1
    FO,F1,FRO,FR1,FWO,FW1,FZX,FZ1,FZX,FZ1,FZX,ZFX,FWX,WFX,FRX,RFX,FX),
  ( U,ZDX,
    ZO,ZX,ZDX,ZX,ZX,
    WO,W1,WZO,ZWX,WDX,ZWX,WZX,ZWX,WX,
    RO,R1,RWO,RW1,RZO,ZRX,RDX,ZRX,RZX,ZRX,RWX,WRX,RX,     --ZDX
    FO,F1,FRO,FR1,FWO,FW1,FZO,ZFX,FDX,ZFX,FZX,ZFX,FWX,WFX,FRX,RFX,FX),
  ( U,DZX,
    ZX,Z1,ZX,DZX,ZX,
    WO,W1,WZX,WZ1,WDX,DWX,WZX,ZWX,WX,
    RO,R1,RWO,RW1,RZX,RZ1,RZX,DRX,RZX,ZRX,RWX,WRX,RX,     --DZX
    FO,F1,FRO,FR1,FWO,FW1,FZX,FZ1,FZX,DFX,FZX,ZFX,FWX,WFX,FRX,RFX,FX),
  ( U,ZX,
    ZX,ZX,ZX,ZX,ZX,
    WO,W1,WZX,ZWX,WZX,ZWX,WZX,ZWX,WX,
    RO,R1,RWO,RW1,RZX,ZRX,RZX,ZRX,RZX,ZRX,RWX,WRX,RX,     -- ZX
    FO,F1,FRO,FR1,FWO,FW1,FZX,ZFX,FZX,ZFX,FZX,ZFX,FWX,WFX,FRX,RFX,FX),
  ( U,WO,
    WO,WO,WO,WO,WO,
    WO,WX,WO,WX,WO,WX,WO,WX,WX,
    RO,R1,RWO,WRX,RWO,WRX,RWO,WRX,RWO,WRX,RWX,WRX,RX,     -- WO
    FO,F1,FRO,FR1,FWO,WFX,FWO,WFX,FWO,WFX,FWO,WFX,FWX,WFX,FRX,RFX,FX),
  ( U,W1,
    W1,W1,W1,W1,W1,
    WX,W1,WX,W1,WX,W1,WX,W1,WX,
    RO,R1,RWX,RW1,RWX,RW1,RWX,RW1,RWX,RW1,RWX,WRX,RX,     -- W1
    FO,F1,FRO,FR1,FWX,FW1,FWX,FW1,FWX,FW1,FWX,FW1,FWX,WFX,FRX,RFX,FX),
```

```
(   U,WZO,
  WZO,WZX,WZO,WZX,WZX,
  WO,WX,WZO,WX,WZO,WX,WZX,WX,WX,
  RO,R1,RWO,WRX,RZO,WRX,RZO,WRX,RZX,WRX,RWX,WRX,RX,        --WZO
  FO,F1,FRO,FR1,FWO,WFX,FZO,WFX,FZO,WFX,FZX,WFX,FWX,WFX,FRX,RFX,FX),
(   U,WZ1,
  ZWX,WZ1,ZWX,WZ1,ZWX,
  WX,W1,WX,WZ1,WX,WZ1,WX,ZWX,WX,
  RO,R1,RWX,RW1,RWX,RZ1,RWX,RZ1,RWX,ZRX,RWX,WRX,RX,        --WZ1
  FO,F1,FRO,FR1,FWX,FW1,FWX,FZ1,FWX,FZ1,FWX,ZFX,FWX,WFX,FRX,RFX,FX),
(   U,WDX,
  WZO,WZX,WDX,WZX,WZX,
  WO,WX,WZO,WX,WDX,WX,WZX,WX,WX,
  RO,R1,RWO,WRX,RZO,WRX,RDX,WRX,RZX,WRX,RWX,WRX,RX,        --WDX
  FO,F1,FRO,FR1,FWO,WFX,FZO,WFX,FDX,WFX,FZX,WFX,FWX,WFX,FRX,RFX,FX),
(   U,DWX,
  ZWX,WZ1,ZWX,DWX,ZWX,
  WX,W1,WX,WZ1,WX,DWX,WX,ZWX,WX,
  RO,R1,RWX,RW1,RWX,RZ1,RWX,DRX,RWX,ZRX,RWX,WRX,RX,        --DWX
  FO,F1,FRO,FR1,FWX,FW1,FWX,FZ1,FWX,DFX,FWX,ZFX,FWX,WFX,FRX,RFX,FX),
(   U,WZX,
  WZX,WZX,WZX,WZX,WZX,
  WO,WX,WZX,WX,WZX,WX,WZX,WX,WX,
  RO,R1,RWO,WRX,RZX,WRX,RZX,WRX,RZX,WRX,RWX,WRX,RX,        --WZX
  FO,F1,FRO,FR1,FWO,WFX,FZX,WFX,FZX,WFX,FZX,WFX,FWX,WFX,FRX,RFX,FX),
(   U,ZWX,
  ZWX,ZWX,ZWX,ZWX,ZWX,
  WX,W1,WX,ZWX,WX,ZWX,WX,ZWX,WX,
  RO,R1,RWX,RW1,RWX,ZRX,RWX,ZRX,RWX,ZRX,RWX,WRX,RX,        --ZWX
  FO,F1,FRO,FR1,FWX,FW1,FWX,ZFX,FWX,ZFX,FWX,ZFX,FWX,WFX,FRX,RFX,FX),
(   U,WX,
  WX,WX,WX,WX,WX,
  WX,WX,WX,WX,WX,WX,WX,WX,WX,
  RO,R1,RWX,WRX,RWX,WRX,RWX,WRX,RWX,WRX,RWX,WRX,RX,        --WX
  FO,F1,FRO,FR1,FWX,WFX,FWX,WFX,FWX,WFX,FWX,WFX,FWX,WFX,FRX,RFX,FX),
(   U,RO,
  RO,RO,RO,RO,RO,
  RO,RO,RO,RO,RO,RO,RO,RO,RO,
  RO,RX,RO,RX,RO,RX,RO,RX,RO,RX,RO,RX,RX,                  --RO
  FO,F1,FRO,RFX,FRO,RFX,FRO,RFX,FRO,RFX,FRO,RFX,FRX,RFX,FX),
(   U,R1,
  R1,R1,R1,R1,R1,
  R1,R1,R1,R1,R1,R1,R1,R1,R1,
  RX,R1,RX,R1,RX,R1,RX,R1,RX,R1,RX,R1,RX,                  --R1
  FO,F1,FRX,FR1,FRX,FR1,FRX,FR1,FRX,FR1,FRX,FR1,FRX,RFX,FX),
(   U,RWO,
  RWO,RWO,RWO,RWO,RWO,
  RWO,RWX,RWO,RWX,RWO,RWX,RWO,RWX,RWX,
  RO,RX,RWO,RX,RWO,RX,RWO,RX,RWO,RX,RWX,RX,RX,             --RWO
  FO,F1,FRO,RFX,FWO,RFX,FWO,RFX,FWO,RFX,FWX,RFX,FRX,RFX,FX),
(   U,RW1,
  RW1,RW1,RW1,RW1,RW1,
```

7.4. BUS RESOLUTION 335

```
    WRX,RW1,WRX,RW1,WRX,RW1,WRX,RW1,WRX,
      RX,R1,RX,RW1,RX,RW1,RX,RW1,RX,RW1,RX,WRX,RX,           --RW1
    F0,F1,FRX,FR1,FRX,FW1,FRX,FW1,FRX,FW1,FRX,FW1,FRX,WFX,FRX,RFX,FX),
(   U,RZ0,
    RZ0,RZX,RZ0,RZX,RZX,
    RW0,RWX,RZ0,RWX,RZ0,RWX,RZX,RWX,RWX,
      R0,RX,RW0,RX,RZ0,RX,RZ0,RX,RZX,RX,RWX,RX,RX,           --RZ0
    F0,F1,FR0,RFX,FW0,RFX,FZ0,RFX,FZ0,RFX,FZX,RFX,FWX,RFX,FRX,RFX,FX),
(   U,RZ1,
    ZRX,RZ1,ZRX,RZ1,ZRX,
    WRX,RW1,WRX,RZ1,WRX,RZ1,WRX,ZRX,WRX,
      RX,R1,RX,RW1,RX,RZ1,RX,RZ1,RX,ZRX,RX,WRX,RX,           --RZ1
    F0,F1,FRX,FR1,FRX,FW1,FRX,FZ1,FRX,FZ1,FRX,ZFX,FRX,WFX,FRX,RFX,FX),
(   U,RDX,
    RZ0,RZX,RDX,RZX,RZX,
    RW0,RWX,RZ0,RWX,RDX,RWX,RZX,RWX,RWX,
      R0,RX,RW0,RX,RZ0,RX,RDX,RX,RZX,RX,RWX,RX,RX,           --RDX
    F0,F1,FR0,RFX,FW0,RFX,FZ0,RFX,FDX,RFX,FZX,RFX,FWX,RFX,FRX,RFX,FX),
(   U,DRX,
    ZRX,RZ1,ZRX,DRX,ZRX,
    WRX,RW1,WRX,RZ1,WRX,DRX,WRX,ZRX,WRX,
      RX,R1,RX,RW1,RX,RZ1,RX,DRX,RX,ZRX,RX,WRX,RX,           --DRX
    F0,F1,FRX,FR1,FRX,FW1,FRX,FZ1,FRX,DFX,FRX,ZFX,FRX,WFX,FRX,RFX,FX),
(   U,RZX,
    RZX,RZX,RZX,RZX,RZX,
    RW0,RWX,RZX,RWX,RZX,RWX,RZX,RWX,RWX,
      R0,RX,RW0,RX,RZX,RX,RZX,RX,RZX,RX,RWX,RX,RX,           --RZX
    F0,F1,FR0,RFX,FW0,RFX,FZX,RFX,FZX,RFX,FZX,RFX,FWX,RFX,FRX,RFX,FX),
(   U,ZRX,
    ZRX,ZRX,ZRX,ZRX,ZRX,
    WRX,RW1,WRX,ZRX,WRX,ZRX,WRX,ZRX,WRX,
      RX,R1,RX,RW1,RX,ZRX,RX,ZRX,RX,ZRX,RX,WRX,RX,           --ZRX
    F0,F1,FRX,FR1,FRX,FW1,FRX,ZFX,FRX,ZFX,FRX,ZFX,FRX,WFX,FRX,RFX,FX),
(   U,RWX,
    RWX,RWX,RWX,RWX,RWX,
    RWX,RWX,RWX,RWX,RWX,RWX,RWX,RWX,RWX,
      R0,RX,RWX,RX,RWX,RX,RWX,RX,RWX,RX,RWX,RX,RX,           --RWX
    F0,F1,FR0,RFX,FWX,RFX,FWX,RFX,FWX,RFX,FWX,RFX,FWX,RFX,FRX,RFX,FX),
(   U,WRX,
    WRX,WRX,WRX,WRX,WRX,
    WRX,WRX,WRX,WRX,WRX,WRX,WRX,WRX,WRX,
      RX,R1,RX,WRX,RX,WRX,RX,WRX,RX,WRX,RX,WRX,RX,           --WRX
    F0,F1,FRX,FR1,FRX,WFX,FRX,WFX,FRX,WFX,FRX,WFX,FRX,WFX,FRX,RFX,FX),
(   U,RX,
    RX,RX, RX,RX,RX,
    RX,RX, RX,RX,RX,RX,RX,RX,RX,
      RX,RX, RX,RX,RX,RX,RX,RX,RX,RX,RX,RX,RX,               --RX
    F0,F1,FRX,RFX,FRX,RFX,FRX,RFX,FRX,RFX,FRX,RFX,FRX,RFX,FRX,RFX,FX),
(   U,F0,
    F0,F0, F0,F0,F0,
    F0,F0, F0,F0,F0,F0,F0,F0,F0,
    F0,F0, F0,F0,F0,F0,F0,F0,F0,F0,F0,F0,F0,
```

          FO,FX, FO,FX,FO,FX,FO,FX,FO,FX,FO,FX,FO,FX,FO,FX,FX),--FO
    (U,F1,
       F1,F1, F1,F1,F1,
       F1,F1, F1,F1,F1,F1,F1,F1,F1,
       F1,F1, F1,F1,F1,F1,F1,F1,F1,F1,F1,F1,
       FX,F1, FX,F1,FX,F1,FX,F1,FX,F1,FX,F1,FX,F1,FX,F1,FX),--F1
    (U,FR0,
       FR0,FR0,FR0,FR0,FR0,
       FR0,FR0,FR0,FR0,FR0,FR0,FR0,FR0,FR0,
       FR0,FRX,FR0,FRX,FR0,FRX,FR0,FRX,FR0,FRX,FR0,FRX,FRX, --FR0
       FO,FX,FR0,FX,FR0,FX,FR0,FX,FR0,FX,FR0,FX,FRX,FX,FX),
    (U,FR1,
       FR1,FR1,FR1,FR1,FR1,
       FR1,FR1,FR1,FR1,FR1,FR1,FR1,FR1,FR1,
       RFX,FR1,RFX,FR1,RFX,FR1,RFX,FR1,RFX,FR1,RFX,FR1,RFX, --FR1
       FX,F1,FX,FR1,FX,FR1,FX,FR1,FX,FR1,FX,FR1,FX,FR1,FX,RFX,FX),
    (U,FW0,
       FW0,FWX,FW0,FW0,FW0,
       FW0,FWX,FW0,FWX,FW0,FWX,FW0,FWX,FWX,
       FR0,FRX,FW0,FRX,FW0,FRX,FW0,FRX,FW0,FRX,FWX,FRX,FRX, --FW0
       FO,FX,FR0,FX,FW0,FX,FW0,FX,FW0,FX,FW0,FX,FWX,FX,FRX,FX,FX),
    (U,FW1,
       FW1,FW1,FW1,FW1,FW1,
       WFX,FW1,WFX,FW1,WFX,FW1,WFX,FW1,WFX,
       RFX,FR1,RFX,FW1,RFX,FW1,RFX,FW1,RFX,FW1,RFX,WFX,RFX, --FW1
       FX,F1,FX,FR1,FX,FW1,FX,FW1,FX,FW1,FX,FW1,FX,WFX,FX,RFX,FX),
    (U,FZ0,
       FZ0,FZX,FZ0,FZX,FZX,
       FW0,FWX,FZ0,FWX,FZ0,FWX,FZX,FWX,FWX,
       FR0,FRX,FW0,FRX,FZ0,FRX,FZ0,FRX,FZX,FRX,FWX,FRX,FRX, --FZ0
       FO,FX,FR0,FX,FW0,FX,FZ0,FX,FZ0,FX,FZX,FX,FWX,FX,FRX,FX,FX),
    (U,FZ1,
       ZFX,FZ1,ZFX,FZ1,ZFX,
       WFX,FW1,WFX,FZ1,WFX,FZ1,WFX,ZFX,WFX,
       RFX,FR1,RFX,FW1,RFX,FZ1,RFX,FZ1,RFX,ZFX,RFX,WFX,RFX, --FZ1
       FX,F1,FX,FR1,FX,FW1,FX,FZ1,FX,FZ1,FX,ZFX,FX,WFX,FX,RFX,FX),
    (U,FDX,
       FZ0,FZX,FDX,FZX,FZX,
       FW0,FWX,FZ0,FWX,FDX,FWX,FZX,FWX,FWX,
       FR0,FRX,FW0,FRX,FZ0,FRX,FDX,FRX,FZX,FRX,FWX,FRX,FRX, --FDX
       FO,FX,FR0,FX,FW0,FX,FZ0,FX,FDX,FX,FZX,FX,FWX,FX,FRX,FX,FX),
    (U,DFX,
       ZFX,FZ1,ZFX,DFX,ZFX,
       WFX,FW1,WFX,FZ1,WFX,DFX,WFX,ZFX,WFX,
       RFX,FR1,RFX,FW1,RFX,FZ1,RFX,DFX,RFX,ZFX,RFX,WFX,RFX, --DFX
       FX,F1,FX,FR1,FX,FW1,FX,FZ1,FX,DFX,FX,ZFX,FX,WFX,FX,RFX,FX),
    (U,FZX,
       FZX,FZX,FZX,FZX,FZX,
       FW0,FWX,FZX,FWX,FZX,FWX,FZX,FWX,FWX,
       FR0,FRX,FW0,FRX,FZX,FRX,FZX,FRX,FZX,FRX,FWX,FRX,FRX, --FZX
       FO,FX,FR0,FX,FW0,FX ,FZX,FX ,FZX,FX ,FZX,FX ,FWX,FX ,FRX,FX ,FX),
    (U,ZFX,

7.5. LOGIC MANIPULATION

```
          ZFX,ZFX,ZFX,ZFX,ZFX,
          WFX,FW1,WFX,ZFX,WFX,ZFX,WFX,ZFX,WFX,
          RFX,FR1,RFX,FW1,RFX,ZFX,RFX,ZFX,RFX,ZFX,RFX,WFX,RFX,   --ZFX
           FX,F1,FX,FR1,FX ,FW1,FX ,ZFX,FX ,ZFX,FX ,ZFX,FX ,WFX,FX,RFX,FX),
       (  U,FWX,
          FWX,FWX,FWX,FWX,FWX,
          FWX,FWX,FWX,FWX,FWX,FWX,FWX,FWX,FWX,
          FRO,FRX,FWX,FRX,FWX,FRX,FWX,FRX,FWX,FRX,FWX,FRX,FRX,   --FWX
           FO,FX,FRO,FX,FWX,FX ,FWX,FX ,FWX,FX ,FWX,FX ,FWX,FX,FRX,FX ,FX),
       (  U,WFX,
          WFX,WFX,WFX,WFX,WFX,
          WFX,WFX,WFX,WFX,WFX,WFX,WFX,WFX,WFX,
          RFX,FR1,RFX,WFX,RFX,WFX,RFX,WFX,RFX,WFX,RFX,WFX,RFX,   --WFX
           FX,F1,FX,FR1,FX,WFX,FX ,WFX,FX ,WFX,FX ,WFX,FX ,WFX,FX,RFX,FX),
       (  U,FRX,
          FRX,FRX,FRX,FRX,FRX,
          FRX,FRX,FRX,FRX,FRX,FRX,FRX,FRX,FRX,
          FRX,FRX,FRX,FRX,FRX,FRX,FRX,FRX,FRX,FRX,FRX,FRX,FRX,   --FRX
           FO,FX,FRX,FX,FRX,FX ,FRX,FX ,FRX,FX ,FRX,FX ,FRX,FX,FRX,FX,FX),
       (  U,RFX,
          RFX,RFX,RFX,RFX,RFX,
          RFX,RFX,RFX,RFX,RFX,RFX,RFX,RFX,RFX,
          RFX,RFX,RFX,RFX,RFX,RFX,RFX,RFX,RFX,RFX,RFX,RFX,RFX,   --RFX
           FX,F1,FX,RFX,FX,RFX,FX ,RFX,FX ,RFX,FX ,RFX,FX ,RFX,FX,RFX,FX),
       (  U,FX,
          FX,FX,FX,FX,FX,
          FX,FX,FX,FX,FX,FX,FX,FX,FX,
          FX,FX,FX,FX,FX,FX,FX,FX,FX,FX,FX,FX,FX,
          FX,FX,FX,FX,FX,FX,FX,FX,FX,FX,FX,FX,FX,FX,FX,FX )--FX
);
```

For every combination of values, the table returns a value which represents the combined interval which covers the component values in the least pessimistic fashion.

7.5 Logic Manipulation

The following sections discuss technology independent facilities which support the manipulation of logic operations.

7.5.1 Overloaded Comparison Operators

The comparison operators in VHDL are used frequently and are useful in manipulating logic values. The standard logic package provides a predefined set of overloaded operators to make modelling readable and easy.

The following code shows the body source for the various overloaded operators.

Overloaded Comparison Operators

```
-- "=" operator.
FUNCTION "="    ( l : t_logic; r : t_state ) RETURN boolean IS
    BEGIN    RETURN( f_state( l ) = r );    END;
FUNCTION "="    ( l : t_state; r : t_logic ) RETURN boolean IS
    BEGIN    RETURN( l = f_state( r ) );    END;

-- "/=" operator.
FUNCTION "/="   ( l : t_logic; r : t_state ) RETURN boolean IS
    BEGIN    RETURN( f_state( l ) /= r );   END;
FUNCTION "/="   ( l : t_state; r : t_logic ) RETURN boolean IS
    BEGIN    RETURN( l /= f_state( r ) );   END;

-- "<" operator.
FUNCTION "<"    ( l : t_logic; r : t_state ) RETURN boolean IS
    BEGIN    RETURN( f_state( l ) < r );    END;
FUNCTION "<"    ( l : t_state; r : t_logic ) RETURN boolean IS
    BEGIN    RETURN( l < f_state( r ) );    END;

-- "<=" operator.
FUNCTION "<="   ( l : t_logic; r : t_state ) RETURN boolean IS
    BEGIN    RETURN( f_state( l ) <= r );   END;
FUNCTION "<="   ( l : t_state; r : t_logic ) RETURN boolean IS
    BEGIN    RETURN( l <= f_state( r ) );   END;

-- ">" operator.
FUNCTION ">"    ( l : t_logic; r : t_state ) RETURN boolean IS
    BEGIN    RETURN( f_state( l ) > r );    END;
FUNCTION ">"    ( l : t_state; r : t_logic ) RETURN boolean IS
    BEGIN    RETURN( l > f_state( r ) );    END;

-- ">=" operator.
FUNCTION ">="   ( l : t_logic; r : t_state ) RETURN boolean IS
    BEGIN    RETURN( f_state( l ) >= r );   END;
FUNCTION ">="   ( l : t_state; r : t_logic ) RETURN boolean IS
    BEGIN    RETURN( l >= f_state( r ) );   END;
```

The following shows an example where overloaded operators are useful.

Logic Value Comparison Example

```
-- Watch clock
IF (clk'EVENT) AND (clk = '1') THEN
    .
    .
    .
END IF;
```

7.5. LOGIC MANIPULATION

The expression **clk** = '1' calls upon the overloaded equality operator allowing the comparison of a signal of type **t_wlogic** with a state of type **t_state**. Without the predefined overloaded operators, this statement would be flagged as an error, and a much more complex VHDL expression would be required to handle all possible values of the **clk** signal.

7.5.2 State/Strength Lookup Tables

As discussed in section 7.2 the logic system as defined is composed of combined state and strength information. Often, extraction of the state or strength from a value is useful. And alternatively, the creation of a value given a state/strength pair is required. The tables described in this section make conversion between logic values and state/strength pairs easy and convenient.

The following shows the source code for the **f_state** table.

State Lookup Table

```
-- Type definition for the table
TYPE f_state_T IS ARRAY (t_logic'low to t_logic'high) OF t_state;

-- Given value return the associated state
CONSTANT f_state : f_state_t := (
        'X',  -- U
        'X',  -- D
        '0',  -- Z0
        '1',  -- Z1
        'X',  -- ZDX
        'X',  -- DZX
        'X',  -- ZX
        '0',  -- W0
        '1',  -- W1
        '0',  -- WZ0
        '1',  -- WZ1
        'X',  -- WDX
        'X',  -- DWX
        'X',  -- WZX
        'X',  -- ZWX
        'X',  -- WX
        '0',  -- R0
        '1',  -- R1
        '0',  -- RW0
        '1',  -- RW1
        '0',  -- RZ0
        '1',  -- RZ1
        'X',  -- RDX
        'X',  -- DRX
        'X',  -- RZX
```

```
        'X',  -- ZRX
        'X',  -- RWX
        'X',  -- WRX
        'X',  -- RX
        '0',  -- F0
        '1',  -- F1
        '0',  -- FR0
        '1',  -- FR1
        '0',  -- FW0
        '1',  -- FW1
        '0',  -- FZ0
        '1',  -- FZ1
        'X',  -- FDX
        'X',  -- DFX
        'X',  -- FZX
        'X',  -- ZFX
        'X',  -- FWX
        'X',  -- WFX
        'X',  -- FRX
        'X',  -- RFX
        'X'   -- FX
);
```

In essence, this table strips off the strength information associated with a value, and returns the appropriate state for the given logic value. The following shows how the **f_state** table can be used in modelling hardware.

Using the State Lookup Table

```
IF f_state( input_signal ) = '1' THEN
    -- Some statements
ELSIF f_state( input_signal ) = '0' THEN
    -- Some statements
ELSE
    -- Some statements
END IF;
```

By utilizing the table lookup in this example, the expression in each if clause is much simpler than if a direct comparison to the signal value were used. In many cases, values which are manipulated in a model can be restricted to pure state information, and only on input and output from a device model are the strengths merged with state information to produce actual signal values. This approach results in models which are easy to read and maintain.

In technology specific applications it is useful to extract the strength of a signal from its value. The following shows the source code for the **f_strength** table.

7.5. LOGIC MANIPULATION

Strength Lookup Table

```
-- Type definition for the table
TYPE f_strength_T IS ARRAY (t_logic'low to t_logic'high) OF t_strength;

-- Given logic value return the strength
CONSTANT f_strength : f_strength_t := (
        'F', -- U
        'U', -- D
        'Z', -- Z0
        'Z', -- Z1
        'Z', -- ZDX
        'Z', -- DZX
        'Z', -- ZX
        'W', -- W0
        'W', -- W1
        'W', -- WZ0
        'W', -- WZ1
        'W', -- WDX
        'W', -- DWX
        'W', -- WZX
        'W', -- ZWX
        'W', -- WX
        'R', -- R0
        'R', -- R1
        'R', -- RW0
        'R', -- RW1
        'R', -- RZ0
        'R', -- RZ1
        'R', -- RDX
        'R', -- DRX
        'R', -- RZX
        'R', -- ZRX
        'R', -- RWX
        'R', -- WRX
        'R', -- RX
        'F', -- F0
        'F', -- F1
        'F', -- FR0
        'F', -- FR1
        'F', -- FW0
        'F', -- FW1
        'F', -- FZ0
        'F', -- FZ1
        'F', -- FDX
        'F', -- DFX
        'F', -- FZX
        'F', -- ZFX
        'F', -- FWX
        'F', -- WFX
        'F', -- FRX
        'F', -- RFX
```

```
        'F',  -- FX
);
```

In essence, this table strips off the state information associated with a value, and returns the appropriate strength for the given logic value. Another related table returns the lowest strength associated with the value interval.

Least Strength Lookup

```
-- Given logic value return the strength
CONSTANT f_strengthL : f_strength_t := (
        'F',  -- U
        'U',  -- D
        'Z',  -- Z0
        'Z',  -- Z1
        'Z',  -- ZDX
        'U',  -- DZX
        'Z',  -- ZX
        'W',  -- W0
        'W',  -- W1
        'Z',  -- WZ0
        'Z',  -- WZ1
        'W',  -- WDX
        'U',  -- DWX
        'W',  -- WZX
        'Z',  -- ZWX
        'W',  -- WX
        'R',  -- R0
        'R',  -- R1
        'W',  -- RW0
        'W',  -- RW1
        'Z',  -- RZ0
        'Z',  -- RZ1
        'R',  -- RDX
        'U',  -- DRX
        'R',  -- RZX
        'Z',  -- ZRX
        'R',  -- RWX
        'W',  -- WRX
        'R',  -- RX
        'F',  -- F0
        'F',  -- F1
        'R',  -- FR0
        'R',  -- FR1
        'W',  -- FW0
        'W',  -- FW1
        'Z',  -- FZ0
        'Z',  -- FZ1
        'F',  -- FDX
```

7.5. LOGIC MANIPULATION

```
            'U',   -- DFX
            'F',   -- FZX
            'Z',   -- ZFX
            'F',   -- FWX
            'W',   -- WFX
            'F',   -- FRX
            'R',   -- RFX
            'F'    -- FX
);
```

The following table returns the highest strength associated with the value interval.

Highest Strength Lookup

```
-- Given logic value return the strength
CONSTANT f_strengthH : f_strength_t := (
            'F',   -- U
            'U',   -- D
            'Z',   -- Z0
            'Z',   -- Z1
            'U',   -- ZDX
            'Z',   -- DZX
            'Z',   -- ZX
            'W',   -- W0
            'W',   -- W1
            'W',   -- WZ0
            'W',   -- WZ1
            'U',   -- WDX
            'W',   -- DWX
            'Z',   -- WZX
            'W',   -- ZWX
            'W',   -- WX
            'R',   -- R0
            'R',   -- R1
            'R',   -- RW0
            'R',   -- RW1
            'R',   -- RZ0
            'R',   -- RZ1
            'U',   -- RDX
            'R',   -- DRX
            'Z',   -- RZX
            'R',   -- ZRX
            'W',   -- RWX
            'R',   -- WRX
            'R',   -- RX
            'F',   -- F0
            'F',   -- F1
            'F',   -- FR0
            'F',   -- FR1
            'F',   -- FW0
```

```
                'F',  -- FW1
                'F',  -- FZ0
                'F',  -- FZ1
                'U',  -- FDX
                'F',  -- DFX
                'Z',  -- FZX
                'F',  -- ZFX
                'W',  -- FWX
                'F',  -- WFX
                'R',  -- FRX
                'F',  -- RFX
                'F'   -- FX
);
```

The **f_strengthL** and **f_strengthH** lookup tables are useful for determining the span of strengths associated with a given value. The following shows how the **f_strength** table can be used in modelling hardware.

Using the Strength Lookup Table

```
IF  (f_strength( input_signal ) = 'U') THEN
    -- Some statements
ELSIF (f_strength( input_signal ) = 'F') THEN
    -- Some statements
ELSE
    -- Some statements
END IF;
```

By utilizing the table lookup in this example, the expression in each if clause is simpler than if a direct comparison to the signal value were used. Extraction of the U strength can be particularly useful in detecting circuit startup problems. For a given technology, knowledge about the strength of a signal is important and this table makes this information easy to access.

The creation of signal values from state/strength pairs is useful. The following shows the source code for the **f_logic** table.

Value from State/Strength Lookup Table

```
-- Type definitions for table
TYPE f_str IS ARRAY (t_strength'low to t_strength'high) OF t_logic;
TYPE f_log_con_T IS ARRAY (t_state'low to t_state'high) OF f_str;

-- Given state/strength, return logic value
CONSTANT f_logic : f_log_con_t := (
        ( D ,      -- '0', 'U'
          Z0,      -- '0', 'Z'
          W0,      -- '0', 'W'
```

7.5. LOGIC MANIPULATION

```
         R0,       -- '0', 'R'
         F0 ),     -- '0', 'F'
       ( D ,       -- '1', 'U'
         Z1,       -- '1', 'Z'
         W1,       -- '1', 'W'
         R1,       -- '1', 'R'
         F1 ),     -- '1', 'F'
       ( D ,       -- 'X', 'U'
         ZX,       -- 'X', 'Z'
         WX,       -- 'X', 'W'
         RX,       -- 'X', 'R'
         FX )      -- 'X', 'F'
);
```

This table returns a logic value given state and strength information as shown below

State/Strength Lookup

Strength	State	Logic Value
Z	0	Z0
W	0	W0
R	0	R0
F	0	F0
Z	1	Z1
W	1	W1
R	1	R1
F	1	F1
Z	X	ZX
W	X	WX
R	X	RX
F	X	FX
U	Z, W, R, or F	D

The following table shows the definition for the **f_logic0** table.

State 0/Interval Lookup

```
-- Type definition for the tables that follow
TYPE f_logs_con_t IS ARRAY (t_strength'LOW TO t_strength'HIGH) OF f_str;

-- Given strength/strength, return logic value
CONSTANT f_logic0 : f_logs_con_t := (
       ( D,        -- 'U', 'U'
         ZDX,      -- 'U', 'Z'
```

```
        WDX,    -- 'U', 'W'
        RDX,    -- 'U', 'R'
        FDX ),  -- 'U', 'F'
      ( ZDX,    -- 'Z', 'U'
        ZO,     -- 'Z', 'Z'
        WZO,    -- 'Z', 'W'
        RZO,    -- 'Z', 'R'
        FZO ),  -- 'Z', 'F'
      ( WDX,    -- 'W', 'U'
        WZO,    -- 'W', 'Z'
        WO,     -- 'W', 'W'
        RWO,    -- 'W', 'R'
        FWO ),  -- 'W', 'F'
      ( RDX,    -- 'R', 'U'
        RZO,    -- 'R', 'Z'
        RWO,    -- 'R', 'W'
        RO,     -- 'R', 'R'
        FRO ),  -- 'R', 'F'
      ( FDX,    -- 'F', 'U'
        FZO,    -- 'F', 'Z'
        FWO,    -- 'F', 'W'
        FRO,    -- 'F', 'R'
        FO )    -- 'F', 'F'
);
```

This table returns the value associated with a given strength interval for the logic 0 state as shown below

7.5. LOGIC MANIPULATION

Logic 0 Lookup

Low Strength	High Strength	Logic Value
U	U	D
U	Z	ZDX
U	W	WDX
U	R	RDX
U	F	FDX
Z	U	ZDX
Z	Z	Z0
Z	W	WZ0
Z	R	RZ0
Z	F	FZ0
W	U	WDX
W	Z	WZ0
W	W	W0
W	R	RW0
W	F	FW0
R	U	RDX
R	Z	RZ0
R	W	RW0
R	R	R0
R	F	FR0
F	U	FDX
F	Z	FZ0
F	W	FW0
F	R	FR0
F	F	F0

The following table shows the definition for the **f_logic1** table.

State 1/Interval Lookup

```
-- Given strength/strength, return logic value
CONSTANT f_logic1 : f_logs_con_t := (
        ( D,        -- 'U', 'U'
          DZX,      -- 'U', 'Z'
          DWX,      -- 'U', 'W'
          DRX,      -- 'U', 'R'
          DFX ),    -- 'U', 'F'
        ( DZX,      -- 'Z', 'U'
          Z1,       -- 'Z', 'Z'
          WZ1,      -- 'Z', 'W'
```

```
             RZ1,      -- 'Z', 'R'
             FZ1 ),    -- 'Z', 'F'
           ( DWX,      -- 'W', 'U'
             WZ1,      -- 'W', 'Z'
             W1,       -- 'W', 'W'
             RW1,      -- 'W', 'R'
             FW1 ),    -- 'W', 'F'
           ( DRX,      -- 'R', 'U'
             RZ1,      -- 'R', 'Z'
             RW1,      -- 'R', 'W'
             R1,       -- 'R', 'R'
             FR1 ),    -- 'R', 'F'
           ( DFX,      -- 'F', 'U'
             FZ1,      -- 'F', 'Z'
             FW1,      -- 'F', 'W'
             FR1,      -- 'F', 'R'
             F1 )      -- 'F', 'F'
   );
```

This table returns the value associated with a given strength interval for the logic 1 state as shown below

7.5. LOGIC MANIPULATION

Logic 0 Lookup

Low Strength	High Strength	Logic Value
U	U	D
U	Z	DZX
U	W	DWX
U	R	DRX
U	F	DFX
Z	U	DZX
Z	Z	Z1
Z	W	WZ1
Z	R	RZ1
Z	F	FZ1
W	U	DWX
W	Z	WZ1
W	W	W1
W	R	RW1
W	F	FW1
R	U	DRX
R	Z	RZ1
R	W	RW1
R	R	R1
R	F	FR1
F	U	DFX
F	Z	FZ1
F	W	FW1
F	R	FR1
F	F	F1

And finally, the following shows the definition for the **f_logicX** table.

State X/Interval Lookup

```
-- Given strength/strength, return logic value
CONSTANT f_logicX : f_logs_con_t := (
        ( D,        -- 'U', 'U'
          DZX,      -- 'U', 'Z'
          DWX,      -- 'U', 'W'
          DRX,      -- 'U', 'R'
          DFX ),    -- 'U', 'F'
        ( ZDX,      -- 'Z', 'U'
          ZX,       -- 'Z', 'Z'
          ZWX,      -- 'Z', 'W'
```

```
            ZRX,      -- 'Z', 'R'
            ZFX ),    -- 'Z', 'F'
          ( WDX,      -- 'W', 'U'
            WZX,      -- 'W', 'Z'
            WX,       -- 'W', 'W'
            WRX,      -- 'W', 'R'
            WFX ),    -- 'W', 'F'
          ( RDX,      -- 'R', 'U'
            RZX,      -- 'R', 'Z'
            RWX,      -- 'R', 'W'
            RX,       -- 'R', 'R'
            RFX ),    -- 'R', 'F'
          ( FDX,      -- 'F', 'U'
            FZX,      -- 'F', 'Z'
            FWX,      -- 'F', 'W'
            FRX,      -- 'F', 'R'
            FX )      -- 'F', 'F'
);
```

This table returns the value associated with a given strength interval for the logic unknown state as shown below

7.5. LOGIC MANIPULATION

Logic 0 Lookup

Low Strength	High Strength	Logic Value
U	U	D
U	Z	DZX
U	W	DWX
U	R	DRX
U	F	DFX
Z	U	ZDX
Z	Z	ZX
Z	W	ZWX
Z	R	ZRX
Z	F	ZFX
W	U	WDX
W	Z	WZX
W	W	WX
W	R	WRX
W	F	WFX
R	U	RDX
R	Z	RZX
R	W	RWX
R	R	RX
R	F	RFX
F	U	FDX
F	Z	FZX
F	W	FWX
F	R	FRX
F	F	FX

The following

Using the Value Lookup Table

```
output_signal <= f_logic( '1','F' ) AFTER delay;   -- F1
```

shows how a logic value can be constructed from a state and strength. In this case the value F1 is generated from the state '1' and strength 'F' and assigned to the signal **output_signal**.

7.5.3 Logic Lookup Tables

The most common operations in a VHDL model will be logic or boolean operations. The tables defined in this section represent a set of primitive logic lookup tables that provide for efficient logic operations on the basic states of the standard logic package. These tables can be combined in VHDL expressions to perform arbitrarily complex logic operations and are suitable for any modelling task which manages combinational logic.

The application of these tables to a specific technology occurs when a given technology package is referenced as described in section 7.3. The boolean operators are overloaded with specific technology dependent expressions as described in sections 7.3.1, 7.3.2, 7.3.3, 7.3.4 and 7.3.5.

The following shows the type definitions which are referenced in the following lookup tables.

Logic Table Types

```
-- Type definition for tables that follow
TYPE f_1_x_1 IS ARRAY (t_state'LOW TO t_state'HIGH) OF t_state;
TYPE f_2_x_1 IS ARRAY (t_state'LOW TO t_state'HIGH) OF f_1_x_1;
```

Type $f\_1\_x\_1$ defines a one dimensional array for one value lookup operations, and $f\_2\_x\_1$ defines a two dimensional array for two value lookup operations.

The following shows the definition of the AND lookup table.

AND Logic Table

```
CONSTANT f_and : f_2_x_1 := (
        ( '0',    -- '0', '0'
          '0',    -- '0', '1',
          '0' ),  -- '0', 'X'
        ( '0',    -- '1', '0'
          '1',    -- '1', '1'
          'X' ),  -- '1', 'X'
        ( '0',    -- 'X', '0'
          'X',    -- 'X', '1'
          'X' )   -- 'X', 'X'
     );
```

The following shows the results this table returns given two states.

7.5. LOGIC MANIPULATION

AND Table Operation

	0	1	X
0	0	0	0
1	0	1	X
X	0	X	X

The results for various combinations of 0 and 1 are straightforward and match the AND boolean operation. With unknown inputs, the table is structured to generate a 0 when at least one of the inputs is a 0, but otherwise an unknown is generated. In general logic tables should be structured to generate the most optimistic but accurate result possible. Generating an X when one of the inputs is a 0 would be accurate but would be overly pessimistic. Clearly generating a 1 when the other input is an unknown is not accurate since the value of the table is driven by the unknown input and inaccurate simulation results are possible if this approach is taken. The uninitialized value is folded into the unknown state when the state is extracted from a logic value and in general should be treated in the same fashion that unknowns are in models. The following shows how the AND table can be used in VHDL expressions.

AND Table Usage

```
next_state := f_and(statea,stateb);
```

The following shows the definition of the OR lookup table.

OR Logic Table

```
CONSTANT f_or : f_2_x_1 := (
        ( '0',    -- '0', '0'
          '1',    -- '0', '1',
          'X' ),  -- '0', 'X'
        ( '1',    -- '1', '0'
          '1',    -- '1', '1'
          '1' ),  -- '1', 'X'
        ( 'X',    -- 'X', '0'
          '1',    -- 'X', '1'
          'X' )   -- 'X', 'X'
);
```

The following shows the results this table returns given two states.

OR Table Operation

	0	1	X
0	0	1	X
1	1	1	1
X	X	1	X

The results for various combinations of 0 and 1 are straightforward and match the OR boolean operation. Similar arguments in the choice of X outputs can be made with this table as with the AND table. The most optimistic but accurate approach to propagating X and U inputs to outputs is taken. The following shows how the OR table can be used in VHDL expressions.

OR Table Usage

```
ASSERT ( f_or( '1', '0' ) = '1' );
```

The following shows the definition of the NAND lookup table.

NAND Logic Table

```
CONSTANT f_nand : f_2_x_1 := (
         ( '1',    -- '0', '0'
           '1',    -- '0', '1',
           '1' ),  -- '0', 'X'
         ( '1',    -- '1', '0'
           '0',    -- '1', '1'
           'X' ),  -- '1', 'X'
         ( '1',    -- 'X', '0'
           'X',    -- 'X', '1'
           'X' )   -- 'X', 'X'
       );
```

The following shows the results this table returns given two states.

NAND Table Operation

	0	1	X
0	1	1	1
1	1	0	X
X	1	X	X

7.5. LOGIC MANIPULATION

The results for various combinations of 0 and 1 are straightforward and match the NAND boolean operation. Similar arguments in the choice of X outputs can be made with this table as with the AND table. The most optimistic but accurate approach to propagating X and U inputs to outputs is taken. The following shows how the NAND table can be used in VHDL expressions.

NAND Table Usage

```
next_state := f_nand(statea,stateb);
```

The following shows the definition of the NOR lookup table.

NOR Logic Table

```
CONSTANT f_nor : f_2_x_1 := (
       ( '1',   -- '0', '0'
         '0',   -- '0', '1',
         'X' ), -- '0', 'X'
       ( '0',   -- '1', '0'
         '0',   -- '1', '1'
         '0' ), -- '1', 'X'
       ( 'X',   -- 'X', '0'
         '0',   -- 'X', '1'
         'X' )  -- 'X', 'X'
);
```

The following shows the results this table returns given two states.

NOR Table Operation

	0	1	X
0	1	0	X
1	0	0	0
X	X	0	X

The results for various combinations of 0 and 1 are straightforward and match the NOR boolean operation. Similar arguments in the choice of X outputs can be made with this table as with the AND table. The most optimistic but accurate approach to propagating X and U inputs to outputs is taken. The following shows how the NOR table can be used in VHDL expressions.

NOR Table Usage

```
ASSERT ( f_nor( '1', '0' ) = '0' );
```

The following shows the definition of the XOR lookup table.

XOR Logic Table

```
CONSTANT f_xor : f_2_x_1 := (
        ( '0',   -- '0', '0'
          '1',   -- '0', '1',
          'X' ), -- '0', 'X'
        ( '1',   -- '1', '0'
          '0',   -- '1', '1'
          'X' ), -- '1', 'X'
        ( 'X',   -- 'X', '0'
          'X',   -- 'X', '1'
          'X' )  -- 'X', 'X'
    );
```

The following shows the results this table returns given two states.

XOR Table Operation

	0	1	X
0	0	1	X
1	1	0	X
X	X	X	X

The results for various combinations of 0 and 1 are straightforward and match the OR boolean operation. Similar arguments in the choice of X outputs can be made with this table as with the AND table. The most optimistic but accurate approach to propagating X and U inputs to outputs is taken. The following shows how the XOR table can be used in VHDL expressions.

XOR Table Usage

```
ASSERT ( f_xor( '1', '1' ) = '0' );
```

The following shows the definition of the NOT lookup table.

NOT Logic Table

```
CONSTANT f_not : f_1_x_1 := (
          '1',    --   '0'
          '0',    --   '1'
          'X'     --   'X'
);
```

The following shows the results this table returns given an input states.

NOT Table Operation

0	1	X
1	0	X

The results for inputs of 0 and 1 are straightforward and match the OR boolean operation. Similar arguments in the choice of X outputs can be made with this table as with the AND table. The most optimistic but accurate approach to propagating X and U inputs to outputs is taken. The following shows how the NOT table can be used in VHDL expressions.

NOT Table Usage

```
ASSERT ( f_not( '1' ) = '0' );
```

7.6 Timing Utilities

For synchronous devices, detection of the rising or falling edge of a clock or input signal is important. The following shows the definition for the f_rising_edge table.

Rising Edge Lookup Table

```
-- Type definition for the following tables
type f_boolean_t is array( t_logic'low to t_logic'high ) of boolean;

-- Given value, return true for rising edge
CONSTANT f_rising_edge : f_boolean_t := (
         false, -- U
         false, -- D
         false, -- Z0
         true,  -- Z1
         false, -- ZDX
         false, -- DZX
```

```
         false, -- ZX
         false, -- W0
         true,  -- W1
         false, -- WZ0
         true,  -- WZ1
         false, -- WDX
         false, -- DWX
         false, -- WZX
         false, -- ZWX
         false, -- WX
         false, -- R0
         true,  -- R1
         false, -- RW0
         true,  -- RW1
         false, -- RZ0
         true,  -- RZ1
         false, -- RDX
         false, -- DRX
         false, -- RZX
         false, -- ZRX
         false, -- RWX
         false, -- WRX
         false, -- RX
         false, -- F0
         true,  -- F1
         false, -- FR0
         true,  -- FR1
         false, -- FW0
         true,  -- FW1
         false, -- FZ0
         true,  -- FZ1
         false, -- FDX
         false, -- DFX
         false, -- FZX
         false, -- ZFX
         false, -- FWX
         false, -- WFX
         false, -- FRX
         false, -- RFX
         false  -- FX
);
```

This table can be utilized in a process which is sensitive to events on a signal to determine whether a rising edge of the signal has just occured as shown in the following.

Detecting Rising Edge of Clock

```
PROCESS (clk)
    BEGIN
```

7.6. TIMING UTILITIES

```
    IF  (f_rising_edge( clk )) THEN
        -- synchronous processing
    END IF;
END PROCESS;
```

Similarly, the **f_falling_edge** table definition is shown in the following.

Falling Edge Lookup Table

```
-- Given value, return true for falling edge
CONSTANT f_falling_edge : f_boolean_t := (
        false,  -- U
        false,  -- D
        true,   -- Z0
        false,  -- Z1
        false,  -- ZDX
        false,  -- DZX
        false,  -- ZX
        true,   -- W0
        false,  -- W1
        true,   -- WZ0
        false,  -- WZ1
        false,  -- WDX
        false,  -- DWX
        false,  -- WZX
        false,  -- ZWX
        false,  -- WX
        true,   -- R0
        false,  -- R1
        true,   -- RW0
        false,  -- RW1
        true,   -- RZ0
        false,  -- RZ1
        false,  -- RDX
        false,  -- DRX
        false,  -- RZX
        false,  -- ZRX
        false,  -- RWX
        false,  -- WRX
        false,  -- RX
        true,   -- F0
        false,  -- F1
        true,   -- FR0
        false,  -- FR1
        true,   -- FW0
        false,  -- FW1
        true,   -- FZ0
        false,  -- FZ1
        false,  -- FDX
        false,  -- DFX
```

```
            false,  -- FZX
            false,  -- ZFX
            false,  -- FWX
            false,  -- WFX
            false,  -- FRX
            false,  -- RFX
            false   -- FX
);
```

This table can be utilized in a process which is sensitive to events on a signal to determine whether a falling edge of the signal has just occured as shown in the following.

Detecting Falling Edge of Clock

```
PROCESS (clk)
    BEGIN
    IF  (f_falling_edge( clk )) THEN
        -- synchronous processing
    END IF;
    END PROCESS;
```

Propagation of hardware devices is often dependent on the value of an output signal. Determination of this propagation is facilitated by the function **f_delay** shown in the following

2 State Delay Calculation

```
FUNCTION f_delay( newlv  : IN t_logic;
                  delay01 : IN time;
                  delay10 : IN time) RETURN time IS
BEGIN
    CASE f_state(newlv) IS
        WHEN '0' => RETURN delay10;
        WHEN '1' => RETURN delay01;
        WHEN 'X' => IF (delay01 > delay10) THEN RETURN delay01;
                    ELSE RETURN delay10;
                    END IF;
    END CASE;
END f_delay;
```

This function uses the new output value **newlv** to determine which delay **delay01** or **delay10** to return. These values represent the 0 to 1 (rising) and 1 to 0 (falling) delays of the device respectively. Notice that the function returns the worst case delay when the output value is unknown as summarized below

7.6. TIMING UTILITIES

2 State Delay Operation

New State	Delay Returned
1	delay01
0	delay10
X	maximum of delay01 and delay10

The following shows how the **f_delay** function can be used in a model.

2 State Delay Example

```
ENTITY nand_twogeneric IS
    GENERIC (tplh,tphl : TIME := 7 ns);
    PORT    (a,b: IN t_wlogic; y: OUT t_wlogic);
END nand_twogeneric;

ARCHITECTURE behavioral OF nand_twogeneric IS
BEGIN
    y <= a NAND b AFTER f_delay(a NAND b, tplh, tphl);
END behavioral;
```

Passing generic parameters to this function is typical and provides for technology independence in the model.

Tristate devices often have an additional Z state which must be handled during timing calculations. The following shows the definition of the **f_zdelay** function.

3 State Delay Calculation

```
FUNCTION f_zdelay( oldlv   : IN t_logic;
                   newlv   : IN t_logic;
                   delay01 : IN time;
                   delay10 : IN time;
                   delayz0 : IN time;
                   delayz1 : IN time;
                   delay0z : IN time;
                   delay1z : IN time) RETURN time IS
    VARIABLE old_strength : t_strength;
    VARIABLE new_strength : t_strength;
BEGIN
    -- Compute the strengths.
    old_strength := f_strength( oldlv );
    new_strength := f_strength( newlv );

    -- If both are 'Z', then take the highest of the four 'z' delays.
```

```
    IF ((old_strength <= 'Z') AND (new_strength <= 'Z')) THEN
       RETURN( f_max_time( delayz0, f_max_time( delayz1,
          f_max_time( delay0z, delay1z ) ) ) );

    -- If the old strength is 'Z', then it must be delayz?.
    ELSIF (old_strength <= 'Z') THEN
       CASE f_state( newlv ) IS
          WHEN '0'       => RETURN delayz0;
          WHEN '1'       => RETURN delayz1;
          WHEN 'X'       => RETURN ( f_max_time( delayz0, delayz1 ) );
       END CASE;

    -- If the new strength is 'Z', then it must be delay?z.
    ELSIF (new_strength <= 'Z') THEN
       CASE f_state( oldlv ) IS
          WHEN '0'       => RETURN delay0z;
          WHEN '1'       => RETURN delay1z;
          WHEN 'X'       => RETURN ( f_max_time( delay0z, delay1z ) );
       END CASE;

    -- Otherwise, use f_delay TO compute the delay.
    ELSE RETURN( f_delay( newlv, delay01, delay10 ) );
    END IF;
END f_zdelay;
```

This function takes as inputs the current value **oldlv** and the new value **newlv** of the output signal along with the various possible delays **delay01**, **delay10**, **delayz0**, **delayz1**, **delay0z** and **delay1z** one for each possible state transition as shown below

2 State Delay Operation

Previous Value	New Value	Delay Returned
Z	0	delayz0
	1	delayz1
	X	maximum of delayz1 and delayz0
0	Z	delay0z
1		delay1z
X		maximum of delay1z and delay0z
0 or 1 or X	0 or 1 or X	f_delay value

The following shows the definition for the **f_setup_check** function.

7.6. TIMING UTILITIES

Setup Constraint Checks

```
FUNCTION f_setup_check( SIGNAL signal_to_check : IN t_logic;
                        setup_constraint : IN time ) RETURN boolean IS
BEGIN
    RETURN (signal_to_check'LAST_EVENT >= setup_constraint);
END f_setup_check;
```

This function takes as inputs a signal **signal_to_check** and a setup delay **setup_constraint** and returns false if a setup violation has occurred, otherwise it returns true. A typical use of this function is shown below

Checking for Setup Errors

```
-- 'Setup' checking on a positive-edge triggered clock.
setup_check: PROCESS( clk )
    BEGIN
    -- Only perform the assertion after the rising edge.
    IF (f_rising_edge( clk )) THEN

        -- Verify that the setup constraint has not been violated.
        ASSERT( f_setup_check( signal_to_check, setup_constraint ) )
            REPORT "Setup check violation on 'signal_to_check'."
            SEVERITY error;
    END IF;
END PROCESS setup_check;
```

A related function **f_hold_check** is shown below.

Hold Constraint Checks

```
FUNCTION f_hold_check( SIGNAL signal_to_check : IN t_logic;
                       hold_time        : IN time ) RETURN boolean IS
BEGIN
    RETURN (signal_to_check'LAST_EVENT >= hold_time );
END f_hold_check;
```

This function takes as inputs a signal **signal_to_check** and a hold delay **setup_constraint** and returns false if a hold violation has occurred, otherwise it returns true. A typical use of this function is shown below

Checking for Hold Errors

```
-- Positive-edge triggered hold time checking process.
hold_check: PROCESS( clk'DELAYED( holdtime ) )
    BEGIN
```

```
        -- Only perform the assertion after the rising edge.
        IF  (f_rising_edge( clk'DELAYED( holdtime ) )) THEN
            -- Verify that the hold constraint has not been violated.
            ASSERT( f_hold_check( signal_to_check, holdtime ) )
                REPORT "Hold check violation on 'signal_to_check'."
                SEVERITY error;
        END IF;
END process hold_check;
```

The following shows the definition for the **f_pulse_check** function.

Pulse Checks

```
FUNCTION f_pulse_check( SIGNAL signal_to_check : IN t_logic;
                              minimum_width    : IN time ) RETURN boolean IS
BEGIN
    RETURN (signal_to_check'LAST_EVENT >= minimum_width);
END f_pulse_check;
```

This function takes as inputs a signal **signal_to_check** and a hold delay **minimum_width** and returns true if a spike has occurred, otherwise it returns false. A typical use of this function is shown below

Checking for Pulses

```
-- Positive pulse checking.
pulse_check: PROCESS( clk )
    BEGIN
    -- Only perform the assertion after the clock falls.
    IF  (f_falling_edge( clk )) THEN
        -- Verify that the pulse width constraint has not been violated.
        ASSERT( f_pulse_check( clk, minimum_width ) )
            REPORT "Pulse check violation on 'clk'."
            SEVERITY error;
    END IF;
END PROCESS pulse_check;
```

7.7 Integer Data Utilities

Often in VHDL models it is more convenient to manipulate integer values rather than the underlying logic value. Counters, decoders, alu's, selectors and other related devices manipulate data which is best represented as integer data. The functions and procedures described in this section make it possible to easily convert arrays of logic values to/from integer values. Overloaded operators are provided which support logical operations on integer data.

The data types used by these functions are show below

7.7. INTEGER DATA UTILITIES

Integer/Logic Conversion Data Types

```
-- Type definition used by integer/logic conversion routines
TYPE t_logarray IS ARRAY (INTEGER RANGE <>) OF t_logic;
```

An array of **t_logic** is defined with the name **t_logarray**. This data type is an unconstrained array making it possible to pass a wide range of array sizes to the functions which follow.

The following shows the **f_logictoint** function which converts an array of logic values into an integer value.

Logic to Integer Conversion

```
PROCEDURE f_logictoint(VARIABLE s : IN t_logarray;
                      VARIABLE unknown : OUT boolean;
                      VARIABLE d : OUT integer) IS
    VARIABLE work : integer := 0;
BEGIN
    work := 0;
    unknown := false;

    FOR i IN s'RANGE LOOP
        work := work * 2;
        IF s(i) = '1' THEN work := work + 1;
        ELSIF s(i) = 'X' THEN unknown := true;
        END IF;
    END LOOP;
    d := work;
END f_logictoint;
```

This function is used primarily in situations where logical operations will be performed using integer data. Each bit is mapped one for one into the integer data type. This function can return both positive and negative integer values depending on the number of logic values passed and the value of the most significant bit. If the logic array contains greater than 32 bits, the most significant bits will be lost. Entry 1 of the logic array is always treated as the most significant bit. If the logic array contains fewer than 32 bits, then the integer value will always be positive and the high order bits of the integer will contain zeros.

The following shows how the **f_logictoint** function can be utilized.

Using Logic to Integer Conversion

```
PROCESS (clk)
    VARIABLE istate : integer;
```

```
    VARIABLE bstate : t_logarray (1 TO 4);
    VARIABLE unknown : boolean;
    SIGNAL a0,a1,a2,a3 : t_wlogic;
BEGIN
    -- parallel load of counter
    IF (clk = '1') THEN
        bstate(1) := a3;      -- MSB
        bstate(2) := a2;
        bstate(3) := a1;
        bstate(4) := a0;      -- LSB
        f_logictoint(bstate,unknown,istate);
    END IF;
```

In this example a counter is loaded on the rising edge of the clock **clk**. Each logic value is copied to an array of logic **bstate** which is then passed to the **f_logictoint** procedure which returns **unknown** and **istate**. istate contains the integer value which represents the loaded data and can now be used for counter operation using integer arithmetic. The **unknown** variable will be true if any of the input data contained an unknown value.

The converse operation is provided as follows which shows the **f_inttologic** function which converts from an integer value into an array of logic values.

Integer to Logic Conversion

```
PROCEDURE f_inttologic(
          VARIABLE d : IN integer;
          VARIABLE s : OUT t_logarray;
          VARIABLE t : IN t_technology) IS
    VARIABLE work : integer := 0;
    VARIABLE offset : integer := 0;
    VARIABLE slen : integer := 0;
    VARIABLE j : integer := 0;
BEGIN
    -- length of bit string?
    IF s'LENGTH > 32 THEN
        slen := 32;
        offset := s'LENGTH - 32;
        IF d >= 0 THEN
            FOR i IN offset-1 DOWNTO 0 LOOP
                s(s'LOW + i) := f_tech('0')(t); END LOOP;
        ELSE
            FOR i IN offset-1 DOWNTO 0 LOOP
                s(s'LOW + i) := f_tech('1')(t); END LOOP;
        END IF;
    ELSE slen := s'LENGTH;
    END IF;

    -- positive value
```

7.7. INTEGER DATA UTILITIES

```
        IF (d >= 0) THEN
            work := d;
            j := slen - 1;
            FOR i IN 1 TO 32 LOOP
                IF j >= 0 THEN
                    IF (work MOD 2) = 0 THEN s(s'LOW+j+offset) := f_tech('0')(t);
                    ELSE s(s'LOW+j+offset) := f_tech('1')(t);
                    END IF;
                END IF;
                work := work / 2;
                j := j - 1;
            END LOOP;

        -- negative value
        ELSE
            work := (-d) - 1;
            j := slen - 1;
            FOR i IN 1 TO 32 LOOP
                IF j >= 0 THEN
                    IF (work MOD 2) = 0 THEN s(s'LOW+j+offset) := f_tech('1')(t);
                    ELSE s(s'LOW+j+offset) := f_tech('0')(t);
                    END IF;
                END IF;
                work := work / 2;
                j := j - 1;
            END LOOP;
        END IF;
END f_inttologic;
```

This function is used primarily in situations where logical operations will be performed using integer data. Each bit is mapped one for one from the integer data type. This function can handle both positive and negative integer values. If the logic array contains greater than 32 bits, the most significant bits will be set to zero. Entry 1 of the logic array is always treated as the most significant bit. If the logic array contains fewer than 32 bits, then the most significant bits of the integer value will be ignored.

The following shows how the **f_inttologic** function can be utilized.

Using Integer to Logic Conversion

```
PROCESS (clk)
    VARIABLE istate : integer;
    VARIABLE bstate : t_logarray (1 TO 4);
    SIGNAL a0,a1,a2,a3 : t_wlogic;
BEGIN
    -- increment counter, generate outputs
    IF (clk = '1') THEN
        istate := istate + 1;
        f_inttologic(istate,bstate,ttl);
```

```
         a0 <= bstate(4);  -- LSB
         a1 <= bstate(3);
         a2 <= bstate(2);
         a3 <= bstate(1);  -- MSB
     END IF;
```

In this example a counter **istate** is incremented using integer arithmetic. The value of the counter is then converted to an array of logic **bstate** using the **f_inttologic** procedure.

For arithmetic operations, two's complement conversion is required. For arrays of 32 logic values the distinction between the previous routines and the following are not noticeable. But in the case where fewer than 32 logic values are required, the representation of each bit will differ between the 32 bit integer two's complement number and the logic array two's complement number. The following routines provide a method for utilizing the integer arithmetic capabilities of VHDL while working with arrays of logic values which contain fewer than 32 bits.

The following shows the **f_logictoint2c** function which converts an array of logic values into an integer value.

Logic to Integer Conversion/2's Complement

```
PROCEDURE f_logictoint2c(VARIABLE s : IN t_logarray;
                         VARIABLE unknown : OUT boolean;
                         VARIABLE d : OUT integer) IS
    VARIABLE work : integer := 0;
BEGIN
    unknown := false;
    work := 0;

    -- negative value?
    IF s(s'LOW) = '1' THEN
        FOR i IN s'RANGE LOOP
            work := work * 2;
            IF s(i) = '0' THEN work := work + 1;
            ELSIF s(i) = 'X' THEN unknown := true;
            END IF;
        END LOOP;
        d := -(work + 1);

    -- positive value
    ELSE
        FOR i IN s'RANGE LOOP
            work := work * 2;
            IF s(i) = '1' THEN work := work + 1;
            ELSIF s(i) = 'X' THEN unknown := true;
            END IF;
        END LOOP;
```

7.7. INTEGER DATA UTILITIES

```
            d := work;
        END IF;
END f_logictoint2c;
```

This function is used primarily in situations where arithmetic operations will be performed using integer data. The two's complement representation of the logic array is converted to the equivalent 32 bit two's complement integer representation. This function can return both positive and negative integer values depending on the sign bit of the logic array. If the logic array contains greater than 32 bits, the most significant bits will be lost. Entry 1 of the logic array is always treated as the most significant bit (sign bit). If the logic array contains fewer than 32 bits, then the integer value will be sign extended.

The following shows how the **f_logictoint2c** function can be utilized.

Using Logic to Integer Conversion/2's Complement

```
PROCESS (clk)
    VARIABLE istate : integer;
    VARIABLE bstate : t_logarray (1 TO 4);
    VARIABLE unknown : boolean;
    SIGNAL a0,a1,a2,a3 : t_wlogic;
BEGIN
    -- parallel load of counter
    IF (clk = '1') THEN
        bstate(1) := a3;    -- MSB
        bstate(2) := a2;
        bstate(3) := a1;
        bstate(4) := a0;    -- LSB
        f_logictoint2c(bstate,unknown,istate);
    END IF;
```

In this example a counter is loaded on the rising edge of the clock **clk**. Each logic value is copied to an array of logic **bstate** which is then passed to the **f_logictoint** procedure which returns **unknown** and **istate**. **istate** contains the integer value which represents the loaded data and can now be used for counter operation using integer arithmetic. The **unknown** variable will be true if any of the input data contained an unknown value. Notice in this example, the counter value **istate** can have a negative as well as positive value since the 4 bits are treated as two's complement numbers.

The converse operation is provided as follows which shows the **f_inttologic2c** function which converts from an integer value into an array of logic values.

Integer to Logic Conversion/2's Complement

```
PROCEDURE f_inttologic2c(
            VARIABLE d : IN integer;
            VARIABLE s : OUT t_logarray;
            VARIABLE t : IN t_technology) IS
    VARIABLE work : integer := 0;
    VARIABLE slen : integer := 0;
    VARIABLE j : integer := 0;
    VARIABLE offset : integer := 0;
BEGIN
    -- length of bit string?
    IF s'LENGTH > 32 THEN
        slen := 32;
        offset := s'LENGTH - 32;
        IF d >= 0 THEN
            FOR i IN offset-1 DOWNTO 0 LOOP
                s(s'LOW + i) := f_tech('0')(t); END LOOP;
        ELSE
            FOR i IN offset-1 DOWNTO 0 LOOP
                s(s'LOW + i) := f_tech('1')(t); END LOOP;
        END IF;
    ELSE slen := s'LENGTH;
    END IF;

    -- positive value
    IF (d >= 0) THEN
        work := d;
        j := slen - 1;
        FOR i IN 1 TO 32 LOOP
            IF j >= 0 THEN
                IF (work MOD 2) = 0 THEN s(s'LOW+j+offset) := f_tech('0')(t);
                ELSE s(s'LOW+j+offset) := f_tech('1')(t);
                END IF;
            END IF;
            work := work / 2;
            j := j - 1;
        END LOOP;
        s(s'LOW) := f_tech('0')(t);

    -- negative value
    ELSE
        work := (-d) - 1;
        j := slen - 1;
        FOR i IN 1 TO 32 LOOP
            IF j >= 0 THEN
                IF (work MOD 2) = 0 THEN s(s'LOW+j+offset) := f_tech('1')(t);
                ELSE s(s'LOW+j+offset) := f_tech('0')(t);
                END IF;
            END IF;
            work := work / 2;
            j := j - 1;
```

7.7. INTEGER DATA UTILITIES

```
        END LOOP;
        s(s'LOW) := f_tech('1')(t);
    END IF;
END f_inttologic2c;
```

This function is used primarily in situations where arithmetic operations will be performed using integer data. The two's complement representation of the integer value is converted to an appropriate two's complement representation in the logic array. This function can handle both positive and negative integer values. If the logic array contains greater than 32 bits, the sign bit is extended for the most significant bits. Entry 1 of the logic array is always treated as the most significant bit (sign bit). If the logic array contains fewer than 32 bits, then the most significant bits of the integer value will be ignored (the most significant bit of the logic array will be forced to the appropriate sign).

The following shows how the **f_inttologic2c** function can be utilized.

Using Integer to Logic Conversion/2's Complement

```
PROCESS (clk)
    VARIABLE istate : integer;
    VARIABLE bstate : t_logarray (1 TO 4);
    SIGNAL a0,a1,a2,a3 : t_wlogic;
BEGIN
    -- increment counter, generate outputs
    IF (clk = '1') THEN
        istate := istate + 1;
        f_inttologic2c(istate,bstate,ttl);
        a0 <= bstate(4); -- LSB
        a1 <= bstate(3);
        a2 <= bstate(2);
        a3 <= bstate(1); -- MSB
    END IF;
```

In this example a counter **istate** is incremented using integer arithmetic. The value of the counter is then converted to an array of logic **bstate** using the **f_inttologic2c** procedure. The **istate** value can be negative in which case the 4 bits will represent the equivalent two's complement number.

The **f_intdecode** function is shown below

Integer Decoding

```
PROCEDURE f_intdecode(
        VARIABLE d : IN integer;
        VARIABLE s : OUT t_logarray;
```

```
                VARIABLE t : IN t_technology) IS
BEGIN
    FOR i IN 0 TO s'LENGTH-1 LOOP
        IF d = i THEN s(s'LOW + i) := f_tech('0')(t);
        ELSE s(s'LOW + i) := f_tech('1')(t);
        END IF;
    END LOOP;
END f_intdecode;
```

This function provides a convenient mechanism for devices which have decoder operation. Given an integer value **d**, the function returns an array of logic values where entry **d** is set to low and all other entries are set to high. The first entry is 0. This function will operate correctly for a maximum of 32 bits.

The following shows a typical use of the **f_intdecode** function.

Decoding an Integer

```
PROCESS (clk)
    VARIABLE istate : integer;
    VARIABLE bstate : t_logarray (1 TO 4);
    SIGNAL a0,a1,a2,a3 : t_wlogic;
BEGIN
    -- clock new address
    IF (clk = '1') THEN
        f_intdecode(istate,bstate,ttl);
        a0 <= bstate(4);
        a1 <= bstate(3);
        a2 <= bstate(2);
        a3 <= bstate(1);
    END IF;
```

A decoder device generates the appropriate decoded outputs on the rising edge of the clock **clk**. The **f_intdecode** function is used to decode the integer value **istate** into the logic array **bstate**. These logic values are then assigned to the output signals **a0**, **a1**, **a2** and **a3**.

In order to facilitate logical operations using integer data, the logical operators are overloaded as follows

Decoding an Integer

```
-- "and" operator.
FUNCTION "and" ( L,R : integer ) RETURN integer IS
    VARIABLE lvalue : t_logarray (1 TO 32);
    VARIABLE rvalue : t_logarray (1 TO 32);
    VARIABLE value  : t_logarray (1 TO 32);
    VARIABLE unknown : boolean;
```

7.7. INTEGER DATA UTILITIES

```
        VARIABLE ivalue : integer;
BEGIN
    f_inttologic(l,lvalue,ttl);
    f_inttologic(r,rvalue,ttl);

    FOR i IN value'RANGE LOOP
        value(i) :=
            f_ttl( f_and(f_state(lvalue(i)))(f_state(rvalue(i))) );
    END LOOP;

    f_logictoint(value,unknown,ivalue);
    RETURN ivalue;
END;

-- "or" operator.
FUNCTION "or" ( L,R : integer ) RETURN integer IS
    VARIABLE lvalue : t_logarray (1 TO 32);
    VARIABLE rvalue : t_logarray (1 TO 32);
    VARIABLE value  : t_logarray (1 TO 32);
    VARIABLE unknown : boolean;
    VARIABLE ivalue : integer;
BEGIN
    f_inttologic(l,lvalue,ttl);
    f_inttologic(r,rvalue,ttl);

    FOR i IN value'RANGE LOOP
        value(i) :=
            f_ttl( f_or(f_state(lvalue(i)))(f_state(rvalue(i))) );
    END LOOP;

    f_logictoint(value,unknown,ivalue);
    RETURN ivalue;
END;

-- "nand" operator.
FUNCTION "nand" ( L,R : integer ) RETURN integer IS
    VARIABLE lvalue : t_logarray (1 TO 32);
    VARIABLE rvalue : t_logarray (1 TO 32);
    VARIABLE value  : t_logarray (1 TO 32);
    VARIABLE unknown : boolean;
    VARIABLE ivalue : integer;
BEGIN
    f_inttologic(l,lvalue,ttl);
    f_inttologic(r,rvalue,ttl);

    FOR i IN value'RANGE LOOP
        value(i) :=
            f_ttl( f_nand(f_state(lvalue(i)))(f_state(rvalue(i))) );
    END LOOP;

    f_logictoint(value,unknown,ivalue);
    RETURN ivalue;
```

```
END;

-- "nor" operator.
FUNCTION "nor" ( L,R : integer ) RETURN integer IS
    VARIABLE lvalue : t_logarray (1 TO 32);
    VARIABLE rvalue : t_logarray (1 TO 32);
    VARIABLE value : t_logarray (1 TO 32);
    VARIABLE unknown : boolean;
    VARIABLE ivalue : integer;
BEGIN
    f_inttologic(l,lvalue,ttl);
    f_inttologic(r,rvalue,ttl);

    FOR i IN value'RANGE LOOP
        value(i) :=
            f_ttl( f_nor(f_state(lvalue(i)))(f_state(rvalue(i))) );
    END LOOP;

    f_logictoint(value,unknown,ivalue);
    RETURN ivalue;
END;

-- "xor" operator.
FUNCTION "xor" ( L,R : integer ) RETURN integer IS
    VARIABLE lvalue : t_logarray (1 TO 32);
    VARIABLE rvalue : t_logarray (1 TO 32);
    VARIABLE value : t_logarray (1 TO 32);
    VARIABLE unknown : boolean;
    VARIABLE ivalue : integer;
BEGIN
    f_inttologic(l,lvalue,ttl);
    f_inttologic(r,rvalue,ttl);

    FOR i IN value'RANGE LOOP
        value(i) :=
            f_ttl( f_xor(f_state(lvalue(i)))(f_state(rvalue(i))) );
    END LOOP;

    f_logictoint(value,unknown,ivalue);
    RETURN ivalue;
END;

-- "not" operator.
FUNCTION "not" ( L    : integer ) RETURN integer IS
    VARIABLE lvalue : t_logarray (1 TO 32);
    VARIABLE value : t_logarray (1 TO 32);
    VARIABLE unknown : boolean;
    VARIABLE ivalue : integer;
BEGIN
    f_inttologic(l,lvalue,ttl);

    FOR i IN value'RANGE LOOP
```

7.7. INTEGER DATA UTILITIES

```
            value(i) := f_ttl( f_not(f_state(lvalue(i))) );
        END LOOP;

        f_logictoint(value,unknown,ivalue);
        RETURN ivalue;
END;
```

This makes it possible to use the logical operators on integer data. The following example shows the use of integer logical operators.

Integer Logical Operations

```
ENTITY test2 IS
    PORT    (a,b: IN integer;
            oand,oor,onand,onor,oxor,onot : OUT integer);
END test2;

ARCHITECTURE full OF test2 IS
BEGIN
    PROCESS (a,b)
    BEGIN
    oand <= a AND b;
    oor <= a OR b;
    onand <= a NAND b;
    onor <= a NOR b;
    oxor <= a XOR b;
    onot <= NOT a;
    END PROCESS;

END full;
```

Two integer ports **a** and **b** are used to generate a number of logical outputs **oand, oor, onand, onor, oxor** and **onot**.

The functions described here handle a maximum of 32 bits of resolution. Algorithms exist for the manipulation of greater than 32 bits for both logical and arithmetic operation in a 32 bit computing environment. Extensions to this package are possible which make it possible to perform arbitrarily complex logical and arithmetic operations.

Bibliography

[ACO88] R.Acosta, M.Alexandre, G.Imken, B.Read, "The Role of VHDL in the MCC CAD System", 25th Design Automation Conference, IEEE 1988, pp 34-39

[ADO85] "Postscript Language Reference Manual", Adobe Systems, Inc, Addison-Wesley, 1985

[AGR80] V. Agrawal, A.Bose, P.Kozak, H.Hham, "The Mixed Mode Simulator", 17th Design Automation Conference, IEEE 1980, pp 626-633

[ARMS88] J.Armstrong, "Chip Level Modelling in VHDL", Addison-Wesley, 1988

[AUG88] L.Augustin, B.Gennart, Y.Huh, D.Luckham, A.Stanculescu, "Verification of VHDL Designs Using VAL", 25th Design Automation Conference, IEEE 1988, pp 48-53

[AUT88] "AUTOCAD Release 9 Reference Manual", Autodesk, 1988

[BOW82] K.Bowden, "Design Goals and Implementation Techniques for Time-based Digital SImulation and Hazard Detection", IEEE Test Conference, Philadelphia, 1982, pp 147-152

[BRE76] M.Breuer, A.Friadman, "Diagnosis and Reliable Design of Digital Sysetms", Computer Science Press, 1976

[CAR88] D.Card, "Atten-shun! You will now design ASICs the military way", Electronic Business, November 1, 1988, pp 98-100

[CHE78] C.Chen, J.Coffman, "Multi-Sim, A Dynamic Multi-Level Simulator", 15th Design Automation Conference, IEEE 1978, pp 386-391

[CHU74] Y. Chu, "Introducing CDL", Computer, December 1974

[CLSI87] CAD Language Systems, "VHDL Tutorial for IEEE Standard 1076 VHDL", Draft, May 1987.

[COEL83] D.Coelho, "HELIX, A Tool for Multi-Level Simulation of VLSI Systems", International Semi-Custom IC Conference, November 1983

[COEL85] D.Coelho, "High–Level Design Using HELIX", ACM Computer-Science Conference, March 1985

[COEL87] D.Coelho, D.Hill, "Multi-Level Simulation for VLSI Design", Kluwer Academic Publishers, 1987

[COELA88] D.Coelho, "VHDL in uniform: mil-spec designs", EE Times, Sept. 26, 1988

[COELB88] D.Coelho, "VHDL: A Call for Standards", 1988 Design Automation Conference, June 1988

[DOD88] "Military Standard 454", 1988 US Government Printing Office

[EIC77] E.Eichelberger, T.Williams, "A Logic Design Structure for LSI Testability", 14th Design Automation Conference, IEEE 1977, pp 462-468

[FAR89] R.Farrow, A.Stanculescu, "A VHDL Compiler Based on Attributed Grammar Methodology", SIGPLAN 1989, ACM 1989

[FLA80] P.Flake, P.Moorby, G.Musgrave, "Logic Simulation of Bi-directional Tri-state Gates", Proc ICCC80, Port Chester, NY pp 594-600

[FLA83] P.Flake, P.Moorby, G.Musgrave, "An Algebra for Logic Strength Simulation", 20th Design Automation Conference, IEEE 1983

[GOEA88] R.Goering, "VHDL toolsets support design verification", Computer Design, July 1988, pp 36-37

[HAR88] T.Harbert, "VHDL challenges industry", EDN News Edition, Sept 1988, pp 1-3

[HAY82] J.Hayes, "A Fault Simulation Methodology for VLSI", 19th Design Automation Conference, IEEE 1982, pp 393-399

[HOR86] E.Horbst, "Logic Design and Simulation", North-Holland 1986

[IEEE87] "IEEE Standard VHDL Language Reference Manual", IEEE standard 1076-1987

[INT88] "Introduction to VHDL", Intermetrics, Inc 1988

[KIM88] K.Kim, J.Tront, "Automatic Insertion of BIST Hardware Using VHDL", 25th Design Automation Conference, IEEE 1988, pp 9-15

[KNU86] D.Knuth, "The T_EXBook", Addison-Wesley 1986

[LAM86] L.Lamport, "$\text{\LaTeX}$, A Document Preparation System", Addison-Wesley 1986

[LEI89] S.Leibson, "VHSIC Hardware Description Language", EDN, March 16, 1989, pp 110-124

[LIP89] "An Introduction to VHDL: Hardware Description and Design", Lipsett, Schaeffer, Ussery, Klewer Academic Publishers 1989

[LOU88] M.Moughzail, M.Cote, M.Aboulhamid, E.Cerny, "Experience with the VHDL Environment", 25th Design Automation Conference, IEEE 1988, pp 28-33

[MCD82] R.McDermott, "Transmission Gate Modelling in an Existing Three-Value Simulator", 19th Design Automation Conference, IEEE 1982, pp 678-681

[MEY89] E.Meyer, "VHDL opens the road to top-down design", Computer Design, February 1, 1989, pp 57-62

[MIL89] B.Milne, "Modeling Efforts Shore up CAE", Electronic Design, February 9, 1989, pp 18

[MOO79] P.Moorby, G.Musgrave, "Simulation of Logic using Wavefronts to give Accurate Timing Analysis", IEEE CADMAT, Sussex, UK, 1989, pp 198-203

[SZY72] S.Szygenda, "TEGAS - Anatomy of a General Purpose Test Generation and Simulation System for Digital Logic", Design Automation

Workshop, 1973, pp 116-127

[SZY77] S.Szygenda, A.Lekkos, "Integrated Techniques for Functional and Gate-Level Digital Logic Simulation", 10th Design Automation Workshop, 1973, pp 159-172

[TI86] The TTL Data Book, Volumes 1 through 4, Texas Instruments, 1986

[TIALS86] ALS/AS Logic Data Book, Texas Instruments, 1986

[TRW84] TDC1028 data sheet, LSI Products Division, TRW Electronic Components Group

[ULR82] E.Ulrich, D.Herbert, "Speed and Accuracy in Digital Network Simulation Based on Structural Modelling", 19th Design Automation Conference, IEEE 1982

[USG88] "Electronic Hardware Description in the VHSIC Hardware Description Language (VHDL)", Data Item Description, 1988 US Government Printing Office

[VAN77] M.vanCleemput, "An Hierarchical Language for the Structural Description of Digital Systems", 14th Design Automation Conference, IEEE 1974, pp 377-385

[VANT89] "The VantageSpreadsheet: System Analyst Tutorial", Vantage Analysis Systems, Inc 1989

[VANU89] "The VantageSpreadsheet: System Analyst Users Guide", Vantage Analysis Systems, Inc 1989

[WEI88] R.Weiss, "VHDL subsets for CAE", EE Times, August 1, 1988, pp 51-52

[WEI89] R.Weiss, "Simulation: critical for VHDL success", EE Times, March 6, 1989, pp 77-80

Index

/=, 338.
<, 338.
<=, 338.
=, 338.
>, 338.
>=, 338.
46 value system, 313.
46 values, 314.
abstraction levels, 4.
    mixing, 4.
accuracy, 25.
accuracy choices, 7.
accuracy continuum, 25.
    2-value unit-delay, 25.
    46-value unit-delay, 28.
    fixed-delay, 30.
    full-delay, 36.
    generic variable-delay, 33.
    variable-delay, 31.
ADA, 2.
ADLIB, 50.
alu, 108.
analog simulation, 8.
anatomy of a vhdl model, 11.
and, 323, 325, 327, 329, 331, 375.
    3-input positive-and, 68.
and gate, 68.
architectural tool, 6.
architecture, 12, 17, 23, 44.
    of VLSI device, 271.
architecture body name, 54.
asic's, 5, 9.
assertion tests, 305.
assertions, 305.
back annotation of delay, 53.

back annotation of delays, 4, 38, 50.
behavior, 12, 17.
    decomposition, 272.
    generic, 46.
    of design, 13.
    of VLSI model, 278.
    technology independent, 33.
behavior transfer function, 37.
bibliography, 377.
bidirectional transmission element, 104.
BIT, 26.
black box, 12.
bottom-up design, 5.
building block, 12.
bus functional model, 270.
bus handlers, 299.
bus resolution, 332.
bxfr, 104.
calculating PAL products, 252.
capacitor, 103.
charge effects, 52.
check_edge, 22.
chip reset, 286.
circuit power-up, 26, 55.
clock pulse check, 41.
cmos, 315, 324.
code.
    modularization, 49.
    sharing, 49.
code coverage, 304.
code documentation, 49.
code quality, 50.
combinational devices, 57.
comparator, 123.
    magnitude 4 bit, 124.
comparison operators, 337.
compatability.
    between models, 49.
    generic parameter, 51.
    model, 55.
compilation.

order, 17.
complementary MOS, 324.
complex devices, 269.
component, 18.
configuration, 14, 45.
    declaration, 17.
    handling timing, 42.
    using, 42.
configuration body, 14, 18.
consistency.
    model, 53.
constraints, 54.
conventions.
    naming, 52, 53.
counter, 196, 202, 208, 214, 220.
    sync 4 bit binary, 214.
    sync 4 bit binary/async clear, 202.
    sync 4 bit decade, 208.
    sync 4 bit decade/async clear, 196.
    sync up/down 4 bit decade, 220.
counters, 195.
coverage of code, 304.
d-type flip flop, 136.
data abstraction, 6.
data item description, 2.
data type.
    BIT, 26.
data types.
    sharing, 23.
decoder, 165, 171.
    2 to 4, 84.
    3 to 8, 79.
    3 to 8/latch, 171.
    3 to 8/register, 165.
defense electronics supply center, 3.
delay, 17, 50.
    average, 31, 32.
    back annotation of, 4, 38, 50, 53.
    calculated, 37.
    calculation, 38.
    calculations, 52.
    formula, 52.
    full, 36.
    generic, 33.
    high to low propagation, 34.
    high to low transition, 32, 34.
    input, 37, 51, 52.
    input pin, 37.
    input to output, 52.
    low to high propagation, 34.
    low to high transition, 32, 34.
    maximum, 43.
    methodology, 50.
    min/max, 290.
    minimum, 43.
    nominal, 31.
    output, 51, 52.
    pessimistic, 32.
    propagation, 34, 37, 51, 52.
    technology independent, 33.
    transition, 52.
    typical, 43.
    unit, 8, 27.
    variable, 31.
    wiring, 36, 52.
    wiring segment, 37.
    zero, 44.
demultiplexer, 165.
    3 to 8/register, 165.
DESC, 3.
design.
    bottom-up, 5.
    building blocks, 6.
    hierarchical, 5.
    logic level, 6.
    middle-out, 5.
    mixing styles, 5.
    multi-level, 5, 49.
    standard component level, 6.
    synchronous, 26.
    top-down, 5.
design documentation, 50.

INDEX

design management, 14.
device.
    dependent data, 291.
    manufacturer specific, 44.
    speeds, 288.
    ssi, 57.
    switch, 99.
    transistor, 104.
device instance, 50.
devices.
    combinational, 57.
    complex, 269.
    memory, 227.
    sequential, 135.
    VLSI, 269.
disconnect, 313.
documentation, 49.
    design, 50.
double blind errors, 302.
drive dependencies, 290.
ecl, 315, 322.
emitter coupled logic, 322.
entity, 12.
    technology dependent, 44.
entity declaration, 17, 22.
entity definition, 273.
error recovery, 55.
errors, 49.
    checking, 38.
    checking section in model, 39.
    clock pulse check, 41.
    double blind, 302.
    handling, 293.
    hold check, 41.
    hold checks, 54.
    invalid control, 54.
    invalid control check, 41.
    invalid data, 54.
    recovery, 55.
    semantic, 54.
    setup check, 41.
    setup checks, 54.
    special timing requirements, 54.
    spike detection, 41, 54.
f_and, 352.
f_busres, 332.
f_cap, 103.
f_ceil, 104.
f_cmos, 325.
f_convu, 320.
f_convx, 322.
f_convz, 317.
f_delay, 34, 43, 360.
f_ecl, 323.
f_falling_edge, 359.
f_hold_check, 363.
f_intdecode, 371.
f_inttologic, 366.
f_inttologic2c, 369.
f_logic, 344.
f_logic0, 345.
f_logic1, 347.
f_logic_bus, 314, 332.
f_logictoint, 365.
f_logictoint2c, 368.
f_logicX, 349.
f_lookup7, 253.
f_lookup8, 254.
f_nand, 354.
f_nmos, 327.
f_nor, 355.
f_not, 356.
f_or, 353.
f_palinit, 229.
f_product, 252.
f_pulse_check, 364.
f_resistor, 101.
f_rising_edge, 357.
f_rominit, 228.
f_setup_check, 362.
f_state, 339.
f_strength, 340.
f_strengthH, 343.
f_strengthL, 342.

f_tech, 315.
f_ttl, 328.
f_ttloc, 330.
f_uxfr, 100.
f_xor, 356.
f_zdelay, 361.
fixed delay, 30.
flip flop, 140, 146, 152.
    d-type, 136.
    JK neg edge, 146.
    JK neg edge/preset, 152.
    JK pos edge, 140.
flip flops, 135.
floating point unit, 269.
forced, 313.
format.
    standardized, 49.
foundation.
    of VLSI model, 278.
full functional model, 270.
functional model.
    bus, 270.
    full, 270.
    partial, 270.
gate.
    2-input positive-or, 73.
    3-input positive-and, 68.
    3-input positive-nand, 71.
    3-input positive-xor, 76.
    and, 68.
    inverter, 64.
    inverter with open collector, 66.
    nand, 57, 60, 71.
    nor, 62.
    or, 73.
    xor, 76.
gates.
    simple, 57.
generic.
    variable delay, 33.
generic behaviors, 46.
generic map, 36.

generic model, 33, 43.
generic parameter, 37.
generic parameter compatibility, 51.
generic parameter instantiation, 35.
generic parameter name, 54.
generic parameters, 33, 34, 42, 50, 52.
getting started.
    on VLSI model, 269.
global relaxation, 99.
guidelines.
    modelling, 50.
hardware.
    describing operation, 50.
hardware implementation, 6.
hardware modelling, 306.
hardware models, 306.
hardware synthesis, 4.
hardware timing, 4.
HHDL, 50.
hierarchical design, 5.
hierarchy, 5.
high impedance, 313.
high level specifications, 5.
high-level specifications, 6.
hold check, 41.
hold checks, 54.
hold time, 4, 295.
IEEE, 2.
    standard 1076-987, 2.
improper simulation results, 55.
index, 381.
industry standard, 5.
initialization, 26.
    of VLSI models, 287.
input.
    PAL, 227.
input delay, 51.
input delay section, 39.
instance of device, 50.
instances.
    binding timing data to, 42.

INDEX 385

instantiation.
    generic parameter, 35.
instruction decoders, 300.
instruction sets, 286, 301.
integer data utilities, 364.
integration of software, 49.
interface specification, 17.
introduction, 1.
invalid control, 54.
invalid control check, 41.
invalid data, 54.
inverter, 64.
    with open collector, 66.
JK flip flop.
    JK neg edge, 146.
    JK neg edge/preset, 152.
    JK pos edge, 140.
knowledge base, 50.
languages.
    programming, 49.
layout, 52.
libraries.
    commercial, 6.
    internal, 6.
    model, 4.
    standard component model, 6.
library.
    modular, 46.
loading dependencies, 290.
local relaxation, 99.
logic designers, 5.
logic lookup tables, 352.
logic manipulation, 337.
logic modelling accuracy continuum, 8.
logic value system, 311.
logic values, 311.
lookup table.
    logic, 352.
    state, 339.
    strength, 339.
magnitude comparator, 124.

    4 bit, 124.
maximum delay, 43.
memory, 227.
    initialization, 227.
    read only, 231.
memory controllers, 269.
memory devices, 227.
messages, 49.
methodology, 49.
microprocessors, 269.
middle-out design, 5.
military.
    documentation requirements, 2.
    procurement, 5.
    standard 454, 2.
min/max timing, 4, 290.
min/typ/max timing package, 43.
minimum delay, 43.
model.
    accuracy continuum, 7.
    anatomy of, 11.
    behavioral, 13.
    compatibility, 49.
    consistency, 53.
    conventions, 48.
    device dependent, 44.
    error checking, 39.
    generic, 33, 43.
    input delay, 39.
    practices, 50, 53.
    pre-compiled, 6.
    referencing technology dependent version, 45.
    separate processes, 39.
    standardized, 48.
    structural, 18.
    structure, 38.
    timing, 288.
    uniform structure, 39.
    version, 14, 45.
model accuracy, 25.
model accuracy continuum, 25.

model compatibility, 55.
model decomposition, 272.
model libraries, 4.
model validation, 304.
modelling.
    conventions, 48.
    guidelines, 50.
    instruction sets, 286.
    microprocessors, 287.
    optimism, 29.
    pessimism, 29.
    practices, 50, 53.
    register sets, 284.
    VLSI devices, 287, 299.
    with overloading, 29.
    without overloading, 29.
modularization of code, 49.
monostable multivibrator, 119.
multi-level design, 5, 49.
multiplexer, 78, 171.
    1 of 2, 96.
    1 of 4, 92.
    1 of 8, 87.
    2 to 4, 84.
    3 to 8, 79.
    3 to 8/latch, 171.
multivibrator, 119.
    monostable, 119.
n-channel MOS, 326.
name.
    architecture body, 54.
    conventions, 52, 53.
    generic parameter, 52, 54.
    port, 54.
    standard, 53.
    uniform, 53.
NAND, 29, 323, 325, 327, 329, 331, 375.
    2-input positive-nand, 57.
    2-input positive-nand with open-collector, 60.
    3-input positive-nand, 71.

true table, 29.
nand gate, 57, 60, 71.
    with open collector, 60.
netlist, 12.
nfet, 107.
nmos, 315, 326.
nor, 323, 325, 327, 329, 331, 375.
    2-input positive-nor, 62.
nor gate, 62.
not, 323, 325, 327, 329, 331, 375.
one shots, 119.
open collector TTL, 330.
operator.
    overloaded, 28, 324, 326, 328, 329, 331, 337.
optimism.
    modelling, 29.
or, 323, 325, 327, 329, 331, 375.
    2-input positive-or, 73.
or gate, 73.
output delay, 51, 52.
overloaded operators, 28, 324, 326, 328, 329, 331, 337.
overloading.
    modelling with, 29.
package.
    min/typ/max timing, 43.
    standard, 50.
    standard logic, 20, 28, 50, 309.
    user defined, 21.
package body, 22.
package declaration, 22.
PAL, 230, 251, 255, 259, 264.
    10 input, 2 output, 6 I/O, 255.
    8 input, 2 I/O, 6 clocked output, 259.
    8 input, 8 clocked output, 264.
    input, 227.
    product, 227.
paldef, 230.
parameter.
    generic, 33, 37, 42, 50, 52.

INDEX

parity checker, 129.
    9 bit odd/even, 129.
parity generator, 129.
    9 bit odd/even, 129.
part.
    manufacturer specific, 44, 45.
pattern recognition, 22.
pessimism.
    modelling, 29.
pfet, 107.
physical models, 306.
PLD, 251, 255, 259, 264.
port, 12.
    input, 12, 17.
    output, 12, 17.
port name, 54.
power-up, 30, 55.
practice.
    modelling, 53.
process statement, 34.
processes.
    using in VLSI model, 278.
product.
    calculating PAL, 252.
    PAL, 227.
productivity, 50.
programmable arrays, 251, 255, 259, 264.
programming languages, 49.
programming methodology, 49.
prom, 236.
    16,384 (4096 by 4), 236.
propagation.
    of unknowns, 29.
propagation delay, 34, 37, 51, 52.
prototypes, 5.
quality.
    code, 50.
quality assurance, 301.
ram, 242, 243.
    64 bit, 243.
random access memory, 242, 243.

read only memory, 231.
register, 158, 177, 182, 188.
    4 bit parallel access, 158.
    8 bit serial shift/parallel out, 177.
    8 bit shift/parallel load, 182.
    8 bit shift/parallel load/clear, 188.
register sets, 284.
registers, 157.
regression test suite, 303.
relaxation.
    global, 99.
    local, 99.
resistor, 101.
results.
    improper simulation, 55.
    simulation, 14, 18.
rom, 228, 231.
    1024 bit (246 by 4), 231.
romdef, 228.
s_t, 43.
schematic, 11.
schematic editor, 12.
selector, 78.
    1 of 2, 96.
    1 of 4, 92.
    1 of 8, 87.
semantic checks, 54.
semiconductors, 6.
sensitivity, 34.
sensitivity list, 34.
separate processes, 39.
sequencers, 301.
sequential devices, 135.
setup check, 41.
setup checks, 54.
setup time, 4, 295.
shell.
    timing, 46.
shift register, 177, 182, 188.
    4 bit parallel access, 158.

8 bit/parallel load/clear, 188.
serial/parallel out, 177.
signal.
    internal, 18.
    strengths, 28.
simulation, 7.
    abstract, 7.
    analog, 8.
software.
    integration of, 49.
special timing requirements, 54.
specification.
    high level, 5.
    high-level, 6.
    of hardware, 5.
spike detection, 4, 41, 54.
spike pass through, 54.
SSI, 6.
ssi devices, 57.
standard.
    industry, 5.
standard components, 5.
standard generic parameter names, 52.
standard logic package, 20, 28, 50, 309.
    accessing, 310.
    using the, 310.
standard names, 53.
standard packages, 50.
standardized coding formats, 49.
standardized models, 48.
startup, 228, 230.
state lookup, 339.
std_cmos, 310, 326.
std_ecl, 310, 324.
std_logic, 310.
std_nmos, 310, 328.
std_ttl, 310, 329.
std_ttloc, 310, 331.
step by step modelling of microprocessors, 287.

stimulus, 13.
    circuit, 19.
strength lookup, 339.
strength output, 30.
strengths, 313.
    disconnect, 313.
    forced, 313.
    high impedance, 313.
    strong resistive, 313.
    weak resistive, 313.
strong resistive, 313.
structure, 11, 18.
    decomposition, 272.
    uniform model, 39.
switch level devices, 99.
switch modelling utilities, 100.
synchronous design, 26.
synthesis, 4.
system architects, 5.
t_choice, 43.
t_logarray, 364.
t_logic, 314.
t_logic_vector, 314.
t_state, 314.
t_strength, 314.
t_technology, 314.
t_time, 43.
t_wlogic, 28, 314.
t_wlogic_vector, 314.
technology, 315.
    independent applications, 315.
    independent signal values, 315.
    rules, 315.
technology dependent entity, 44.
technology dependent model.
    referencing, 45.
technology dependent timing, 45.
technology independence, 42.
test and development, 304.
test plan, 302.
timing.
    calculations, 52.

INDEX 389

handling, 42.
  technology dependent, 45.
  technology independent, 33.
timing characteristics, 50.
timing data, 42.
timing methodology, 50.
timing model, 288.
timing shell, 46.
timing utilities, 357.
top down design, 4.
top-down design, 5.
transfer function, 17.
transistor.
  output calculation, 100.
transistor transistor logic, 328.
transition delay, 52.
transmission element, 104.
transmission gate, 107.
  basic, 107.
  complementary, 107.
truth table.
  nand, 29.
ttl, 315, 328.
ttloc, 315, 330.
typical delay, 43.
U, 27, 313.
uniform approach.
  to device dependent data, 291.
uniform names, 53.
uninitialized, 313.
uninitialized value, 27, 28, 30, 55.
unit delay, 8, 25, 27, 28.
unknown.
  suppression, 30.
  versus uninitialized, 30.
unknown handling, 54.
unknown propagation, 29.
unknown values, 28.
unknowns, 293.
  handling, 54.
use statement, 310.
user defined packages, 21.

utilities.
  integer data, 364.
  switch modelling, 100.
validation, 304.
validation of model, 304.
value.
  unknown, 28.
value system, 21, 311.
version, 14.
VHDL.
  as a design tool, 4.
  contracts awarded, 3.
  describing electronic hardware
in, 11.
  file, 16.
  history of, 1.
  introduction, 1.
  language reference manual, 50.
  layered language features, 4.
  learning, 4.
  stimulus, 14.
  structure in, 12.
  subsets, 4.
  tools, 3.
  using in design, 5.
  vendors, 3.
  version 7.2, 1.
VLSI, 6.
  modelling techniques, 299.
  using processes in model, 278.
VLSI devices, 269.
waveform checking, 296.
weak resistive, 313.
wiring.
  input, 37.
wiring delay, 36, 52.
wiring segment delays, 37.
xor, 323, 325, 327, 329, 331, 375.
  3-input positive-xor, 76.
xor gate, 76.
zero delay, 44.